Advance Praise for *The War on Science*

"Higher education isn't what it used to be. Cancel Culture and DEI have caused many to keep their mouths shut. Not so the authors of this book. This collection of essays tells of threats to open inquiry, free speech, and the scientific process itself. A much-needed book."

–Sabine Hossenfelder, Physicist and Author of *Existential Physics: A Scientist's Guide to Life's Biggest Questions*

"Provocative, impassioned, and often personal, these essays by a starry cast of scholars offer vivid views of the most disturbing trend in academia today: the distortion of science by activists who put their idiosyncratic notions of social justice ahead of truth. Biology, medicine, and even math—not just the humanities and social sciences—are falling prey to DEI, decolonization, and de-sexed language. Not only does *The War on Science* elucidate the problem, it also models the solution: recommitment to scientific objectivity, scholarly integrity, and—above all—humility in the search for knowledge."

–Jonathan Rauch, Author of *The Constitution of Knowledge*

"The encroachment of ideology and groupthink on the production of knowledge may sound abstract, but it has a profound effect on every single aspect of life in the world today. *The War on Science* is an absolutely crucial response to this dangerous trend."

—Greg Lukianoff, President of the Foundation for Individual Rights and Expression, Author of *Unlearning Liberty* and Co-Author of *The Coddling of the American Mind* and *The Canceling of the American Mind*

"There may be no more perverse irony of history: The civilization that created the scientific method and that shared its fruits with all of humanity is now hell bent on destroying its creation. The destruction is coming from within science itself—perpetrated by ideologues, and abetted by administrators and scientists too cowardly to defend this greatest of intellectual accomplishments. With the publication of *The War on Science*, no university bureaucrat will be able to dismiss the anti-science crusade as just a 'right-wing' chimera. *The War on Science* marshals evidence from every corner of the scientific world to document the corruption of truth-seeking in favor of political fads. Its illustrious authors have defied censorship and vilification. If the authors' peers gain the courage from this book to join the rebellion against enforced lies, there will yet be hope to salvage the scientific enterprise in the place of its birth."

—Heather Mac Donald, Thomas W. Smith Fellow at the Manhattan Institute, Author of *When Race Trumps Merit*

"This book is a clarion call. Woke imperialism, the new oppressor, steamrolls over religion, culture, customs, language, and reason, constructing a Tower of Babel where science isn't just sidelined—it's sent to DIE. We're at DEFCON 1, alarms blaring as ideological spaghetti soldiers emerge from our universities, infiltrating all industries from aviation to medical and dismantling excellence with copypasta in the name of equity. The stakes are high: our future is on the line."

–Cyan Banister, Co-Founder and
General Partner of Long Journey Ventures,
Tech Investor (Uber, SpaceX, Anduril)

Also by Lawrence M. Krauss

The Fifth Essence: The Search for Dark Matter in the Universe
Fear of Physics: A Guide for the Perplexed
The Physics of Star Trek
Beyond Star Trek: From Alien Invasions to the End of Time
Atom: A Single Oxygen Atom's Journey from the
Big Bang to Life on Earth..and Beyond
Quintessence: The Mystery of Missing Mass in the Universe
Hiding in the Mirror: The Mysterious Allure of Extra
Dimensions, from Plato to String Theory and Beyond
Quantum Man: Richard Feynman's Life in Science
A Universe from Nothing: Why is there Something Rather than Nothing.
The Greatest Story Ever Told…So Far
The Edge of Knowledge: Unsolved Mysteries of the Cosmos
The Physics of Climate Change

THE WAR ON SCIENCE

THE WAR ON SCIENCE

Thirty-Nine Renowned Scientists and Scholars Speak Out About Current Threats to Free Speech, Open Inquiry, and the Scientific Process

EDITED BY
LAWRENCE M. KRAUSS

A POST HILL PRESS BOOK
ISBN: 979-8-88845-756-6
ISBN (eBook): 979-8-88845-757-3

The War on Science:
Thirty-Nine Renowned Scientists and Scholars Speak Out About Current Threats to Free Speech, Open Inquiry, and the Scientific Process
© 2025 by Lawrence M. Krauss
All Rights Reserved

Cover design by Jim Villaflores

This book, as well as any other Post Hill Press publications, may be purchased in bulk quantities at a special discounted rate. Contact orders@posthillpress.com for more information.

This is a work of nonfiction. All people, locations, events, and situations are portrayed to the best of the author's memory.

No part of this book may be reproduced, stored in a retrieval system, or transmitted by any means without the written permission of the author and publisher.

Post Hill Press
New York • Nashville
posthillpress.com

Published in the United States of America
1 2 3 4 5 6 7 8 9 10

To Nancy, whose integrity and bravery serve as a constant inspiration.

"Universities need to abandon the concept that they have a central role in Moral Education."

Lawrence Summers

Former President, Harvard University

"A university has no capacity to eliminate 'hate,' nor should that be its mission."

Heather Mac Donald

"Our apparatniks will continue making the usual squalid mess called History: all we can pray for is that artists, chefs and saints may still appear to blithe it."

W. H. Auden

Table of Contents

Introduction and Overview xxi
    Lawrence M. Krauss

Section 1: Free Speech, Victimhood, and Ideology

Scientific Truth Stands Above Human Feelings and Politics 3
    Richard Dawkins

How Ideology Threatens to Corrupt Science 36
    Alan Sokal

The Treason of the Intellectuals 46
    Niall Ferguson

Excerpts from a Conversation with Sam Harris in October 2017 About Free Speech, Universities, and Cancel Culture 56
    Nicholas A. Christakis

Universities as Dispensers of Parasitic Ideas 69
    Gad Saad

Spotlight on Scientific Censorship 78
    Anna Krylov and Jay Tanzman

The Tyranny of Victimhood: How Progressives Turned Regressive 93
    Maarten Boudry

The Two Universities Redux 108
    Abigail Thompson

Section 2: Ideological Corruption of Academic Disciplines

Radical Equalitarianism and Mathematics 117
    Sergiu Klainerman

How Do You Decolonize Mathematics? 129
    John Armstrong

How the Law Went Woke 138
    Bruce Pardy

The Ideological Subversion of Biology 146
    Jerry A. Coyne and Luana S. Maroja

Social Justice, MD—Medicine under Threat 171
    Sally Satel

Burying Science Under Indigenous Religion 181
    Elizabeth Weiss

An Apology for Philology 192
    Solveig Lucia Gold and Joshua T. Katz

Section 3: Cancel Culture

Indigenization Destroys Academic Values: The University of
Lethbridge Case 203
    Frances Widdowson

Why I Left Harvard 214
    Carole Hooven

Delicious Moral Treats 222
    Christian D. Ott

Section 4: The Impact of Diversity, Equity, and Inclusion Bureaucracies on Scholarship in Academia and Beyond

Equity Hiring, by an Equity Hire 245
    Janice Fiamengo

Equity, Diversity, and Inclusion: The Dismal Pseudoscience Threatening Science 256
    Geoff Horsman

The Leaning Ivory Tower 267
    Alessandro Strumia

DEI in Science and Medicine: Missing Metrics and Measures 280
    Roger B. Cohen and Amy L. Wax

Understanding Why Plagiarists Are Not Fired: From Veritas to DEI 294
    Peter Boghossian

Section 5: Gender, Race, Ideology, Science, and Scholarship

Gender-Affirming Care: The Abandonment of Medical and Academic Standards 301
    Lauren H. Schwartz and Arthur W. Rousseau

A Deafening Silence: Bioethics and Gender-Affirming Health Care 317
    Alex Byrne and Moti Gorin

The Gender Wars and Academic Freedom 335
    Judith Suissa and Alice Sullivan

The Desexing of Language in Women's Health Research and Care: A Story of Marginalization of Science, Cultural Imperialism, and Abuse of Power 354
    Karleen Gribble

Section 6: What Can Be Done?

Science and Politics: Three Principles, Three Fables 377
   Dorian Abbot

A Five-Point Plan to Save Universities from Themselves 390
   Steven Pinker

Teaching Inclusion in a Divided World 396
   Nicholas A. Christakis

Depoliticizing Science Requires Revolutionary Reform of the Professoriate 398
   Richard E. Redding

The Archetypal Fate of the Universities 404
   Jordan B. Peterson

Epilogue 417
References and Footnotes 419
Authors 421

Introduction and Overview

Universities and science institutions in the West are unfortunately no longer guaranteed to be places where the free and open exchange of ideas is encouraged, nor where scientific progress can be carried out unhindered by ideology.

Faculty, even tenured faculty, have lost their jobs or been otherwise censured for their speech or writing, and students are afraid to ask questions in class and are even often encouraged by their universities to report language they find offensive.

Universities have created vast bureaucracies under the guise of what is generally called diversity, equity, and inclusion. They are accountable to no one and police both behavior and language. Questions ranging from the difference between sex and gender to the question of whether Indigenous science warrants the name are often forbidden to even be raised in the classroom or in academic meetings.

Merit, at the heart of academic progress, is also now being questioned in academia, and in scientific institutions more broadly. Identity-based quotas—both gender and racially based—are being imposed on academic hiring and promotion around the world, independent of both the questions of how this might impact academic excellence and to what extent such measures are actually required or even useful. From a fundamental perspective, is it essential that the composition of researchers and teachers in every discipline reflect the underlying demographics of society as a whole, independent of the available pool of emerging talent?

As university and government DEI programs have restricted hiring and promotion to those whose ideological predilections conform to DEI manifestos, faculty and administrators are becoming more prone to impose critical social justice (CSJ) ideology throughout their academic activities. The notion—critiqued by former Harvard University President Lawrence Summers in the epigraph of this book—that universities have a role to play in moral education, has resurfaced with a vengeance.

Dubious postmodern notions regarding objective, evidence-based inquiry and epistemology, and, more recently, CSJ, while once restricted to fringe departments, have now become endemic, making their way into the mainstream, even in hard science curricula. Those hired under the banner that racism and sexism are rampant in academia tend to echo that in their research and teaching. As a result, the debate about scientific issues often becomes stifled. For example, the question of whether gender-based differences between fields reflect underlying sexism or simply deeper psychological or sociologically based differences in interests between the sexes cannot safely even be raised in numerous universities, as various faculty who have lost their jobs or affiliations can attest.

One of the more pernicious facets of this new incursion of ideology into scholarship is the notion that language conflates with violence and that personal offense gives one special rights. The use of certain words is proscribed in academic environments, government, and the media as if they were sacred mantras. Scholars and journalists have been sacked for using the wrong word, independent of its context. And even more worrisome, various academic journals now advise their editors to not accept papers that might cause offense, regardless of their validity or potential importance.

A vigorous dialectic is essential for academic progress, especially in the sciences. That means active questioning and open debate, and even the possibility of provocation or offense must not only be tolerated, it should be encouraged. Science is a social activity, unlike popular perception, and it cannot properly function unless ideas can be dissected and even attacked. Only in this way can weak or incorrect ideas be filtered out.

Students are also suffering in this environment. Not only are they missing out on necessary experiences of critical reasoning, but there are also important subject areas that faculty simply avoid at the risk of stirring up controversy. The notion that language is somehow a form of violence has created additional tensions in classroom education. The creation of "safe spaces," where students will not have to confront ideas or words that make them uncomfortable, hasn't helped: students need to be protected from hearing them at all costs, and most importantly, if anyone is particularly sensitive to any word or idea, too often, all those around them must be protected too.

In fact, numerous studies have suggested that this escalation has caused students to feel less safe, increased their anxiety levels, and encouraged them to hold back from open discussions in classrooms. At the same time, one is witnessing a generation of students, including those educated at supposedly elite institutions, who have not been trained to think critically. The current conflict in the Middle East is something that generates strong opinions and passions, but watching videos of some Columbia student protestors made it clear that they had little or no critical perception of what they were protesting for or against.

In effect, universities have often become hostile work and learning environments, ironically in large part because university bureaucracies—in a misplaced effort to identify and root out hostile environments—have created hypervigilant policing that encourages a sense of victimhood and creates a climate of conflict and tension.

Years ago, I wrote that one of the central purposes of education is to make you uncomfortable. Only when you move outside of your comfort zone are you likely to examine your own preexisting biases and discover truths about the world that you might never have otherwise allowed to enter your consciousness. Unfortunately, in the current climate, intellectual comfort has become a standby at universities rather than an anathema.

The emerging threat to science and reason has percolated outside ivy-covered walls to scientific and government institutions and to industry. Formerly prestigious science journals are not only devoting increasing space to parroting critical social justice ideology; they are, in some cases, censoring scientific articles that editors feel may not be in accord with them, even if the results of these articles are not in dispute! As I reported recently in the *Wall Street Journal*, an analysis of grants demonstrated that US science funding agencies like the National Science Foundation and the National Institutes of Health spent hundreds of millions of dollars on social justice initiatives instead of fulfilling their mandates of supporting scientific research. And the heads of these organizations often almost proudly claimed, without evidence and purely on the basis of demographic data, that their own institutions are systemically racist. At the same time, a number of these organizations have required prospective grant recipients to conform to certain political and social justice ideals as a part of the granting process. This intrusion of ideology into the process of science funding is particularly egregious.

Beyond the direct threat to science, there was the opportunity cost of drawing key funding away from the scientific enterprise and of forcing researchers to take time out of doing science to explicitly prepare proposals conforming to certain CSJ dictates. And parochially, for the West at least, the issue of scientific competition becomes relevant. For countries to compete economically in the twenty-first century, support for science and engineering is vital. China, India, Singapore, Korea, and numerous other countries are building up their research enterprises, sometimes beginning to outstrip the West, unencumbered by these ideological strictures.

Recent events, following the election of Donald Trump, have changed the landscape of federal funding. Executive orders issued require funding agencies to remove numerous extraneous DEI-related programs as well as requirements for grant proposals. This is a very welcome change, and while some agencies have already complied, it remains to be seen how this will percolate down through the academic hierarchy.

While this volume focuses, for the most part, on scientific disciplines, it is worth pointing out that academic policing of language and ideas is occurring in the humanities perhaps even more severely and is dramatically impacting hiring and scholarship in other fields. In this regard, as Joshua Katz and Solveig Gold describe later in this volume, in the field of classics—which, after all, traces a civilization closely tied to the emergence of modern science and philosophy in the West—there is new critical social justice resistance to even studying Latin or Greek and their connection to the emergence of modern Western philosophical and scientific thought. In areas like music and art, masterpieces are being dismissed, and in the name of equity, merit, too, has become an anathema. As Heather MacDonald, in her book *When Race Trumps Merit*, poignantly argued, "when museum and other cultural organizations declare their mission to be anti-racism…the final result, if unchecked, will be the cancellation of a civilization."

It is vitally important, for the progress of scholarship and the scientific health of the nation and the world, for these trends to be opposed. The first step in this process is to make the public more fully aware of just how worrisome the situation has become. While many faculty and administrators recognize the problems in higher education and in institutions supporting scientific and scholarly research, many are afraid of speaking out because of the possible consequences for their own careers.

Nevertheless, the potential problems are so grave, and the stakes so serious, that a stellar group of thirty-nine distinguished scholars from a wide variety of disciplines and a wide variety of experiences—including a full spectrum of political leanings—have agreed to contribute to this volume describing the existing and emerging threats. All of them have a common interest in combating the current trends hindering higher education, scholarship, science, and reason. Some of these scholars have been canceled as a result of their statements or actions, and others have watched from the sidelines. Some have successfully fought for change within their own institutions, and others have not been so successful. All share, however, the recognition that something needs to be done. They have agreed to brave the minefield to describe the current challenges and, in some cases, point out possible solutions to the current dilemma of ideology trumping science and reason in our society. I applaud them for their clarity and bravery in speaking out at the current time, and I thank them for the time and energy they devoted to writing about these important issues. I have given each of them carte blanche to write on subjects that most concern them based on areas of their own interest and expertise and have at most lightly edited the pieces. The views expressed therein are their own. As a result, this compilation is not monolithic. It reflects a diversity of opinion on certain issues, different concerns, and different perspectives. That is precisely what the authors of this volume are advocating should be reflected in the functioning of our universities and research support institutions.

You, the reader, are encouraged to draw your own conclusions.

LAWRENCE M. KRAUSS

***

This volume is organized into sections corresponding to major issues, including: free speech, victimhood, and ideology; ideological corruption of academic disciplines; cancel culture; the impact of DEI policies; gender, race, ideology, science, and scholarship; and finally, what can be done? Together, they broadly encompass deep concerns about the current war on science, scholarship, and reason. In the rest of this introduction, I want to provide a brief overview of these issues, with selected examples, to provide some perspective for what is to follow.

I have found myself recently reflecting on the contrast between the situation today and that of when I began my career as a faculty member at Yale University in the mid-1980s. The science departments at Yale were located on what is called Science Hill; however, the separation between science and the rest of campus at the time wasn't just geographic. Those of us in the science departments sometimes felt like poor cousins at a university dominated by very well-known English and history departments where the bulk of Yale undergraduate students studied.

At that time, something called "deconstructionism" was in vogue in several of the humanities departments, notably the English department. An outgrowth of postmodern ideas associated with Michel Foucault, deconstructionism was based on the notion that there was no ultimate truth or meaning in any scholarly work or text that could be understood without first deconstructing the ideological biases (gender, racial, political, cultural, etc.) of the authors and the society in which they wrote. Equally important was the idea that human social and intellectual activity derived from forms of domination, where notions of power and oppression completely governed all relations and writing.

We in the sciences scoffed at what we interpreted to be the lack of intellectual standards in a movement that argued against objective truth and claimed that all knowledge was tainted by ideological biases. After all, stars emit frequencies of light that are not only independent of human biases but were also emitted before humans even evolved!

In 1996, the US mathematical physicist Alan Sokal published a famous spoof paper in the social science journal *Social Text*. The paper was entitled "Transgressing the Boundaries: Towards a Transformative Hermeneutics of Quantum Gravity." The paper made no sense scientifically, but his purpose at the time was to demonstrate how low the scholarly standards of postmodern scholarship in the humanities had become. Here is a typical quote from the paper:

> ...most recently, feminist and poststructuralist critiques have demystified the substantive content of mainstream Western scientific practice, revealing the ideology of domination concealed behind the façade of 'objectivity'...all 'reality', no less than social 'reality' is at bottom a social and linguistic construct;...the scientific community...cannot assert a privileged epistemological status with respect to counter-hegemonic narratives emanating from dissident or marginalized communities.

At the time, scientists laughed when reading this during coffee discussions when the spoof was announced, and they congratulated themselves with the statement: It can't happen here!

Fast-forward twenty-seven years. In 2023, the *Journal of Chemical Education* published an article entitled "A Special Topic Class in Chemistry on Feminism and Science as a Tool to Disrupt the Dysconcious Racism in STEM." The abstract and body of the paper contain phrases that are eerily familiar and equally nonsensical.

> This article presents an argument on the importance of teaching science with a feminist framework and defines it by acknowledging that all knowledge is historically situated and is influenced by social power and politics.... [The course will] explore the development and interrelationship between quantum mechanics, Marxist materialism, Afro-futurism/pessimism, and postcolonial nationalism. To problematize time as a linear social construct, the

Copenhagen interpretation of the collapse of wave-particle duality was utilized.

In the same year, the *Physical Review*, the preeminent journal of physics in the United States, published in its Physics Education section a paper entitled "Observing Whiteness in Introductory Physics: A Case Study." Here is some of the text:

> Within whiteness, the organization of social life is in terms of a center and margins that are based on dominance, control, and a transcendent figure that is consistently and structurally ascribed value over and above other figures.... Entangled with the above is the use of whiteboards as a primary pedagogical tool...they also play a role in reconstituting whiteness as social organization.... They collaborate with white organizational culture, where ideas and experiences gain value (become more central) when written down.

This incursion of postmodern gobbledygook was not limited to the physical sciences alone. Here is the title of a paper presented that same year at the Joint Mathematics Meeting, the largest math congress in the world: "Undergraduate Mathematics Education as a White Cisheteropatriarchal Space and Opportunities for Structural Disruptions to Advance Queer of Color Justice."

Lest it seem that these are merely fringe occurrences, laughed off by the bulk of the science community, it is worth noting that the prestigious National Academy of Sciences (NAS) awarded their highest prize in scientific communication to a physicist who wrote a paper entitled "Making Black Women Scientists Under White Empiricism: The Racialization of Epistemology in Physics." Among the many gems in the paper is the phrase: "Black women must, according to Einstein's principle of covariance, have an equal claim to objectivity regardless of their simultaneously experiencing intersecting axes of oppression."

The misuse of the physics principle of covariance here is virtually straight out of Sokal. Interestingly, later that year, Sokal himself published an in-depth analysis of this paper, tearing apart almost every word of it. Why, you might ask, would he waste his time on such nonsense? Because it turned out that that paper was number fifty-six in the Altmetric ranking of the top one hundred most discussed scholarly articles for 2020, cited thirty-seven times in the literature, including fourteen citations in the science education literature! Moreover, the only place where Sokal could publish his counter-critique was a new journal entitled the *Journal of Controversial Ideas*.

Attacks on science, reason, and scholarship are occurring in at least four major areas: the sociology of science, the infrastructure of science, academic freedom and free speech, and finally, the scientific and scholarly enterprise itself.

(1) Sociology: Science Is Sexist and Racist

In June 2020, after the murder by police in Minneapolis of George Floyd, in my own field of physics, the American Physical Society (APS)—the major US physics society comprising over fifty-five thousand members—endorsed a "strike for black lives," organized in part by a group of minority physicists and social justice advocates with the rather quixotic name, "Particles for Justice." This strike was meant to "shut down STEM" in academia. The APS closed down its office not to protest police violence or racism but to "commit to eradicating systemic racism and discrimination, especially in academia, and science," stating that "physics is not an exception" to the suffocating effects of racism in American life.

Leaving aside the question of whether shutting down STEM programs for a day is a good response to perceived social inequalities, the APS gave no direct empirical justification for their claims. The presumption of systemic racism was all that seemed necessary.

Science works hard to distinguish between correlation and causation in the analysis of empirical data. It is, therefore, particularly disheartening when this distinction vanishes in public pronouncements about science and in subsequent policy recommendations. For example, demographic data correlating gender, race, and sexual orientation with the numerical fraction of physicists are generally assumed to arise from underlying systemic racism without an explicit causal connection being demonstrated.

The APS decided to act on its perception of systemic discrimination in two different ways. The first, involving affirmative action programs, has become increasingly common among scientific societies and universities. Programs created uniquely for women and selected minorities have begun to proliferate. The APS also wants to change the public image of scientists. With rare exceptions, no longer are white men shown in public photographs representing the field.

In December 2020, the APS sent out a letter to their entire membership arguing that a recent Presidential Executive Order 13950 on Combating Race and Sex Stereotyping was "in direct opposition to the core values of the American Physical Society" and that the order needed to be rescinded in order to "strengthen America's scientific enterprise." The executive order quoted Martin Luther King, stating that in government-supported scientific institutions, people should "not be judged by the color of their skin, but the content of their character." The order further argued that materials from places like Argonne National Laboratory that equate "color blindness" and "meritocracy" with "actions of bias" or from Sandia National Laboratories that state that an emphasis on "rationality over emotionality" is a characteristic of "white male[s]" were inappropriate training materials for government-supported science institutions.

The presumption of racism as endemic to the sociology of academia is not restricted to physics, of course. At Princeton, again in 2020, and again in response to the George Floyd murder, more than one hundred faculty members, including more than forty in the sciences and engineering, wrote an open letter to the president with proposals to "disrupt the institutional hierarchies perpetuating inequity and harm." This included the creation of a policing committee that would "oversee the investigation and discipline of racist behaviors, incidents, research, and publication on the part of faculty," with "racism" to be defined by another faculty committee, and requiring every department, including math, physics, astronomy, and other sciences, to establish a senior thesis prize for research that somehow "is actively anti-racist or expands our sense of how race is constructed in our society."

The presumption of inherent racism and sexism in science is sufficiently widespread that even a renowned scientific institution like the University of California San Francisco has created a course for all basic science first-year PhD students entitled "Racism in Science," which explores "the relationship between notions of race and science and how scientific research has been informed by and perpetuates anti-Black racism."

The connection between demographics and presumed racism and sexism in science also led the NAS, purported to be the most distinguished honorific organization in the sciences in the United States, to change its own policies shortly after the first female president of the academy, Marcia McNutt, took office. They now assign "slots based on the diversity of the lists of nominees" that have been forwarded. Classes that have a more diverse list get more slots. The next year, the council reviews how these slots have been filled and adjusts the distribution based on performance. As the home secretary put it, "If [the selectors] used them to pick a bunch of white guys from Harvard, they get penalized." Not surprisingly, one-half of the members of the incoming 2021 class were women. As a result, this academy, designed to be the pinnacle of scientific meritocracy, now filters its membership based on racial and gender identity.

(2) Administration, Staffing, and Students in Science and Scholarship

After the George Floyd incident, all of the major US federal funding sources for science and technology, the National Institutes of Health (NIH), the National Aeronautics and Space Administration (NASA), the Department of Energy (DOE), and the National Science Foundation (NSF), committed themselves to enacting policies of so-called anti-racism and gender equity in their respective funding decisions.

While the leadership of each of these agencies produced language adhering to the accepted critical social justice narrative of systemic bias, perhaps the most extreme statement was made by Francis Collins, who was the director of by far the largest US science funding agency, the NIH. In a speech, Collins apologized for the existing "structural racism in biomedical research."

These agency policies diffused down to universities in two different ways. First, through the use of ideological requirements imposed on grantees for federal funding of research, and secondly, through the exponential rise of diversity, equity, and inclusion (DEI) bureaucracies at most US institutions of higher learning.

All of the major US funding agencies included boilerplate language about their commitment to both DEI and anti-racism. The DOE required that a section of all research proposals include a section that "describes the activities and strategies that investigators and research personnel will incorporate to promote diversity, equity, inclusion, and accessibility in their research projects." NASA required research proposals to elaborate how proposals will "further NASA's inclusion goals." The NIH major BRAIN initiative required applicants to submit a "Plan for Enhancing Diverse Perspectives," where "perspectives" is defined to mean people, not ideas, and other programs required DEI plans and reporting on strategies for affirmative action recruiting.

As I have described, this landscape for federal funding of science, in the US at least, has been completely altered following the election of Donald Trump. The DOE has already ended its requirements for DEI sections in research proposals, for example. This is an evolving situation, but it is nevertheless remarkable how quickly this long established set of DEI requirements for federal funding has now been removed.

Nevertheless, as went government funding, so went university administrations. The fastest growing area of academic administration at US universities has involved DEI officers and offices. For example, the University of California Berkeley's Division of Equity and Inclusion has 152 staffers and a $36 million budget, and the University of Michigan paid more than $30 million last year for 241 DEI staff. These offices do not passively monitor diversity initiatives but actively insert themselves in the governance of everything from hiring to speech and even the academic curriculum, as a number of authors in this book describe.

Starting in 2019 or so, one began to see additional criteria in advertisements for faculty openings. As a recent Cornell ad put it, "Also required is a statement of diversity, equity and inclusion describing the applicant's efforts and aspirations to promote equity, inclusion and diversity through teaching, research, and service."

In 2021, I examined twenty-five advertisements for new faculty in my own field of physics, from research institutions like Caltech to liberal arts colleges like Bryn Mawr, and even in areas as esoteric as quantum engineering and theoretical astrophysics, and twenty-four of them required applicants to demonstrate an explicit, active commitment to the DEI agenda.

Signaling a commitment to racial diversity and inclusion may seem like an innocuous requirement, and it would be if that were all that was required. However, merely stressing that one encourages and supports education and research independent of race, gender, sexual orientation, religion, and so on is deemed inconsistent with the required *active* anti-racism needed to pass the DEI tests. Instead, as several authors here describe in detail, prospective faculty in fields like, say, theoretical particle physics have to demonstrate how they and their research have actively promoted diversity goals throughout their careers, even if those careers have required them to spend most of their time thinking about eleven-dimensional universes.

In many institutions, DEI offices and officers have filtered applications before they got to the faculty. In 2018–19, for example, the life-sciences department at Berkeley reported that 76 percent of applicants were rejected based on their diversity statements without looking at their research records. This is not lost on students and postdoctoral researchers. As one colleague of mine at a major research institution wrote, "I have a student on the market this year, agonizing more on the diversity statement than on their research proposal."

John Sailer at the National Association of Scholars got access under a Freedom of Information Act inquiry to DEI assessments at Ohio State University. Here are some examples:

- A committee searching for a professor of freshwater biology used a rubric that cited several "problematic approaches" for which a

candidate can receive a zero score—for example, if he "solely acknowledges that racism, classism, etc. are issues in the academy."
- For a search in astrophysics, "the DEI statement was given equal weight to the research and teaching statements."
- In a search for a professor of chemistry, the report notes that one candidate's "experiences as a queer, neurodivergent Latinx woman in STEM has provided her with an important motivation to expand DEI efforts beyond simply representation and instead toward social justice."

Once again, it is worth emphasizing that all of these requirements stem from the assumption of systemic racism and/or sexism in departments at universities and scientific and scholarly institutions. As Efimov et al. argue in their masterful review, published in *Frontiers of Research Metrics and Analytics*, Volume 9 (2024), "*the demand to provide an inclusion plan without evidence that there is a need for one is compelled speech and an intrusion of ideology into the conduct of science.*"

Besides the imposition of ideology, DEI bureaucracies also tend to enforce identity constraints on hiring, a.k.a. affirmative action. This conflates equality of opportunity with equality of outcome. People vary, and those variations supersede concerns about race or gender. If the incoming applicant pool doesn't reflect the background demographics of society, then discrimination based on race or gender is fundamentally unfair to both applicants who are excluded on these grounds and to successful applicants, whose subsequent careers will, in the eyes of their colleagues, always be marked with an asterisk.

Canada has fiercely taken up the notion that appointments in the sciences must match background demographic indices to the nearest decimal point. The most prestigious government-supported research positions are called Canada Research Chairs (CRCs). They were originally created to help recruit back to the country ex-patriot Canadian scientists who, like me, had left the country because the opportunities to pursue their studies within Canada had earlier been limited.

Recently new quotas were announced so CRCs would match the background population demographics. By 2029, 50.9 percent of CRCs must be "women and gender minorities," 22 percent must be racialized minorities, 7.5 percent must be people with disabilities, and 4.9 percent must be Indigenous. It doesn't take a rocket scientist to realize that with such discriminatory hiring based on identity, merit will no longer be the chief determinant for successful candidates.

In Canada, discrimination based on sex or race in academic hiring is legal as long as it can be argued to right past wrongs. And as a result, current university recruitment efforts—including in fields that have little to do with identity, such as quantum computing and computational biology—now flat-out exclude white males who don't self-identify as disabled or LGBT+. Canada's leading university, the University of Toronto, recently announced that *all* recruiting for Canada Research Chairs will be restricted to the "designated" groups in regard to engineering, dentistry, medicine, and various other disciplines. It is not unique. The University of Guelph advertised a CRC in experimental physics, searching for a world-class researcher, as long as they were either "a woman, person with disabilities, Indigenous person, or a racialized person." Queen's University advertised for a CRC in geotechnical studies open to applicants of any race, so long as candidates "self-identify as women."

Social justice adherence is being explicitly presented as a prerequisite for hiring at all levels. McGill University, for example, advertised a tenure track position with the following proviso: "*a demonstrated relevance of the candidate's work to addressing anti-Black racism or systemic inequities…will be regarded as an important asset.*" The hiring faculty was a computer science department.

Quotas and discrimination based on sex or gender are not unique to North America. In Australia, the National Medical and Health Research Council announced they would award half of their research grants for mid-career and senior-level faculty to women and non-binary applicants, in spite of the fact that only 20 percent of the grant proposals come from women, and the number from non-binary applicants is so small as to be unreported. In the Netherlands, Eindhoven University of Technology, which specializes in engineering, announced a radical new policy in 2019, in which for the first six months of the recruiting season for permanent positions, only female applicants would be considered.

All of these efforts continue in spite of little evidence that any underlying root causes of demographic disparities in STEM are actually addressed by them. Affirmative action hiring at the university faculty level does little to alleviate what may be deep systemic disparities in education or economic opportunities in the public at large. Moreover, it is not even clear that they work.

There is, in fact, a potentially more glaring demographic disparity at US universities. The undergraduate population at US universities is now overwhelmingly female, with a ratio of almost 60 percent female to 40 percent male. A significant demographic imbalance favoring females persists in graduate school at both the master's level and PhD level. But this disparity receives very little attention, and as far as I know, there are no systemic efforts at a national level to address it.

The religious fervor that accompanies what has become an almost sacred claim that science is systemically racist and sexist—a claim that cannot be questioned without the risk of being canceled as a heretic—is perhaps what is most disheartening about the current movement to impose identity-based restrictions on science. While in healthy times, science fosters debate and discussion, any such potential provocation today is taken as an attack that needs to be quashed. To give a sense of the depth of feeling here, I reproduce the content of an email thread from a colleague who questioned a recipient of a recent $10 million grant to get more women into computer science. His questions may have been intended to provoke. But it is the answers to the questions that reflect, to me, the real problem:

> Dear XXX:
>
> As a (sort of) social scientist, may I suggest a couple of questions…: 1. How do we know the correct percentage of women vs. men in CS (or any other subject)? …2. Does racial diversity imply idea diversity, as many believe? … Anyway, sorry to intrude….

The Response:

> Given you're neither a computer scientist, Black person, Black woman, or a woman at all…I'm going to be brief because your extremely offensive email didn't warrant the time I took to respond…. I know that, as a white man in academia and America, you believe that what you think is all that matters…. However, I can assure you that you and your ideas will never be centered or a priority in anything I do, believe, or teach….

In the worst case, "offensive" questions like these can lead to censure or dismissal. So much for informed debate and discussion, which is the hallmark of good science and good scholarship.

In its most extreme form, quotas for senior positions can be sufficiently embarrassing as to backfire because they put a stigma on individuals who eventually are hired. Consider, for example, MIT, which, as an engineering school, has a faculty that has been typically male-dominated. However, the head of the corporation, the president, the provost, the director of research, the chancellor, the dean of science, and five of the eight chairs of engineering departments are all women. In 2015, a study published by the National Academy of Sciences (the same academy that now effectively introduces quotas in its membership) found a two-to-one preference for hiring women in STEM positions at that time. Yet, in spite of this, governments and institutions still feel it is necessary to impose quotas on hiring.

(3) Academic Freedom and Free Speech

Perhaps no other area currently inhibits scholarship and education more than the recent spate of attacks on faculty, staff, and students, not because of proscribed behaviors, but rather proscribed ideas. This has produced an atmosphere of fear and recrimination, and it needs to stop.

There are a host of egregious examples of late. To give a general sense of the variety, however, I will explicitly list a few here, including two authors in this volume:

- *Stephen Hsu*: Hsu, a physicist and former colleague of mine at Harvard, became vice president for research at Michigan State University. During the strike for black lives initiated by Particles for Justice that I alluded to earlier, activists at MSU took advantage of that day to launch a protest campaign against Hsu. His crime? His research involved computational genomics to study how genetics might be related to cognitive ability—something which, to the protesters, smacked of eugenics. He was also accused of supporting psychology research at MSU (published in the *Proceedings of the National Academy of Sciences*) on the national statistics of police shootings that didn't support the ongoing narrative regarding racial bias. Within a week, the university president requested Hsu's resignation. (Note: Shortly after, the authors of the study that the NAS had published asked the journal or retract their article, not because of flaws in the analysis, but rather their concerns over media "misuse" of their results.)
- *Frances Widdowson*: In 2019, Frances Widdowson, then a professor at Mount Royal University in Alberta, attended a lecture by Dr. Gregory Cajete, a professor of Native American studies and education at the University of New Mexico. In this talk, Cajete was discussing indigenous "star knowledge" and its importance for educational institutions. Widdowson asked about the relevance of this for introductory astronomy classes. In October 2020, this became the subject of a harassment complaint. At an arts faculty council meeting in March 2021, a colleague asserted that a number of indigenous scholars had left MRU, and he wondered how faculty members could make the university a "safe place" for them. Widdowson directed a question to the dean of arts as to whether saying that indigenous science was not science made the university "unsafe." This resulted in Widdowson being castigated by a number of colleagues and another unsuccessful harassment complaint filed against her. MRU listed the incident as one of the "examples" for why she should be fired: "Your conduct at a meeting in the spring of 2021 was so disruptive that it was a significant contributing

factor to the development of new procedures at arts faculty council meetings as well as the disabling of the chat function during these meetings." Widdowson was subsequently fired from her tenured position in December 2021.

- *John Kormendy*: Eminent astronomer John Kormendy retracted an article intended for publication in the *Proceedings of the National Academy of Sciences*. His article focused on statistical results relating to the evaluation of the "future impact" of astronomers' research as a means to "inform decisions on resource allocation such as job hires and tenure decisions." The response was speedy and vicious. Online critics attacked Kormendy's use of quantitative metrics, which they felt cast doubt on the application of diversity criteria in personnel decisions. The online protest was too much for Kormendy, who retracted his article and released an abject apology. Publication of a book he wrote on the same subject was then stopped, and all copies destroyed.

- *Elizabeth Weiss*: In 2022, Weiss sued San José State after being removed as curator of the university's collection of remains and locked out of the research facility because she is opposed to the repatriation of ancient bones, based on arguments from indigenous religious creation myths, instead of keeping them for scientific study. Later, in 2023, she and four colleagues had a paper removed from a session at the American Anthropological Association because it discussed how biological sex remains a necessary analytic category in anthropology. The AAA stated it would cause "harm to members represented by the Trans and LGBTQI of the anthropological community as well as the community at large." To suggest that a spirited discussion of the importance of sex in the context of archeology threatened the scientific integrity of the program is to misunderstand the nature of science.

LAWRENCE M. KRAUSS

(4) The Scientific and Scholarly Enterprise

Science is based on explorations that study reality and on developing models that predict the future accurately. Often, science unveils inconvenient truths that cause us to rethink previous assumptions. To me, that is one of the greatest benefits of science for humanity and something that makes it endlessly beautiful.

While discovering one's previous assumptions might be wrong is a remarkably liberating experience, that experience is now too often viewed in academic institutions and scientific journals as too liberating.

Consider, for example, the Royal Society of Chemistry. Last year, a directive went out to editors of the journals of the Royal Society of Chemistry with a new set of guidelines. The letter read, in part:

> A set of guidelines has been produced by Royal Society of Chemistry staff to help us minimise] the risk of publishing inappropriate or otherwise offensive content. Offence is a subjective matter and sensitivity to it spans a considerable range; however, we bear in mind that it is the perception of the recipient that we should consider, regardless of the author's intention…. Please consider whether or not any content (words, depictions or imagery) might have the potential to cause offence, referring to the guidelines as needed.

One might be willing to forgive this remarkable abuse of scientific editorial privilege if an offense was particularly harsh or inappropriate. However, the document later defined offensive content as: "Any content that could reasonably offend someone on the basis of their age, gender, race, sexual orientation, religious or political beliefs, marital or parental status, physical features, national origin, social status or disability." It is hard to imagine anything left!

Two of what have been the most prestigious scientific journals in print, *Science* and *Nature*, have recently jumped on the critical social justice bandwagon in their editorials and features, often taking it as self-evident that science is both racist and sexist. One of the *Nature* journals, *Nature Human Behaviour*, recently allowed their concerns about potential harm to minorities to impact their decision to publish scientific articles. In an editorial entitled, "Science Must Respect the Dignity and Rights of All Humans," the editors established a new editorial policy: "The journal will reject articles that might potentially harm (even inadvertently) those individuals or groups most vulnerable to 'racism, sexism, ableism, or homophobia.'" In short, if the results of scientific investigation, while true, might cause offense, they must not be published!

An equally worrisome change involved the conviction of federal science agencies that supporting cutting-edge science is in some sense ancillary to their social justice mission—a conviction that is now being addressed by the new administration. Before the Presidential executive order requiring this mandate to be removed, grantees were required to "describe the activities and strategies of the applicant to promote equity and inclusion *as an intrinsic element to advancing scientific excellence* [italics mine]." The presumption here was that promoting equity and inclusion—a potentially laudable political objective—is intrinsic to good science. However much one might like that to be the case, there is no evidence that it is.

The NSF proudly proclaimed their "Big Ideas" programs that "identify areas for future investment at the frontiers of science and engineering… to position our Nation at the cutting edge of global science and engineering leadership." Among the cutting-edge programs is the most awkwardly named NSF program I know: Inclusion across the Nation of Communities of Learners of Underrepresented Discoverers in Engineering and Science, so named to allow the acronym INCLUDES. Over $40 million has already been devoted "to catalyze the STEM enterprise to work collaboratively for inclusive change…to advance equity and systemic change." One can hope that with the new administration's orders, this program will disappear.

A new program called New Frontiers in Research in Canada recently sponsored an academic initiative entitled "Decolonizing Light," which studies "the reproduction of colonialism in and through physics" and "how colonial scientific knowledge authority was and is still reproduced in the context of light." The program makes it clear that physics is fundamentally bad: "Physics is considered 'hard' and objective science, disconnected from social life and geopolitical history. This narrative both constitutes and reproduces inequality."

The Royal College of Physicians and Surgeons of Canada (RCPSC) developed a new project called CanMEDS 2025 for medical education. While most people who go to a doctor would probably prefer if the chief emphasis of that doctor's education involved developing their medical expertise, not so says the RCPSC:

> CanMEDS 2025 affords us the opportunity to think critically and propose a vision for the practice of medicine which is rooted in social justice, anti-racism, anti-oppression, and cultural safety, promoting a broader cultural shift which is necessary for the profession. A new model of CanMEDS would seek to centre values such as anti-oppression, anti-racism and social justice, rather than medical expertise.

Fields as unrelated to the vicissitudes of modern political life as mathematics are not safe from this kind of recalibration. A state-proposed new "Pathway to Equitable Math Instruction Dismantling Racism in Mathematics Education" in California, funded in part by the Bill and Melinda Gates Foundation, the Lawrence Hall of Science, and UC Berkeley, includes the following remarkable claims:

- The concept of mathematics being purely objective is unequivocally false.
- White supremacy culture shows up when:
- There is a greater focus on getting the "right" answer....
- Students are required to "show their work" in standardized ways.

An emerging field in astronomy is called astrobiology, promoted initially by NASA director Dan Goldin in the 1990s. The goal is to try and understand whether conditions exist elsewhere in the universe for the development of life and what kinds of life might exist in the universe. It has become particularly susceptible to the incursion of postmodern ideas. A recent *Scientific American* (another once-great journal now fixated on CSJ ideas) article appeared last year entitled "Cultural Bias Distorts the Search for Alien Life," arguing that "decolonizing" the search for extraterrestrial intelligence (SETI) could boost its chances of success. An astrobiology meeting at Penn State decided to forbid using the word "intelligence" in SETI as it is a "white" construct.

The effort to put "indigenous science" on the same par as Western science in schools achieved its most significant advance in New Zealand in 2021. In that year, the Ministry of Education implemented a policy that Maori "ways of knowing" would have equal standing with Western science in science classes. Two members of a group of scientists who publicly questioned the scientific rationale for this policy were investigated for removal by the Royal Society of New Zealand, and two members were removed from teaching evolution classes at the University of Auckland. There is no doubt that indigenous peoples practiced science at some level. Determining by experimentation which fruits are edible, what plants might prove beneficial for treating wounds, and so on, were of vital importance for their health and well-being. In this sense, there is nothing that separates such indigenous science from any other kind of science. As the comic songwriter Tim Minchin said, "You know what they call alternative medicine that works? Medicine."

Indigenous myths are also treated as if they are to be respected as factual in places such as the American Museum of Natural History. In its Northwest Coast Hall, which reopened in 2022, there is a case with the warning label: "This display case contains items used in the practices of traditional Tlingit doctor. Some people may wish to avoid this area, as Tlingit tradition holds that such belongings contain powerful spirits."

Finally, one more emotionally charged change in the scientific enterprise has to do with who has a right to publish scientific papers. While it is conventional that all authors who make significant contributions to a work intended for publication should be listed as authors, the American Astronomical Society has recently stated that "sanctioned" behavior (unrelated to the research in question) can be grounds for removing authorship even though such a change violated the AAS's own code of ethics, as a former president of the society pointed out in a published note.

The imposition of political or ideological "moral" constraints on the scientific enterprise has a long and checkered history. Historical examples include Lysenko and the Stalinist campaign against genetics (described in various contexts later in this volume) and the Nazi condemnation of "Jewish science." As Niall Ferguson points out here, these provide fair warning.

Science is a process that has resulted in remarkable progress for the human species for one reason and one reason alone: it works. More generally, reason and scholarship enhance the human experience. Science and reason work because they are based on the freedom to question, the freedom to debate, the primacy of empirical evidence, and the requirement of a willingness to change one's mind in the face of such evidence. Imposing any other set of moral rules that curtail any of these features diminishes science, scholarship, and the human experience.

Many of the current threats to science and reason that I have discussed here, and that will be discussed in the rest of this book, may stem from an earnest desire to do good. But good intentions are just that, and the results can be devastating. With this in mind, it is sometimes useful to retain one's historical perspective. I find comfort in an ancient example that I learned of from one of the contributors to this book, Anna Krylov:

The Dutch scientist Antonie van Leeuwenhoek's pioneering development of microscopy led to his 1677 discovery of spermatozoa in semen. He was concerned that communicating his new results might cause offense. As he put it when communicating his results to the president of the Royal Society for publication in its *Philosophical Transactions*, "If your Lordship should consider these observations may disgust or scandalize the learned, I earnestly beg your Lordship to regard them as private and to publish them or destroy them as your Lordship sees fit." Fortunately for the progress of biology, his lordship wasn't concerned about causing offense, and van Leeuwenhoek's results were published. For the good of science, scholarship, and society, as the numerous scholars who have contributed to this volume reiterate through their essays, this is the example we need to emulate today.

It is not lost on this author, nor I expect on many of my colleagues, who wrote their contributions as did I before the election of the current US administration, that another very different war on science, and more generally, on other enlightenment institutions, appears to be emerging. Initially motivated, at least in part, to counter some of the trends we discuss in this book, this external attack has appeared to morph into the conviction that institutions that are not directly tied to political goals are either dispensable, or a direct threat to those goals and must be brought under control if possible. This means that much of the current US infrastructure supporting advanced scientific research may be under siege. Paradoxically it also means that authoritarian countermeasures are sometimes being introduced on the basis of constraining various incursions of ideology into science, that nevertheless appear to threaten free speech. While it is vitally important to expose and resist any such efforts in order for science and reason to progress, these new perceived threats, unlike the ones described in this book, are largely external. They need to be fought at the ballot box and in legislatures and the courts. The War on Science being addressed here needs to be fought on another front. It involves largely internal threats and is thus largely a war for the hearts and minds of the academic community itself. To resist such an attack, we need to expose its nature and encourage the scholarly community to stand up for science and reason, and for the public to support and encourage such efforts. That is the purpose of this anthology

SECTION 1
Free Speech, Victimhood, and Ideology

The imposition of ideological constraints in academia, in general, and science, in particular, has a long and ignoble history. In this section, we hear from a diverse group of authors describing historical precedents to our current situation and disturbing current examples of both the dangers of such ideological corruption and its various forms.

The section begins with a wide-ranging essay by Richard Dawkins describing the nature of science and some classic examples of ideology negatively impacting science and society in the former Soviet Union under the combined tyranny of the biologist Lysenko and Stalin, and then more recent examples involving modern biology and gender. His arguments presage some of the specific examples discussed by others later in this book.

Next, Alan Sokal, whose famous spoof paper in the journal *Social Text* in 1996 raised concerns about incursions of postmodernist thought into social science, writes about more modern incursions of ideology into science, recalling the wisdom of John Stuart Mill.

The historian Niall Ferguson gives us a chilling historical reminder of how easily academia can be perverted by ideology.

Nicholas Christakis, whose own work spans social science, biology, statistics, and public health, describes, in extracts from an interview he did with Sam Harris, both his general deeply considered views about free expression at universities as well as his own less-than-optimal experience at Yale.

This is followed by Gad Saad, who argues, from the point of view of an evolutionary psychologist, that universities are not merely suffering from the effects of ideology being imposed on scholarship but that they are the source of that ideology.

Chemist Anna Krylov and her writing partner Jay Tanzman have explored in a number of written pieces various aspects of how ideological fixations harm academia. In the essay, they discuss the growing incursion of censorship in the scientific enterprise.

The culture of victimhood, which now appears to be ubiquitous in society, and especially in academic environments, is critiqued by philosopher Maarten Boudry.

Finally, mathematician Abigail Thompson discusses the increasing chasm between the university of our idealizations and the university of today's realities.

Scientific Truth Stands Above Human Feelings and Politics

RICHARD DAWKINS

"There once was a man who said 'God
Must think it exceedingly odd
If he finds that this tree
Continues to be
When there's no one about in the Quad'"

Ronald Knox, satirizing Bishop Berkeley

If intelligent extraterrestrial beings[1] ever visit us, what common ground shall we find for conversation?[2] Science, of course. Overwhelmingly science, very probably nothing but science and mathematics. Our other preoccupations will be too alien to them, or too parochial, to arouse their interest. And vice versa. The aliens will revere their equivalents of Newton and Einstein, of Planck and Heisenberg. They'll have the Pythagorean theorem. They'll have computed (in their own notation, of course) π and the other great constants of mathematics and physics. They'll have the atomic theory. They'll recognize the same list of elements as we do and will group them into their equivalent of our periodic table. It's less obvious, but I stick my neck out and maintain that they'll also revere their Darwin, for I have argued[3] that some version of evolution by natural selection is the only way intelligence can come into being anywhere in the universe.

Let's bend over backward to concede that their biological nature may be more strange than we can imagine. They may be clouds of gas, pulsating amorphous lumps of jelly, or swarms of distributed mentality—rather than solid bodies such as we animals would recognize as living. But they'll still be products of a form of Darwinian evolution (or artifacts ultimately made by products of Darwinian evolution). And their science will be recognizable as (or will include, though probably surpassing) the science of Newton, Maxwell, Einstein, Schrödinger, Mendeleev, and others of our scientific pantheon. If they didn't have at least that level of physics and mathematics, they wouldn't be capable of getting here. Over the entrance to Plato's academy was inscribed, "Let no one ignorant of geometry enter here." A rather deplorable slogan, a snobbish barrier to learning, an education filter. But there is a cosmic filter and a slogan that needs no displaying at the threshold of any planet. "No one ignorant of science and mathematics can enter our gravity well because they'd be incapable of climbing out of their own."

Would our alien visitors revere their own Freud or Marx, to name two Terrans sometimes touted as Darwin's equals? To say the least, you'd have an uphill struggle arguing that they would. As for honoring their own Foucault or Derrida, their own Judith Butler or "Ibram X Kendi," the idea is a mirthless joke. Parochial ephemera such as "systemic racism," "decolonizing the curriculum," or "cultural appropriation" will be beneath their notice: as trivial, as meaningless, as futile as the proverbial angels pirouetting on a pinhead.

All this is just to say that science is, or incrementally approaches, universal truth *sub specie aeternitatis*, truth for all time and all space. Unlike gender studies, media studies, women's studies, black studies, white studies, Mickey Mouse studies, science is not parochial or ephemeral, not limited to one dot in the universe, one species, or one meager span of history. Moreover, science advances as the centuries (decades, years, weeks) go by in ways that cannot be said of theology, philosophy, sociology, or, I think, any other academic discipline. Even the glories of Shakespeare, Beethoven, and Michelangelo, profound though they may be, soul-enriching as they are, to put it mildly, are like a fine wine that doesn't travel and doesn't age well, at least on the cosmic scale of time and space.

LIGO (detecting gravitational waves smaller than the width of a proton), cracking the DNA code, the prediction and then detection of the Higgs boson, vaccination, the soft landing on a comet, the mission to Pluto, the periodic table, the recognition that evolution by nonrandom survival is the process that made us, these are achievements for which every member of that extraordinary species *Homo sapiens* can stand proud. Not just in our knowledge of science, but also in the methods by which we acquire it. Those methods, too, are likely to be universal, not confined to our one planet. I shall not here revisit the great works of scientific epistemology and the scientific method. Instead, I'll come down to earth with the humble double-blind trial as an exemplar and embodiment of the spirit of science, in particular, science's respect for objective evidence and its suspicion of subjective or ideological bias.

Medical researchers, when testing a new drug, insist that neither the doctors handing out the pills nor the patients receiving them are allowed an inkling of which pills are the placebo and which are the drug under trial. If a patient knew that his pills were (or were not) the experimental drug, it could influence his health psychosomatically. If a doctor knew which pill she was handing out, her manner could betray it to the patient.[4] If a researcher could keep track of the way the results of the experiment were turning out, he might be tempted to terminate it prematurely when it seemed to be going his way. Even with no cheating, the outside world can never be totally persuaded by the experiment unless it is scrupulously controlled. The double-blind control trial is the gold standard, even if, in particular cases, you can't implement it.[5] It stands for the exclusion of feelings, subjective impressions, "lived experience," and the pernicious notion that "your truth is different from my truth, and everyone's truth is equally valid." The only fully reliable truth is the truth supported by objective, publicly inspectable evidence.

A young woman of my acquaintance gave up her social science course in despair and switched to another university where she could study medicine and science. The last straw was the social anthropology lecturer who said, "The beauty of our subject is that when two of us look at the same data we come to opposite conclusions." I once had an argument with a social anthropologist who said, approximately, "As a Western scientist you look at the moon and see a lump of rock obeying Newton's Laws, some 240,000 miles away. But if there are a tribal people who believe the moon is an old cooking pot tossed into the sky not far above the treetops, their belief is equally true. You believe what you believe because you are a Western scientist. But your truth is no more true than their truth."

No, a thousand times no. An anthropologist may need to respect how the tribe's beliefs about the moon fit the rest of their culture. They have a coherent worldview in which everything hangs together, including the belief that the moon is not far above the trees. But however mutually coherent and consistent their beliefs, they may still be simply false. The "moon is a cooking pot" theory is not "a different kind of truth," not just "a different way of knowing."[6] It's false. By all means, sympathize with different cultural beliefs; by all means, study them as an anthropologist or a historian. But where facts are concerned, there is only one way to determine their truth or falsehood, and that's to examine the evidence. The best way we know to evaluate evidence is the scientific method.[7] If a better method comes along, science will adopt it.

Science is the jewel in humanity's crown. We have every right to take pride in it. But pride as we uncover nature's truths should be tempered by humility in their contemplation. The facts of science were facts long before there were humans to discover them. And they will continue to be facts long after our species has gone extinct, long after any brains exist to remember us. The facts to which science aspires are true, not just here and now for us to see, but through all time and through every distant reach of the universe where there are not, nor ever will be, eyes to see them.

This essay discusses two examples of vain hubris in the face of the eternal and universal verities of science. Both elevate humanity in a kind of species-level solipsism, updated echoes of the philosophy satirized in my epigraph on Bishop Berkeley. First, a disgraceful Russian episode in which mere politics was elevated over scientific truth. In the Soviet Union in the 1930s and '40s, an ignorant science-denier managed to gain the ear of one supremely powerful man, with the result that Russian (and later Chinese) food production was set back half a century, with tragic consequences for millions of people. Marxist Leninist theory was dictatorially assumed to outrank genetics. And second, there's the postmodern hubris, which presumptuously, and falsely, dismisses science as a mere "social construct." I'll come to this in the second half of my essay—specifically, the ridiculous claim that we can override science and choose our own sex from a spectrum. The human conceit here is the idea that personal feelings can change reality.

I shall not deal with religious faith in detail—I have done so in *The God Delusion* and *Outgrowing God*—but I'll just mention in passing that it provides a third significant example of the elevation of human feelings over scientific truth. The exciting hypothesis that the universe was designed by a massive creative intelligence deserves to be considered on its scientific merits (and, I think, rejected). But more relevant to this essay is the contention that religious faith is validated by its power to give comfort, psychological succor, or consolation to depressed, bereaved, or frightened humans. I recently had a public discussion with a dear friend, previously a stalwart atheist, who had suddenly announced her conversion to Christianity. She gained the sympathy of the audience, including me, by her revelation that Christian prayer had pulled her out of an episode of deep, near-suicidal depression. When asked whether she really believed Christian doctrines, her reply was, "I choose to believe…" But, while sympathetic, I couldn't help pointing out that "I am comforted by believing X is true" does not logically imply "X is true." To paraphrase Steven Pinker, you may choose to believe that the tiger pursuing you is a rabbit, but your chosen belief, no matter how fervent, cannot alter the brute fact that the brute is a tiger. As you will soon painfully discover. But I'll say no more on this and will switch to the first of my two major examples of human hubris in the face of objective reality: the elevation of politics over scientific truth.

The Lysenko Disaster

> "'Do you remember writing in your diary, "Freedom is the freedom to say that two plus two make four"?'
> 'How many fingers am I holding up, Winston?'
> 'Four.'
> 'And if the Party says there are not four but five—then how many?'
> 'Four.'
> The word ended in a gasp of pain."
>
> George Orwell, *1984*

Hitler and Stalin both professed respect for science, but both were profoundly anti-scientific in their outlook. Hitler's theory of race was scientifically ridiculous as well as being unspeakably cruel, while Stalin threw his weight behind the spiteful chicanery of the charlatan Trofim Denisovitch Lysenko (1898–1976), with disastrous consequences for Soviet and later Chinese agriculture. In what follows, I shall be quoting from a book that Lysenko produced in 1949. Its bland and pedestrian title, *The Situation in Biological Science*, belies its sinister content. In English translation, it's the proceedings of a 1948 Moscow conference on Soviet agricultural science. Its pages are innocent of scientific evidence but heavy on ideology and political slogans, filled with praise for neo-Lamarckian "Michurinism" while denouncing "Mendelism-Morganism-Weismannism." Mendelism-Morganism-Weismannism was Lysenko's disparaging name for what we would call genetics. Gregor Mendel (1822–1884) showed that genes are particulate. Thomas Morgan (1866–1945) showed that they are linearly arranged in linkage groups, which physically correspond to the chromosomes. August Weismann (1934–1914) argued for the separation of the germline, which passes longitudinally down the generations, from the mortal soma, which does not. This view, the very antithesis of Lamarck's inheritance of acquired characteristics, is supported by all available evidence.

And "Michurinism"? Ivan Vladimirovitch Michurin (1855–1935) was a gifted gardener who successfully bred a large number of new fruit varieties. He practiced, and improved, artificial selection and hybridization techniques. His methods didn't violate any principles of Mendelian genetics, although the scientifically illiterate Michurin wouldn't have known one way or the other. It is not clear to me why Lysenko used Michurin as the patron saint of his peculiar neo-Lamarckian ideas. No theorist, he just had green fingers and loved his plants. We should probably think of "Michurinism" as nothing more than a sloganeering label, of the kind that ideologues wave about in the absence of any coherent meaning. Like "dialectical materialism" or "postmodernism." Perhaps Lysenko felt he couldn't get away with "Lysenkoism," and so grafted his theories to Michurin's name, much as Michurin himself might have grafted a new cultivar to an existing apple stock without really understanding what he was doing.

Lysenko was an ambitious ignoramus who rejected genes and chromosomes while knowing nothing about them. A British botanist, Sydney Harland (1891–1982), spent several hours with Lysenko and found him "completely ignorant of the elementary principles of genetics and plant physiology…to talk to Lysenko was like trying to explain the differential calculus to a man who did not know the twelve times table."[8] Not a smidgen of scientific evidence could be adduced to support Lysenko in his Lamarckian campaign against genetics, chromosomes, and Weismann, and his vilification of the great plant geneticist Nikolai Ivanovitch Vavilov (1887–1943). But he was Stalin's pet, and Stalin was not interested in evidence, only political ideology.

Lysenko owed his early ascendancy to his discovery (actually rediscovery, for the phenomenon was already known) that plants can change in response to environmental stress. For example, cold can induce them to flower, a process whose English name is vernalization. Where Lysenko went wrong—and it was a very big wrong—was his firm belief that such vernalization was inherited by future generations. Some kind of memory of their old experience was supposed to be passed on. This neo-Lamarckian theory directly contradicts everything we know—and everything that was known at the time—about genetics. Indeed, Lysenko vocally despised genetics and what he called the "chromosome theory." His nonsense would have died a natural death had it not conformed to Soviet ideology and the quasi-religious Marxist faith in the indefinite improvability and malleability of man, which goes with a general suspicion of genetics. If science contradicts political ideology, science has to be wrong. Weismannian inviolability of the germplasm must not be allowed to impede society's upward march to the glorious proletarian future. With such impeccable Marxist-Leninist credentials, Lysenkoism caught the attention of Joseph Stalin. Consequently, Lysenko became a star of Russian science. He was eventually granted supreme power over Soviet agronomy and agricultural policy—with disastrous results.

> Although it's impossible to say for sure, Trofim Lysenko probably killed more human beings than any individual scientist in history. Other dubious scientific achievements have cut thousands upon thousands of lives short: dynamite,

poison gas, atomic bombs. But Lysenko, a Soviet biologist, condemned perhaps millions of people to starvation through bogus agricultural research—and did so without hesitation. Only guns and gunpowder, the collective product of many researchers over several centuries, can match such carnage.[9]

The infection of Lysenkoism spread to China, again presumably because of its resonance with Marxist theology:

> As a part of Mao's Great Leap Forward, an effort to jump-start every sector of China's economy and industry, the mistakes of collectivization and Lysenkoism were mimicked to the letter. The failure of these efforts was predictable, but the scale of the destruction wrought in China surpassed that of even the Holocaust. Between 1959 and 1961, as many as 45 million Chinese people died as a result of starvation, malnutrition, illness, and injury after the country's farms were wrecked by Lysenko's crackpot ideas. This period of time would become known as The Great Chinese Famine.[10]

Lysenko was not just unscientific. He was overbearingly dictatorial with it. His hectoring approach to science, and his toadying to Stalin and Lenin, is evident in almost every sentence of his opening speech to the 1948 conference (said to have been edited in advance by Stalin himself):[11]

> Thus, Comrades, as regards the theoretical line in biology, Soviet biologists hold that the Michurin principles are the only scientific principles. The Weismannists and their followers, who deny the heritability of acquired characters, are not worth dwelling on at too great length. The future belongs to Michurin (Applause).
>
> VI Lenin and JV Stalin discovered IV Michurin and made his teaching the possession of the Soviet people. By their great paternal attention to his work they saved for biology

the remarkable Michurin teaching.... Our Academy must work to develop the Michurin teaching. In this it ought to follow the personal example of concern for the work of IV Michurin shown by our great teachers—VI Lenin and JV Stalin" (Loud applause).

Lysenko's speech contains ominous swipes at Russian scientists who still dared to follow what we would think of as genetics. Speeches by others sycophantically echoed Lysenko's wanton intrusion of political ideology into science, offering political slogans in place of scientific data:

> The only correct theory, the one capable of illuminating the path of practical agronomy, is the Michurin-Lysenko theory.... I cannot refrain from telling you of a curious incident which occurred in our college. Docent Platonov, who was originally recommended by professor Zhebrak, delivers his course to his students in a manner which caused the students to protest. Thereupon, the Dean of the faculty, professor Orgulnik, asked Platonov to come and see him and requested him to revise his position. In reply, the director received a lengthy missive from Platonov, in which he wrote in substance as follows: I shall give my course as I gave it hitherto, and just try to stop me!" (Speech by NG Belenky, p99).

Well done, Platonov! I hope he came to no harm. This anecdote chillingly foreshadows contemporary incidents where American students complain, often successfully, to their dean that a professor should be suspended because his lectures make them feel "unsafe."

> The roar of guns on the battlefields had not yet ended, the loyal sons of the Soviet people were still shedding their blood in defence of our country's honour, freedom and independence, and the toilers at home, while helping the front, were at the same time rebuilding wrecked towns and villages, factories and mills; but representatives of the

Mendel-Morgan trend in biology, like Professor Dubinin, were busy solving so extremely 'important' a problem as ascertaining the number and ratio of the fruit fly populations....| (Speech by PF Plesetsky).

The gratuitous sarcasm in that last sentence was born partly from disdain for orthodox geneticists (notably including Morgan himself), for whom fruit flies were a favorite experimental subject, and partly from the view that all genetic research should be directed towards agricultural productivity, not frivolous academic pursuits. *Pravda* had said something similar in 1927 when praising Lysenko earlier in his career. As paraphrased in Peter Pringle's book, here was

> a young "barefoot scientist", a practical researcher who had not attended a university, who did not toil in a laboratory away from the land studying "the hairy legs of flies" but "went to the root of things."[13]

Counting hairs on *Drosophila* legs was what geneticists outside Russia were well known to do. Seen through the eyes of an ignorant ideologue, what could be more academic, less likely to lift Soviet productivity to the utopian heights of a workers' paradise? Even Vavilov's renowned collection of plants and seeds from all over the world was seen as too academic. In 1931, a rival agronomist, Alexander Kol, publicly attacked Vavilov on the grounds that his work "on the centers of origin of plants has now replaced VI Lenin's revolutionary task of invigorating the Soviet farmland with new plants." The obsession with immediate practical results dominated Soviet agricultural science through Stalin's imperious demands for dramatic improvements in food production in an impossibly short time. Lysenko cashed in on this by making promises that couldn't possibly be fulfilled but which appealed to Stalin: "Bravo, Comrade Lysenko, Bravo!" (Joseph Stalin, Feb. 14, 1935).[14]

One after another, the delegates rose to cheer on Lysenko and denounce the "incorrect" heresy of Mendelism-Morganism-Weismanism. And so to Lysenko's concluding statement for the whole conference. One can almost hear his yells echoing through the hall, inviting thunderous applause and, of course, the inevitable standing ovation.

> Progressive biological science owes it to the geniuses of mankind, Lenin and Stalin, that the teaching of IV Michurin has been added to the treasure house of our knowledge, has become part of the gold fund of our science." (Applause) "Long live the Michurin teaching, which shows how to transform living nature for the benefit of the Soviet people!" (Applause). Long live the party of Lenin and Stalin, which discovered Michurin for the world" (Applause) "and created all the conditions for the progress of advanced materialist biology in our country." (Applause) "Glory to the great friend and protagonist of science, our leader and teacher, Comrade Stalin!" (All rise. Prolonged applause.)

This, I repeat, is not how one does science. Science proceeds through the dispassionate evaluation of evidence, not by yelling political slogans, slavishly proclaiming political loyalty, and denouncing heresies.

At the end of the book are a few sad little statements from heretics, groveling apologies reminiscent of Galileo's forced retraction under threat from the Inquisition.

> When I leave this session, the first thing I must do is to review not only my attitude towards the new, Michurinian science, but my entire earlier activity. I call upon my comrades to do likewise.... From tomorrow on I shall not only myself, in all my scientific activity, try to emancipate myself from the old reactionary Weismann-Morganian views, but shall try to reform and convince all my pupils and comrades (by SI Alikhanian).

> I admit that the position I held was wrong.... A sleepless night helped me to think over my behaviour.... Let the past which divided me from TD Lysenko (although not always, it is true) be forgotten. Believe me, that I take this step today

> as a party member, as a sincere member of our party—that is honestly (Applause). (Statement by PM Zhukovsky).

We must hope that these abject penitents got away with it and were spared the fate of many colleagues who fell afoul of Lysenko's vindictive zeal, mirroring, in a small way, that of his patron Stalin. The most famous victim was Nikolai Vavilov, already mentioned: a genuinely distinguished scientist, highly educated where Lysenko was little more than an ignorant peasant. After initially going out of his way to mentor Lysenko despite well-founded misgivings, Vavilov eventually stood up to him, and—an example to others—he paid the ultimate price. Arrested in 1941, his death sentence on a trumped-up charge of espionage was commuted to life imprisonment, and he died in prison in 1943 under conditions of terrible cruelty. He was posthumously rehabilitated in 1955, and a 1987 Russian stamp bears his portrait.

Today, the name of Trofim Lysenko is universally reviled, a synonym in biology for a charlatan. Those Western scientists such as J. D. Bernal, and even J. B. S. Haldane for a while, who took him seriously, stand ignominiously as a warning against placing mere human politics above scientific truth.[15] Lysenko's particular brand of pseudoscience is not a problem we have to deal with now, thank goodness. But the still disturbing aspect of the Lysenko affair is not the (now dead) bad science so much as the bossy, despotic manner with which it was put across, the elevation of politics over scientific truth, and the intolerance summarily displayed towards dissent.

This brings me to the second of my specimens of science denial: the presumptuous elevation of personal feelings, in particular, the hubristic assertion that the scientific reality of the sex of our birth can be overruled by human psychology and legislation. Behind it lurks "postmodernism,"[16] a school of thought that seems bereft of sense or precise definition, even in the minds of those who profess it. It includes a disdain for objective scientific truth, replacing it with subjective preference or political fiat.

Is Sex a Social Construct?

> "I accept the Universe."
> Margaret Fuller (attrib.)

> "Gad, she'd better."
> Thomas Carlyle (attrib.)

The physicist Alan Sokal, one of the most effective critics of the postmodernist attitude to science, has summarized it as the belief that

> So-called scientific knowledge does not in fact constitute objective knowledge of a reality external to ourselves, but is a mere social construction, on a par with myths and religions, which therefore have an equal claim to validity.

He illustrates it with the following remarkable quotation from the sociologist Harry Collins: "The natural world has a small or non-existent role in the construction of scientific knowledge." And there's this, equally astonishing, from another social scientist, Kenneth Gergen: "The validity of theoretical propositions in the sciences is in no way affected by factual evidence." Science, according to these social scientists, is no more than a social construct.

What is a social construct? The perfect example is money. A rectangle of green paper with "100 dollars" written on it is seen by most people as a desirable object, and a hundred times more desirable than a piece of green paper with "One dollar" written on it. A rectangle of green paper is not useful in the same sense as a rectangle of chocolate, bread, or integrated circuit board is useful. A dollar is useful only because society deems it to be exchangeable into things like chocolate or bread. "One hundred dollars" has no meaning outside the meaning that society chooses to accord it. Money, unlike science, really is a social construct.

The number of days in a year is real, determined by the rotation of Earth around the sun in relation to Earth's rotation on its own axis. But the calendar, with its division of the year into twelve months, is a social construct. When it became apparent that the Julian calendar had drifted out of phase with celestial reality, it was replaced by the Gregorian calendar, more in sync with Earth's orbital cycle. In 1582, a papal bull from Pope Gregory XII removed ten days from the calendar to bring it into phase. A few Catholic countries obeyed, but it wasn't until 1752 that the British Empire finally fell into line, and September 3rd jumped straight to 14th. That's the kind of thing you can decide to do to a social construct like a calendar. Human power extends that far. It may or may not be a myth that people protested in the so-called "Calendar Riots" because they thought the government had deprived them of eleven days of their lives. In any case, it's a good joke. The date of your death will be determined by implacable facts of nature, not by a mere social construct like the Gregorian versus the Julian calendar.

Another of the quotations Alan Sokal used to illustrate postmodernism was from Stanley Aronowitz: "Science legitimates itself by linking its discoveries with power, a connection which determines (not merely influences) what counts as reliable knowledge…." By power, he meant political power. This attitude exemplifies the hubristic aggrandizement of human power, arrogating to mere humans a quasi-divine dominion over nature itself. As with Lysenko's determination to overrule science if science conflicted with Marxist/Leninist theory, if reality is a mere social construct, society has the power to change reality. Like the joke about legally repealing the laws of thermodynamics so that we can have perpetual motion machines. I would argue that legally declaring a man to be a woman, just because he wants to be a woman, or vice versa, has much in common with the perpetual motion joke and the calendar riots joke. But unfortunately, it is no joke. It's the law in several countries.

There are not just males and females, so the claim goes. They are but the extremes of a spectrum. Where you place yourself in the spectrum, man or woman or somewhere in between, it's all a matter of personal choice. As with Lysenkoism, this involves a denial of genetic reality, and a Marxist-like faith in the malleability of nature. Where Lysenko thought wheat strains were malleable to vernalization, a politically powerful lobby today thinks your sex is not genetically determined but is malleable under your personal whim, sometimes backed up by law. If you feel you are a woman, you are a woman. Never mind if you have a Y-chromosome, testes, and a penis, no matter if you have breasts and ovaries, your male or female identity is something you get to decide for yourself, as easily as you might choose your political party or favorite football team.

It is a doctrine that has become highly influential. The American Medical Association in 2023 laid down some "Best practices for sex and gender diversity in medical education."[17] Medical students are to be taught that both sex and gender are "[socially] constructed." And "it is appropriate to affirm each individual's self-determination regarding both sex and gender labels." But are "male" and "female" really social constructs like money or like our twelve-month calendar? Is that really the considered view of the American Medical Association? Are we seriously training a generation of young doctors to think that the sex of a patient is a matter of individual choice, not objective anatomical and physiological reality? Let us at least hope that, when treating patients, they forget to ignore biological sex. In which case, what exactly is the point of the AMA's "Best Practice" document? It can only be virtue signaling, and rather silly virtue signaling at that. If I ever encounter a doctor who gives me a full medical examination and then asks me what sex I am, I'd be inclined to ask for a more thorough examination, if not another doctor. There are welcome auguries that the fashion is finally on the wane, at least in Britain.[18] It is to be hoped that, in America, it will soon go the way of McCarthyism. The otherwise loathsome President Trump made the upholding of biological maleness and femaleness the subject of an Executive Order, as one of his first actions after taking office (perhaps the only good thing he has ever done). I could imagine future lawsuits against surgeons who, in violation of the first clause of the Hippocratic oath, have cut off the breasts of girls below the age of consent for no better reason than a subsequently regretted claim to have been "assigned" the wrong sex. What, after all, does "below the age of consent" mean if not too young to make permanent, life-changing decisions?

Two and Only Two Sexes

> "No practical biologist interested in sexual reproduction would be led to work out the detailed consequences experienced by organisms having three or more sexes; yet what should he do if he wishes to understand why the sexes are, in fact, always two?"
>
> R. A. Fisher

How can I be so sure that there are only two sexes? Isn't it just a matter of opinion? Sir Ed Davey, leader of the British Liberal Democrat party, is of the opinion that women "quite clearly" can have a penis.[19] Words are our servants, not our masters, and he might claim that it's open to him to say, "I define a woman as anybody who self-identifies as a woman; therefore, a woman can have a penis." That is logically unassailable in the same way as "I define 'flat' to mean what you call 'round'; therefore, the world is flat."[20] I think it's clear that if we all descended to that level of sophistry, rational discourse would soon dig itself into the desert sand. The question is whether a particular redefinition is helpful for any constructive purpose. The flat Earth example clearly is not. It's simply perverse. How about the redefinition of woman (or man)? I shall make the case that the redefinition of a woman as capable of having a penis, if not downright perverse, is worse than unhelpful. I shall advocate instead what I shall call the universal biological definition (UBD) based on gamete size. Biologists use the UBD because it's the only definition that applies all the way across the animal and plant kingdoms and all the way through evolutionary history.

Gametes come in two radically different sizes, the phenomenon of anisogamy. Female gametes are very much larger than male gametes, and that is how biologists define female and male. A human egg contains at least ten thousand times as much matter as a human sperm. In ostriches, the discrepancy is obviously even greater, by a very large amount. The UBD is universal in the sense that it applies to all animals, vertebrate and invertebrate. All plants, too, unless you count algae as plants. Admittedly, not all individuals produce gametes at all, or throughout their life. Worker bees are sterile females. We call them female because they have the *potential* to produce macrogametes. Every worker would have turned out as a queen if she had been fed differently as a larva. That's "potential." A human male baby or fetus has the potential to produce microgametes, for all that he doesn't produce any yet. An older woman remains female, though she has ceased to produce ova.

The UBD has the virtue that, in addition to being universally applicable, it explains a diverse load of facts. And it's grounded in a body of powerful and widely illuminating theory. Here's how. It's an argument that should appeal to economists. In the words of R. A. Fisher (1930), "In organisms of all kinds, the young are launched upon their careers endowed with a certain amount of biological capital derived from their parents." When two gametes unite to make a zygote, they must, between them, provide this requisite quantity of expensive nourishment. In a fair and equitable world, you might expect the two parents to contribute equally, each bearing half the necessary costs. Such a system is known as isogamy. It doesn't exist in animals and plants but can be found in some microorganisms and algae. Clever mathematical modeling by various scientists, including Geoffrey Parker[21] of the University of Liverpool, indicates that, under plausible conditions, isogamy is unstable. It tends to be replaced, in evolutionary time, by its opposite, anisogamy: two different kinds of gamete, radically different from one another.

Here's a verbal version of the Parker type of mathematical model of anisogamy. Imagine you are an individual in an isogamous system. If you produce slightly larger than average isogametes, each zygote will be better endowed and therefore more likely to survive. On the other hand, since there's no such thing as a free lunch, you can only afford to make fewer zygotes. Conversely, by stinting on gamete size, you could contribute to making a larger number of zygotes, but they'd be poorly endowed and less likely to survive. Unless, that is, your smaller-than-average isogametes could somehow seek out larger-than-average isogametes to partner with. Parker and others developed plausible models whereby, over evolutionary time, half the individuals produce gametes in ever-decreasing numbers but ever-increasing size. These gametes eventually evolve into eggs. The other half goes in the other direction. They evolve smaller and smaller gametes in larger and larger numbers, which eventually become sperms.[22] You could, if you wish, say that the sperm producers exploit the egg producers. Or you could say that, being more valuable, eggs don't have to go out of their way to seek sperms. They can just sit and wait to be approached. Sperms, therefore, evolved miniature outboard motors (waving tails) with which to actively seek out eggs. Both types, the macrogamete producers and the microgamete producers, flourish in the presence of the other.

The fundamental economic inequality of anisogamy illuminates a large number of biological phenomena, thereby justifying my claim that the UBD does lots of explanatory work. If you define females as macrogamete producers and males as microgamete producers, you can immediately account for the following facts (see any recent textbook on ethology, sociobiology, behavioral ecology, or evolutionary psychology):

1. In mammals, it's the females that gestate the young and secrete milk.
2. In those bird species where only one sex incubates the eggs, or only one sex feeds the young, it is nearly always the females.
3. In those fish that bear live young, it is nearly always the females that bear them.
4. In those animals where one sex advertises to the other with bright colors, it is nearly always the males.
5. In those bird species where one sex sings elaborate or beautiful songs, it is always the male who does so.
6. In those animals where one sex fights over possession of the other, it is nearly always the males who fight.
7. In those animals where one sex has more promiscuous tendencies than the other, it is nearly always the males.
8. In those animals where one sex is fussier about avoiding miscegenation, it is usually the females.
9. In those animals where one sex tries to force the other into copulation, it is nearly always the males who do the forcing.
10. When one sex guards the other against copulation with others, it is nearly always the males that guard females.
11. In those animals where one sex is gathered into a harem, it is nearly always the females.

12. Polygyny is far more common than polyandry.
13. When one sex tends to die younger than the other, it is usually the males.
14. Where one sex is larger than the other, it is usually the males.

That's quite a lot of explanatory heavy lifting,[23] although, admittedly, not all the fourteen are independent of each other. In all cases, the key is economics: large gametes cost more than small ones. In various ways, this inequality plays out. Large gametes are more precious, more worth guarding, more worth fighting for, more worth protecting against wastage through mating with the wrong species or wrong individual. This predictive power is an additional virtue of the UBD, to set alongside its universality. Rather than "set alongside," it's more a case of *explaining* the universality. The "nearly always" or "usually" with which I cautioned the fourteen generalizations actually amount to "exceptions that prove the rule" when looked at more closely.

It is no idle whim, no mere personal preference, that leads biologists to define the sexes by the UBD. It is rooted deep in evolutionary history. The instability of isogamy, leading to extreme anisogamy, is what brought males and females into the world in the first place. Anisogamy has dominated reproduction, mating systems, and social systems for probably two billion years. All other ways to define the sexes fall afoul of numerous exceptions. Sex chromosomes come and go through evolutionary time. Profligate gamete-spewing into the sea gives over to paired-off copulation, and vice versa. Sex organs grow and shrink and grow again as the eons go by, or as we jump from phylum to phylum across the animal kingdom. Sometimes one sex cares for the young, sometimes the other, often both, often neither. Harem systems change places with faithful monogamy or rampant promiscuity. Psychological concomitants of sexuality change like the wind. Amid a rainbow of sexual habits, parental practices, and role reversals, the one thing that remains steadfastly constant is anisogamy. One sex produces gametes that are much smaller, and much more numerous, than the other. That is all ye know of sex differences and all ye need to know, as Keats might have only slightly exaggerated if he'd been an evolutionary biologist.

Here are some apparently anomalous examples that test (the true meaning of "prove" in the proverb) the rule. Unlike most mammals, spotted hyena females are larger than males and socially dominant over them. They have a hugely enlarged clitoris, scarcely distinguishable from a penis. They can get erections. They have false testes made of fatty tissue. The sight of apparently male hyenas giving birth has given rise to numerous myths of hermaphroditism. Given that so many roles and signals are reversed or ambiguous, how can we even know what we're talking about when we use the words "male" and "female" in describing the anomalies of hyenas? By the UBD, of course.

Many species of fish are livebearers. As listed above, it is usually the female who gets pregnant. But in seahorses, it's the male, in the sense that he has a pouch for holding the fertilized eggs, and he gives birth from his pouch. How do we know it's the male? Couldn't we *define* the female as the one that gets pregnant? We could, but then, "In seahorses, it's the female who gets pregnant" becomes a tautology that leads nowhere. The UBD leads on to interesting further questions. It is the male, as defined by gamete size, who gets pregnant. Now that's interesting. It's unusual. It generates questions for further research. What is it about seahorses that leads to the microgametic sex getting pregnant, in a departure from the usual pattern? I won't discuss possible answers here, but the question is obviously worth asking. It would make for an interesting PhD thesis.

Some worms and snails, and many plants, are simultaneous hermaphrodites. They are capable of producing both micro- and macro-gametes. Not a problem: the UBD is easily applied, and sex remains binary. The earthworm has organs appropriate to both sexes defined by gamete size. Enthusiasts for fluidity of "gender" love anemone fishes, also known as clown fishes. They, along with many other creatures, are sequential hermaphrodites. The largest, most dominant fish in a group of clownfish is female. If she dies, the dominant male becomes female. But what does that even mean? By what definition of male and female? On the UBD, it's very simple. When the dominant egg-producer dies, the largest sperm producer starts to produce eggs instead.

"Non-binary" advocates are very fond of "intersexes," and they often quote a figure of 1.7 percent as the frequency of intersexes in the human population. Even if that figure were true, it is still a remarkably low percentage on which to build an ideology. And it isn't true. It's wrong by a huge margin. This false figure originated with Anne Fausto-Sterling, an expert in "gender studies." As pointed out by Leonard Sax,[24] Fausto-Sterling inflated her figure by including such conditions as Klinefelter syndrome (XXY, with male genitalia), Turner syndrome (one X chromosome and no Y, with female genitalia), and late-onset congenital adrenal hyperplasia. These are not intersexes by any sensible definition. Sax estimated the true percentage of intersexes based on genital anatomy as 0.018, two orders of magnitude lower than Fausto-Sterling's bogus figure.

Fausto-Sterling had an agenda. Her idea of a "Sexual Continuum" had "profound implications…our current notions of masculinity and femininity are cultural conceits." She cheerfully offered a cultural conceit of her own. Some intersexual individuals might "become the most desirable of all possible mates able to pleasure their sexual partners in a variety of ways." But "continuum" implies that the frequency distribution, even if bimodal, should at least have noticeable intermediates between the two peaks. In a true continuum, it should be possible to plot a frequency bar chart on a normal sheet of paper. You can't do it. It wouldn't fit. If we represent the frequency of unambiguous male newborns by one of the twin towers of New York's World Trade Center, and the frequency of unambiguous female newborns by the other tower, the frequency of intersexes would be represented by a medium-sized molehill between them. Some continuum! It gets worse. My molehill was based on Sax's estimate of 0.018. If we take the UBD seriously, even that is an overestimate. The true figure is zero, not even a molehill, for nobody produces middle-sized gametes, intermediate between egg and sperm.

Relative gamete size is the only way in which the male/female distinction is defined universally across all animal phyla. All other ways to define maleness versus femaleness are bedeviled by numerous exceptions. Especially with respect to the sex chromosomes, where you can't even speak of a rule, let alone exceptions to it. In mammals, many insects, and other animals, sex is determined in development by the XX XY chromosome system, the male sex having unequal sex chromosomes. Birds and Lepidoptera have the same system but in the opposite direction and, therefore, presumably evolved independently. It's the females who have unequal chromosomes. How do we know? Couldn't you *define* males as the sex with unequal chromosomes? Well, you could, but then you'd have to say it's the male that lays the eggs, the females that fight over males, etc. You'd lose every one of the fourteen explanations I discussed earlier. Far better to stick with the UBD and say birds use sex chromosomes to determine sex, but it evolved independently of the mammal system. Birds are descended from dinosaur reptiles, and most modern reptiles don't have sex chromosomes at all. Often, sex is determined by incubation temperature. In some cases, higher temperatures favor males. In other cases, higher temperature favors females. In yet other cases, extremes of temperature, high or low, favor females, males developing at intermediate temperatures. Many snakes, some lizards, and a few terrapins use sex chromosomes, but which sex has unequal sex chromosomes varies. Amidst all this variation, the only reliable discriminator is gamete size.

The way the sexes are *defined* (the UBD, universal and without exception) is separate from the way an individual's sex is *determined* during development (variable and far from universal). How we, in practice, *recognize* the sex of an individual is yet a third question, distinct from the other two. In humans, one look at a newborn baby is nearly always enough to clinch it. Even if it occasionally isn't, the UBD remains unshaken.

Gender

A watered-down version of the ideology concedes that sex may be binary, but "gender" is not. The word gender enters the discourse trailing clouds of confusion. To grammarians, gender is clear. It is a classification of nouns by how adjectives and pronouns agree with them. French nouns fall into two genders, English and German nouns into three. Kivunjo, according to Steven Pinker, has fifteen genders. French genders could have been named A and B (*la table, le tapis*), English and German genders A, B, and C. As it happens, all males belong in gender B, all females in gender A, and that same neat separation occurs in most languages. It is, therefore, convenient to use "feminine" and "masculine" as names for two of the genders, rather than A and B. This perfect correlation permits the use of "gender" as a coy euphemism for sex. Alex Byrne, in *Trouble with Gender*, and Kathleen Stock, in *Material* Girls, both make valiant attempts to sort out the confusion. In the mind of this reader, accustomed to scientific standards of rigor, the confusion remains. Stock herself sensibly tries to avoid the term, replacing it with "concrete, clearer terms that do whatever jobs I want them to at the time." For the same reason, gender is not a word I normally use except in the grammarian's sense. If you want to speak French properly, you really do need to respect every noun's preferred pronoun.

The current fashion for transsexualism belongs in a cluster of interrelated vogues, sometimes called "woke," partly stemming from a sincere concern for social justice, largely well-meaning (which could not be said of Lysenko) but misguided and scientifically ill-informed (which could be said of Lysenko in spades). The cluster includes "identitarianism"[25] and the view that alternative "ways of knowing" (women's ways of knowing, indigenous ways of knowing, personal lived experience) are just as valid as objective science in understanding nature. The various strands have been helpfully listed and persuasively criticized by Jerry Coyne and Luana Maroja, originally published in *Skeptical Inquirer*[26] and reproduced in this volume.

Transsexualism has strictly no necessary connection to whether sex is "binary" or a continuously varying spectrum, although, in practice, the same people are often partisan with respect to both. Those of us who argue that there is no spectrum of intermediates between male and female—that sex is "binary"—should not be seen as threatening to transsexualism. Whether or not there are "intersexes" with ambiguous genitalia, or abnormal sex chromosomes, is irrelevant to transsexualism because no trans person claims to be intersex. A trans woman insists that she actually is a woman; a trans man, that he actually is a man. Neither claims to be hermaphrodite. Rather, the claim is a psychological one. There's a disjunct, or so it is claimed, between a person's biological sex and the gender that they feel themselves to be.

There are many dimensions along which human personality can be measured. They might include assertiveness, ambition, empathy, aggressiveness, selfishness, methodicalness, volatility, perseverance, affection, bossiness. A mathematician might see each person as situated in a multidimensional space defined by these dimensions. We are all amateur psychologists who gossip about each other. Without being mathematicians, we implicitly classify each other along dimensions such as those I have listed. Perhaps there is a psychological dimension of masculinity/femininity, which is more or less correlated with some of the other dimensions listed. In defiance of much hard-won feminist progress, you might invoke sexual stereotypes in an implicit invocation of such a dimension. "Cecil is effeminate. Ros is butch. Lizzy is a tomboy; she doesn't like dolls, loves climbing trees, and plays with wheeled toys."

We can situate ourselves along such personality dimensions, including the perceived dimension of masculinity/femininity. We may even go so far as to wish we had been born the opposite sex. We might phrase it as being trapped inside the wrong body. It's a version of dualism, a belief in a kind of disembodied soul, the real you, who is of a different sex or gender from the body in which the "real" you lurks. If that is how you are inclined, it might take little encouragement from the surrounding culture to push you over the edge into full-fledged belief. And today's surrounding culture—doctors, psychiatrists, teachers, political leaders, lawyers, and perhaps above all schoolfriends—gives more than a little push. Sex "assigned at birth" is arbitrary, we are told, and you only discover whether the real you is male or female by introspection.

An early account of what it is like to feel you are trapped in the wrong body is Jan Morris's *Conundrum*. As what she called a "true transsexual," she had little time for "the poor castaways of intersex, the misguided homosexuals, the transvestites, the psychotic exhibitionists, who tumble through this half-world like painted clowns, pitiful to others and often horrible to themselves."[27]

A feeling of being in a body of the wrong sex seems to be a real psychological condition. Such "dysphorics" can feel genuine distress. When anorexics look in the mirror, they see an emaciated body that they think is too fat. "Gender" dysphorics look in the mirror and see the wrong genitals. Both deserve sympathy and understanding. Nobody is phobic about anorexics. Why should anyone be phobic about gender dysphorics? "Transphobia" is a pernicious fiction.

Partly influenced by Jan Morris and partly out of normal politeness, it is my custom to refer to people by their preferred pronouns. But I draw the line at the belligerent slogan, "Trans women *are* women," because it is scientifically false, a debauching of language, and because, when taken literally, it can infringe the rights of other people, especially women. It logically entails the right to enter women's sporting events, women's changing rooms, women's prisons, and so on. So powerful has this "postmodern" counterfactualism become that newspapers refer to "her penis" as a matter of unremarked routine.

> *She is accused of four counts of engaging in sexual activity in the presence of a child and two counts of exposure where she "intentionally exposed her penis intending someone would see it and cause alarm or distress."*
>
> Bournemouth Daily Echo, Jan. 23, 2023

> *Even the* Times, *Britain's traditional newspaper of record, could begin an article (Jan. 18, 2023) with these words: "A transgender woman has denied raping two woman with her penis as she went on trial at the High Court in Glasgow."*

If the journalist had said "with his penis," the *Times* could have been in trouble with the police for "misgendering." In 2020, Humberside police descended on the workplace of Harry Miller to warn him that one of his tweets "was being recorded as a hate incident." What did the offending tweet say? "I was assigned Mammal at Birth, but my orientation is Fish. Don't mis-species me." A neat joke, in my opinion, and pretty gentle when compared with the satire of, say, Evelyn Waugh, Tom Lehrer, Ricky Gervais, Tim Minchin, Monty Python, W. S. Gilbert, or Jonathan Swift. Evincing an almost superhuman inability to take a joke, the police recorded it as a "hate incident" and threatened the satirist. Have the British police become George Orwell's Thought Police?[28] Will it come to that? The resemblance occurred to Mr. Justice Knowles, before whom the Harry Miller case came up. "In this country we have never had a Cheka, a Gestapo or a Stasi. We have never lived in an Orwellian society." Well said, M'Lud! I hope the Humberside police officers have learned their lesson. Perhaps somebody might take them aside and patiently explain about this thing called satire.

On the day that I wrote this paragraph, J. K. Rowling[29] called attention to a remarkable feat of doubletalk by another of Britain's leading newspapers, the left-wing *Guardian*. The report said that a woman called Scarlet Blake had been convicted of murdering a man called Jorge Carreno as he walked home by the peaceful River Cherwell in Oxford. Blake had earlier filmed and live streamed herself killing her neighbor's cat and putting it through a blender. She explained to the court that she identified as a cat, and she meowed to them in support of this claim. The *Guardian* reported all this, using the pronouns "she" and "her" throughout, never once mentioning that this murderer of an innocent stranger was actually a man. The headline used the word "woman" of the murderer. When Louise Tickle, a former *Guardian* writer, complained, the paper belatedly changed the website version of the story. Louise Tickle says she has evidence that Blake has not even legally transitioned. This murderer and cat-homogenizer is a man through and through, legally a man as well as biologically. The *Guardian* reported the murder as the work of a woman for no better reason than that Alan Blake chose to call himself a woman, as easily as one might choose to call oneself a socialist or a Manchester United supporter. I suppose they were terrified of being accused of transphobic misgendering. Or maybe they've sincerely bought into the superstition that uttering the magic incantation, "I am a woman," turns you into one, like a pumpkin turning into Cinderella's coach.

What is especially galling is that the violent deeds of the Scarlet Blakes of this world will swell the official statistics of crimes by women. The journalist Josephine Bartosch forcefully made this point,[30] quoting Richard Garside, director of the Centre for Crime and Justice Studies. More than 90 percent of convicted murderers are male. This means that if a murder by a trans woman is added to the female side of the ledger, the percentage effect on female/male murder statistics is much more dramatic than if the murder is clocked up to a male. I plugged in the actual figures, and it turned out that the change in the ratio is fifteen times as great if a trans murderer is counted as female than if the same murderer is counted as male. This huge and misleading statistical effect is a direct consequence of taking "Trans women are women" seriously. That slogan doesn't just harmlessly satisfy an individual's private emotional needs. If taken seriously, it dramatically distorts official statistics that might be used to guide public policy.

I am sorry to say it looks horribly as though otherwise sensible and certainly well-meaning political leaders are pandering to an intimidating lobby, militant activists ever ready to pounce, Lysenko-like, on what they see as heresy. Such activism is especially dominant among young people. Several senior publishers have confided to me that they are under strong pressure from young employees to censor, or even suppress, books that they perceive as "transphobic." I hereby place on record my regretful suspicion that some otherwise respected scientists, too, are betraying science in a desperate attempt to curry favor with "the kids," perhaps especially their own.

Sex and Race: A Double Standard

If you think about it, it's rather surprising that the current craze for a spectrum of non-binary self-identifications should have hit sexual identity rather than racial identity. Modern American culture is obsessed with race[31] and racial identity at the same time as being obsessed with pronouns and sexual identity. But whereas sex is clearly binary (or at least you have a hard time arguing that it isn't), race is manifestly non-binary (or at least far closer to being a continuum than sex). When a male mates with a female, each offspring is either male or female, not intermediate. Like Mendel's wrinkled versus smooth peas. When a black-skinned person mates with a white-skinned person, the offspring are usually of an intermediate color. Unlike Mendel's peas. This is because skin color is polygenically inherited. But when Americans *speak* of a person with one white parent and one black parent, they almost invariably call them "Black."

Because skin color is polygenically inherited, you get a complete range of intermediates among the population of African Americans, but the cultural label "Black" is inherited as if it were a Mendelian dominant. If either of your parents is Black, society identifies you as Black. If any one of your four grandparents is Black, society identifies you as Black. If any one of your eight great-grandparents is Black, society identifies you as Black. In the case of sex, you know that exactly 50 percent of your great-grandparents, exactly 50 percent of your ancestors in any previous generation, were male, exactly 50 percent female. In the case of race, there's a continuous gradation in the American population. You'd think it would, therefore, be relatively easy to indulge individuals who self-identify as whatever "race" they choose. Why, I repeat, did the craze for non-biological self-identification hit sex and not race? Why the double standard? An individual who chooses to identify with the sex opposite to their biology is treated with sympathy and respect. But if they try the analogous self-identification where race is concerned, what happens? They are ostracized, placed in the modern equivalent of the medieval stocks, and pelted with metaphorical tomatoes.

Let's find names for two responses to a controversy. Vavilov patiently tried to explain genetics to Lysenko, so in his honor, let's use "Vavilov Style" as our name for rational argument: a reasoned response to a question. The "Lysenko Style" is the opposite: no rational argument at all, just summary denunciation and name-calling, followed by such punitive sanctions as are available. In an example of the "Vavilov" style of calm, rational discussion, the philosopher Rebecca Tuvel published an article, "In Defense of Transracialism," in the 2017 volume of the feminist philosophy journal *Hypatia*.[32] She compared the case of Rachel Dolezal, who identified as a member of a different race, with Caitlin Jenner, an American athlete (formerly William Bruce Jenner) who identified as a member of a different sex. The response to her article was pure Lysenko and hysterical in tone. A majority of the journal's associate editors issued an apology. The editor resigned. Tuvel herself was publicly accused of "epistemic violence," was described as crazy, racist, transphobic, and stupid, and threatened with the loss of her career. Others may attempt an explanation for the hysteria. I shan't even try. The double standard itself is quite beyond me for the reasons given in the previous paragraph. Admittedly, there is the additional fact that Dolezal concealed her true parentage, but that's just her as an individual and has nothing to do with the principle. The principle is that modern American culture allows you to choose your sex at will, but not your race. You'd think the freedom offered, even encouraged, in the one case should extend to the other. The more so since race really is a continuum. The double standard, ruthlessly enforced, may have provided an understandable motive for Dolezal's deception. Perhaps she lied in realistic anticipation of the kind of treatment meted out to Rebecca Tuvel. "Transsexual women are women" is all the rage. But "Transracial Blacks are Black" will see you ostracized as a pariah.

In 2021, I invited Twitter readers to discuss the weird double standard:

> In 2015, Rachel Dolezal, a white chapter president of NAACP, was vilified for identifying as Black. Some men choose to identify as women, and some women choose to identify as men. You will be vilified if you deny that they literally are what they identify as. Discuss.

As a lifelong Oxford teacher, I have become accustomed to inviting my tutorial pupils to discuss controversial issues, counterfactual hypotheses, thought experiments, and interesting paradoxes. When I wrote "Discuss" at the end of my tweet, I was obviously signaling my hope for a Vavilov style of response. What I got was pure Lysenko, complete with miniature, and rather pathetic, gulag-surrogate. The details aren't worth spelling out. If you're curious, just Google my name together with "American Humanist Association."

The Theology of Woke

Those twin American obsessions, with race and with sexual self-identification, have something else in common. Each has its own specific and detailed analogy with Christian theology. First race—and white guilt over slavery and colonialism. Original sin is one of the central ideas in Christian theology. Christianity is obsessed with sin, sin as a kind of abstract entity that accounts for all that's negative, including even disease. In a biology class at my school, the teacher asked us about the causes of disease. Before any of us could suggest viruses or bacteria, autoimmunity, or cancer, one boy put his hand up and volunteered, "sin." His contribution received predictably short shrift from the biology teacher. It's easy to see where it came from. But the Christian obsession with sin goes beyond individual misdemeanors. It's not enough to feel guilt at our own transgressions. That would be understandable in any religion or none. More specifically, in Christianity, we are each held responsible for the sin of an alleged remote ancestor (who never existed, as even theologians now admit, but let that pass), going much further back than Exodus's "visiting the iniquity of the fathers upon the children, and upon the children's children, unto the third and to the fourth generation." We are all born in sin. Straight out of the womb, we are already guilty of the sin of Adam. Augustine, the chief architect of this rather nasty idea, even thought that Adam's sin passes down the generations in semen. Jesus, conceived without semen, was, therefore, without sin. It was theologically necessary that his mother, Mary, too, should be without sin, so she too had to have been "immaculately conceived" rather than by the taint of semen, and this was conveniently revealed to Pope Pius IX in 1854.

The theology of atavistic guilt is directly carried over into the fashionable idea that all white people are automatically racist; all white people should feel guilty because they are white, guilty because of the appalling behavior of slave-owning white people of past centuries. And it really was unspeakably appalling. A picture of the economically driven packing of a slave ship, chained men lying like sardines in a tin, conveys the horror. The scale of the suffering is beyond imagining. Our forebears who perpetrated such massive and hideous wrongs should have felt guilt on a massive scale. Today, we cast around in desperation for someone to blame for such depraved cruelty. But they are all dead and unavailable. As are the African chieftains who sold other Africans into slavery. As are the Arab slave traders in East Africa. Whether your ancestors happen to be among the guilty ones (statistically, they probably are), if your skin is the same color as theirs, you are expected to share their guilt. But we are not our ancestors. Their genes have come down to us in sperm and eggs, but their sins have not. We should feel appalled at their monstrous evil, and this should drive us to a determination not to perpetrate whatever might be its modern equivalent. But there's no reason to feel collective guilt specifically because we are the same color as those old slavers, just as no German under the age of ninety should feel guilt for Hitler any more than the rest of us who belong to the same species as that monster. Treating people as members of a race, whether for blame or credit, rather than as individuals, was precisely what Martin Luther King hoped we'd grow out of.

> I have a dream that my four little children will one day live in a nation where they will not be judged by the color of their skin but by the content of their character. I have a dream today.

Not only is Dr. King's noble hope unfulfilled.[33] Even the aspiration itself, if without mentioning his name, is coming under ideological attack.[34]

So to my second theological comparison. "Transubstantiation" and "transgender" have more in common than just their first five letters. Protestants see the eucharist bread and wine as merely symbolic of the body and blood of Christ. But the Roman Catholic (and, less clearly, the Eastern Orthodox) church teaches that the bread and wine actually become the body and blood. The sense in which they mean this comes, like so much else in Medieval Christendom, from Aristotle, albeit he antedated Jesus by centuries. Aristotle made a distinction between true "substance" and "accidentals." In the theology of Thomas Aquinas and others, the bread and wine retain the *accidental* properties of starchy food and alcoholic liquid, respectively, but on being blessed by a priest their Aristotelian *substance* becomes body and blood. Hence the word tran*substanti*ation. If that makes no sense to you, never mind. Theology is not noted for making sense.[35] My point is that the same style of mystery pervades extreme transgenderism. A person may be biologically male, but his penis, testes, Y-chromosome, masculine physique, and assignment by inspection at birth are mere Aristotelian accidentals. Her true Aristotelian substance is her female gender, revealed by introspection. It is not a mere symbolic change of "gender" (Aristotelian accidental) but an actual revelation of true sex (Aristotelian substance) that the magic achieves.

Those two analogies are with Christian theology in particular. There is a more general analogy with religions in general, especially in the zealous hunting down of heretics, followed by vicious punishment. Which brings us back to Lysenko and his relentless persecution of the Mendelist-Morganist heresy. I think it is fair to compare the militant "sex is non-binary" activists to Lysenko in two main respects. First, both subscribe to anti-genetic, scientifically untenable dogma. And, second, they advance their case less by rational argument than by pejorative name-calling and attempts to "cancel" dissenters or pitilessly destroy their careers.

Concluding Remarks

If your science is so weak that the best you can do is yell that your opponent is a "Mendelist-Morganist" or a "Weismannist," a "Transphobic bigot," a "TERF," or a "full-on MAGA alt-right Trump-supporter," you've already lost the argument.[36] Sometimes the name-calling goes further and becomes overtly threatening. In Lysenko's case, very obviously and tragically so. At a London Pride demonstration in 2023, "Sarah Jane" Baker (previously Alan Baker) told a cheering crowd, "If you see a TERF, punch them in the fucking face." I don't think I'm unduly guilty of sexist stereotyping if I say that such language is more typical of the sex that "Sarah Jane" claims to have left than the one she aspires to join.

Sky News (January 23, 2023) had a picture of two Scottish Nationalist Party politicians, members of the British Parliament and the Scottish Parliament, respectively, at a transgender demonstration in Glasgow, grinning inanely in front of a large, colorful sign depicting a guillotine and the slogan "Decapitate TERFS." Threats of violence have no place in decent society. But I would make two points here. The first is that we need to be clear about what we mean by violence. The "postmodern" tendency to redefine existing words, giving them new politically inspired meanings in terms of "oppressors" and "oppressed," doesn't apply only to sex. It has been extended to redefining "hate" ("hate speech" can mean "anything I disagree with") and the very word "violence" itself. The *Oxford English Dictionary* defines "violence" as "the deliberate exercise of physical force against a person, property, etc." "Physical force" is the key phrase. At the very least, "violence" should include the threat of physical force. Yet you will find claims on the Internet that "misgendering is an act of violence." If you think it's an act of "violence" to call somebody "he" rather than "they," you are entitled to your private redefinition. But you may get short shrift from somebody who has suffered real physical violence and knows firsthand what violence means. If the word "violence" is so devalued as to include the mere uttering of a pronoun, what word is left for brutal wife-beating and rape, for murder with knife, machete, or baseball bat?

The second thing I'd say is that "Decapitate TERFS" is not only horrible. It's pathetic. So is an exhortation to punch a TERF "in the fucking face." A position should be supported, or refuted, by rational discussion informed by evidence. People who terminate an argument by resorting to threats or name-calling are ignominiously signaling that they've lost the argument. Like Lysenko. It took decades, and the death of Stalin, for Russia to see through Lysenko. Let us hope we, in our century, don't take quite so long to come to our senses.

On March 19th, 2025, as this book was going to print, the Sullivan Report commissioned by the British Government was published https://assets.publishing.service.gov.uk/media/67d98b8a4ba412c67701ed92/review-of-data-statistics-research-on-sex-and-gender.pdf and summarised in The Times the same day.

Briefly, the report shows that official UK statistics on health, crime and education have used people's declared "gender" rather than their biological sex in classifying data. The information about biological sex has been lost with, in some cases, serious consequences. For example, patients have not been called up for cancer tests because their biological sex had been erased from their medical records.

How Ideology Threatens to Corrupt Science

ALAN SOKAL

Science—and that includes both the natural and the social sciences—is, or at least is supposed to be, a truth-seeking enterprise. The phenomena that one decides to study may be chosen for their conceptual significance, for their social or economic importance, or simply out of personal curiosity. But whatever topic a scientist decides to investigate, she is intellectually and morally obliged to follow the evidence wherever it leads—even (or especially) if that evidence conflicts with her preconceptions or her desires.

Science doesn't always work this way, of course—scientists are, after all, human—but that is anyway the ideal towards which we strive. And if there is freedom of debate within the scientific community—freedom to hold each other's ideas to stringent conceptual and empirical scrutiny—then the scientific community collectively is more likely to reach objectively true conclusions than any of its members could do alone.

A scientist's political and social values may, of course, influence her selection of topics to study—that is perfectly legitimate. But those values should be carefully put to the side when evaluating the evidence. The goal of the scientific endeavor is to find out how things really are, not to confirm how we wish they were.

Many decisions that we must make collectively—about anything from educational methods to pandemics to climate change—need to be based on scientific knowledge: we require detailed factual evidence about how children learn to read, how viruses spread, and how the earth's oceans and atmosphere behave. But although this scientific information forms the essential background for public policy, it doesn't *determine* that policy, since policy decisions also involve values and tradeoffs between competing values (Association Française pour l'Information Scientifique n.d.). But whatever your values, it still behooves you to have as accurate an understanding as possible of reality to inform your policy choices (Sokal 2021). (If you don't, you risk implementing policies that are counterproductive as assessed by *your own* values.) And in a democracy, every citizen has the right, and should have the opportunity, to do the same.

One important social mechanism within science is peer review: proposed scientific contributions are evaluated for their correctness and importance by experts in the field (ideally double-blind); and depending on that evaluation, the article may be accepted for publication, accepted subject to revision, or rejected entirely. That system isn't perfect—it can be compromised by personal rivalries, competing research programs, and simple reviewer sloppiness—but it is the best that we have been able, thus far, to devise. The key desideratum is that submissions should be evaluated for their conceptual rigor, their methodological soundness, their empirical thoroughness, and their importance to the scientific field. Social and political values may play a role in this last aspect—telling us which topics are most important to investigate—but they should play no role in the evaluation of which contributions on that subject are fit to publish. That evaluation should be based solely on the scientific quality of the research, not on whether we find its results congenial.

This, anyway, has been the official policy of the scientific community for the past three centuries—implemented imperfectly, to be sure, but nevertheless functioning as an important regulative ideal. But times have changed: now ideology threatens openly to corrupt the truth-seeking enterprise that we call science.

Two years ago, the prestigious journal *Nature* issued a new "ethics guidance" concerning proposed submissions (Unsigned 2022). But the guidance does not pertain simply to the protection of human research subjects; that issue has been strictly regulated for decades. Nor is it about restricting the publication of information that poses serious material dangers, such as facilitating the production of nuclear or biological weapons. Rather, the guidance purports to address other forms of "harm" that could be caused by a scientific publication. And on these grounds, the editors arrogate to themselves an astoundingly broad power:

> Regardless of content type (research, review or opinion) and, for research, regardless of whether a research project was reviewed and approved by an appropriate institutional ethics committee, editors reserve the right to request modifications to (or correct or otherwise amend post-publication), and in severe cases refuse publication of (or retract post-publication):…Content that undermines—or could reasonably be perceived to undermine—the rights and dignities of an individual or human group on the basis of socially constructed or socially relevant human groupings.

That vague and subjective language is an open door to ideological censorship of valid scientific contributions—a censorship that the editors do not even attempt to disguise. It is, therefore, imperative to evaluate the justifications that the editors of *Nature* have offered in support of this brave new policy.

The document starts ominously: "Although academic freedom is fundamental, it is not unbounded."

(Vague assertions of this kind are always a bad sign: one knows what is coming next [Sokal (2024)].) The guidance purports to apply "ethical principles" analogous to those used to protect human research subjects, but now concerning other types of "harms":

> For example, research may—inadvertently—stigmatize individuals or human groups. It may be discriminatory, racist, sexist, ableist or homophobic. It may provide justification for undermining the human rights of specific groups, simply because of their social characteristics.

Let's slowly unpack these claims.

(1) What Could It Mean for Scientific Research to "Stigmatize" Individuals or Human Groups? And to Do So "Inadvertently"?

Suppose research finds that obesity can cause cancer (it can [See the references cited in Wikipedia 2024]). Does that "stigmatize" overweight people? Some people would argue that it does (Knapton 2019, Varshney 2021); but that is shooting the messenger because we don't like the message. In fact, suppressing this research would do harm, above all, *to overweight people* by denying them information that they could use—if they wish, and only if they wish—to protect their health.

Or suppose research finds that gay men have more sexual partners, on average, than heterosexual men (they do [Mercer et al. 2016, Table 2]). Does that "stigmatize" gay men? Maybe it does, at least in the eyes of people who disdain sexual promiscuity. But it is also important information in planning interventions to reduce the risk of sexually transmitted diseases—interventions that would disproportionately benefit *gay men*.

The editors of *Nature* have thus assigned to themselves the purely subjective task of judging which scientific research "stigmatizes" some social group and have empowered themselves to suppress valid scientific contributions—information that is likely to be *true* and important—on that sole basis.

(2) What Could It Mean for Scientific Research to Be "Discriminatory, Racist, Sexist, Ableist or Homophobic"?

If the research incorporated racist or sexist *presuppositions*, that would be an *epistemic* defect that would undermine the quality of the research and perhaps invalidate it entirely, purely on traditional scientific criteria; no new "ethics guidance" is needed on that score. Clearly what the editors are getting at is not racist or sexist presuppositions, but rather *conclusions* from the research that the editors, in their infinite wisdom, judge to be racist or sexist. But that is again shooting the messenger.

Suppose, for instance, that research finds (as it seems to [Johnson, Carothers, and Deary 2008; Archer 2019; Stevens and Haidt 2017, Stevens

2017a, 2017b]) that men show larger variation than women over a range of cognitive and psychological traits, including various types of intelligence—so that men are overrepresented at both the low and high ends of the scale, even when the means (i.e., averages) are equal. Surely this is not the only reason why women are underrepresented among scientists—sexist stereotypes, influencing girls and young women, must also be a major contributing factor (Nosek et al. 2009, Schmader 2023), and there are undoubtedly other factors as well—but it might form part of the explanation (Summers 2005, Halpern et al. 2007); it might mean that even in a future non-sexist society the majority of scientists (and also of people with intellectual disabilities) will be men. Should this information be suppressed? If that happens, then our ignorance of relevant facts will interfere with our ability to determine accurately the extent to which sexist discrimination persists in different fields; and it will also impede us from distinguishing between ameliorative policies that are effective and those that are not.

(3) And What, Finally, Can It Mean to "Provide Justification for Undermining the Human Rights of Specific Groups"?

Consider again the research about sex differences in the variation of mathematical ability. Would this research provide a "justification" for discriminating against women scientists? Absolutely not! It might provide a lame *excuse* for such discrimination, but not a justification. Since each individual's work can be evaluated on its own merits, the statistical properties of the groups to which that individual belongs are completely irrelevant.

So what the editors seem to have in mind is not research that could *justify* undermining the human rights of specific groups—indeed, it's hard to see how *any* scientific research could do that, simply because one cannot derive an "ought" from an "is"—but research that some people might attempt to *misuse* as a supposed justification for undermining human rights. But valid ideas should not be suppressed because some people might misuse them; rather, it is the misuse that should be criticized instead.

The bottom line is that the editors of *Nature* have arrogated to themselves the right to suppress valid scientific work—work that is both correct and important—purely because it allegedly "undermines—or could reasonably be perceived to undermine—the rights and dignities of an individual or human group."

But what could it mean for a scientific contribution—that is, information about reality—to undermine anyone's rights or dignities? Once again, the editors are perpetrating a severe confusion between "is" and "ought"; indeed, the policy is entirely founded on that confusion.

But then the editors cover their tracks by introducing, in astute lawyer-like fashion, a new element: the scientific work need not *actually* undermine anyone's rights or dignities; rather, it suffices that some unnamed people (note the editors' strategic use of the passive voice) could reasonably *perceive* the work to undermine someone's rights or dignities. But this is an extraordinarily broad criterion: it is likely that *any* controversial scientific work that has public-policy implications will cause *some* people to perceive it as undermining someone's rights or dignities. For instance, an article reviewing the neuropsychological effects of puberty blockers (Baxendale 2024a) will likely be labeled by advocates of gender-identity ideology as undermining the rights and dignities of transgender people ("stigmatising an already stigmatised group," as one of this article's anonymous peer reviewers explicitly put it [Baxendale 2024b]); others will reply that this research helps to *protect* the rights of gender-nonconforming teenagers by offering them accurate information about the benefits and risks of proposed medical interventions (Baxendale 2024b; Cohn 2023; Barnes 2023; Smith 2023; Malone, Wright, and Robertson 2019; Cass 2024).

Admittedly, the editors require that the research slated for suppression could *reasonably* be perceived to undermine the rights and dignities of an individual or group. But who gets to decide which perceptions are reasonable and which are not? The editors themselves, of course. And these are the same editors who insist, among other things, that sex as defined by gametes and chromosomes—the well-established biological understanding—"has no foundation in science," that "sex [is] more complex than male and female," and that the now-outdated (according to them) biological view "would undermine efforts to reduce discrimination against transgender people and those who do not fall into the binary categories of male or female" (Unsigned 2018).

Consequently, any scientific article that employs the standard biological concept of sex now risks being characterized by the *Nature* editors as undermining the rights and dignities of transgender people—and ipso facto as being reasonably perceived as doing so. Since that criterion would exclude a huge chunk of work in biology and medicine, the editors cannot apply it consistently without sabotaging their own journal. So they will, of necessity, apply it selectively: to suppress those studies whose conclusions they dislike (Stavroula 2022a, 2022b). As psychologist Bo Winegard has perceptively pointed out:

> Imagine for a moment that this editorial were written, not by political progressives, but by conservative Catholics, who announced that any research promoting (even "inadvertently") promiscuous sex, the breakdown of the nuclear family, agnosticism and atheism, or the decline of the nation state would be suppressed or rejected lest it inflict unspecified "harm" on vaguely defined groups or individuals. Many of those presently nodding along with Nature's editors would have no difficulty identifying the subordination of science to a political agenda (Winegard 2022).

The *Nature* editors attempt to soften the blow of their brazen announcement of future censorship by declaring that

> There is a fine balance between academic freedom and the protection of the dignity and rights of individuals and human groups. We commit to using this guidance cautiously and judiciously, consulting with ethics experts and advocacy groups where needed.

As Winegard comments:

> This is not at all reassuring. Asking ethicists to assess the wisdom of publishing a [scientific] journal article is as

antithetical to the spirit of science as soliciting publication advice from a religious scholar. Who are these "ethics experts" and "advocacy groups" anyway?

Imagine the outcry on the Left if a journal announced it would be consulting pro-life advocates before publishing an article about the effects of abortion on wellbeing. Or if it decided to consult conservative evangelicals when evaluating an article about the effects of adoption by homosexual couples.

In practice,

> The journal is effectively announcing the employment of sensitivity readers, who it can safely be assumed, will invariably recommend the risk-averse option of suppression whenever the possibility of controversy arises.

Further information on the perils of politicizing science can be found in eloquent articles by chemist Anna Krylov and statistician Jay Tanzman, biologists Jerry Coyne and Luana Maroja, geophysicist Dorian Abbot, social psychologist Lee Jussim, sociologist Musa al-Gharbi and social psychologist Cory Clark, sociologist Yves Gingras, and journalist Jonathan Rauch (Krylov 2021, 2022; Krylov and Tanzman 2021, 2023; Coyne and Maroja 2023; Abbot 2021, 2023; Jussim 2019; al-Gharbi and Clark 2023; Clark et al. 2023; Gingras 2022; Rauch 2021, 2022).

There is one further danger that the advocates of ideological censorship in science would do well to ponder.

As John Stuart Mill observed long ago in his celebrated essay *On Liberty*,

> The peculiar evil of silencing the expression of an opinion is, that it is robbing the human race; posterity as well as the existing generation; those who dissent from the opinion, still more than those who hold it. If the opinion is right, they are deprived of the opportunity of exchanging error for truth: if wrong, they lose, what is almost as great a benefit,

the clearer perception and livelier impression of truth, produced by its collision with error (Mill 1859/2003, 87).

The first side of this bifurcation is clear: though we all naturally think that our current opinions are correct (otherwise, they wouldn't be our opinions), we still ought to be willing to admit that we are not infallible. And that means that if you really care about truth, you ought to be open to hearing arguments against your current opinions and open to changing those opinions whenever the counterarguments turn out to be cogent. Perhaps the *Nature* editors are so utterly certain that their views—on a huge variety of disparate subjects—are all 100 percent correct that they are unable to imagine learning even a tiny bit from listening to reasoned criticisms; if that is the case, then they themselves are the losers (as are their readers who are prevented from hearing relevant evidence).

But the other side of Mill's bifurcation is less obvious, so let me quote Mill again:

> He [sic] who knows only his own side of the case, knows little of that. His reasons may be good, and no one may have been able to refute them. But if he is equally unable to refute the reasons on the opposite side; if he does not so much as know what they are, he has no ground for preferring either opinion.
>
> Nor is it enough that he should hear the arguments of adversaries from his own teachers, presented as they state them, and accompanied by what they offer as refutations. That is not the way to do justice to the arguments, or bring them into real contact with his own mind. He must be able to hear them from persons who actually believe them; who defend them in earnest, and do their very utmost for them. He must know them in their most plausible and persuasive form....
>
> Ninety-nine in a hundred of what are called educated men are in this condition; even of those who can argue fluently

for their opinions. Their conclusion may be true, but it might be false for anything they know: they have never thrown themselves into the mental position of those who think differently from them, and considered what such persons may have to say; and consequently they do not, in any proper sense of the word, know the doctrine which they themselves profess (Mill 1859/2003, 104–105).

Mill's two-pronged argument in favor of the freedom of debate is, in fact, a crucial ingredient in legitimizing knowledge in general, and scientific knowledge in particular; and it is striking that Mill himself used an example from science—namely, Newtonian mechanics—to explain why. Isaac Newton published his celebrated laws of motion in 1687; and by the time Mill was writing in 1859, scientists had accumulated overwhelming evidence, from both terrestrial and astronomical observations, that Newtonian physics is correct (even to the point of predicting accurately, in 1846, the existence and precise location of the hitherto-unknown planet Neptune). But, Mill points out, if at some point the government (or even just the scientific societies) had decided that, in view of the overwhelming evidence of the correctness of Newtonian mechanics, it would henceforth be forbidden to dispute it, then we would now have *much less reason* to believe in the correctness of Newtonian mechanics! It is precisely the fact that Newtonian mechanics has held up in the face of free and open debate that gives us such justified confidence in its correctness.

So even if the "progressives" are 100 percent correct on every subject and have nothing whatsoever to learn from their thoughtful critics, censorship of opposing views is *still* harmful *to their own cause*, as it undermines the good reasons for anyone else to adopt their ideas.

It would be a real positive step if the *Nature* editors were to reflect on this argument—which is, after all, Mill's, not mine—and respond to it. But people with power are unfortunately not accustomed to acknowledging (much less addressing) reasoned critiques from lesser mortals. So don't hold your breath.

The Treason of the Intellectuals

NIALL FERGUSON

Anyone who has a naive belief in the power of higher education to instill morality has not studied the history of German universities in the Third Reich.

In 1927, the French philosopher Julien Benda published *La trahison des clercs*—"The Treason of the Intellectuals"—which condemned the descent of European intellectuals into extreme nationalism and racism. By that point, although Benito Mussolini had been in power in Italy for five years, Adolf Hitler was still six years away from power in Germany and thirteen years away from victory over France. But already, Benda could see the pernicious role that many European academics were playing in politics.

Those who were meant to pursue the life of the mind, he wrote, had ushered in "the age of the intellectual organization of political hatreds." And those hatreds were already moving from the realm of ideas into the realm of violence—with catastrophic results for all of Europe.

A century later, American academia has gone in the opposite political direction—leftward instead of rightward—but has ended up in much the same place. The question is whether we—unlike the Germans—can do something about it.

For nearly ten years, rather like Benda, I have marveled at the treason of my fellow intellectuals. I have also witnessed the willingness of trustees, donors, and alumni to tolerate the politicization of American universities by an illiberal coalition of "woke" progressives, adherents of "critical race theory," and apologists for Islamist extremism.

Throughout that period, friends assured me that I was exaggerating. Who could possibly object to more diversity, equity, and inclusion on campus? In any case, weren't American universities always left-leaning? Were my concerns perhaps just another sign that I was the kind of conservative who had no real future in the academy?

Such arguments fell apart after October 7, as the response of "radical" students and professors to the Hamas atrocities against Israel revealed the realities of contemporary campus life. That hostility to Israeli policy in Gaza regularly slides into antisemitism is now impossible to deny.

I cannot stop thinking of the son of a Jewish friend of mine, who is a graduate student at one of the Ivy League colleges. After October 7, he went to his assigned desk to find, carefully placed under his computer keyboard, a note with the words "ZIONIST KIKE!!!" in red and green letters.

Just as disturbing as such incidents—and there are too many to recount—has been the dismally confused responses of university leaders.

Testifying before the House Committee on Education and the Workforce on December 5, 2023, Harvard President Claudine Gay, MIT President Sally Kornbluth, and University of Pennsylvania President Elizabeth Magill showed that they had been well-briefed by the lawyers their universities retain for such occasions.

They gave technically correct explanations[1] of how First Amendment rules apply on their campuses—if they did apply. Yes, context matters. If all students did was chant "From the river to the sea," that speech is protected, so long as there was no threat of violence or "discriminatory harassment."

But the reason Claudine Gay's carefully phrased answers infuriated her critics is not that they were technically incorrect, but that they were so clearly at odds with her record—specifically her record as dean of the Faculty of Arts and Sciences in the years 2018–2022, when Harvard was sliding to the very bottom of the rankings for free speech at colleges.

The killing of George Floyd happened when Gay was dean. Six days after Floyd's death, she published a statement[2] on the subject that suggested she felt personally threatened by events in distant Minneapolis. Floyd's death, she wrote, illustrated "the brutality of racist violence in this country" and gave her an "acute sense of vulnerability." She was "reminded, again, how even our [i.e., black Americans'] most mundane activities, like running...can carry inordinate risk. At a moment when all I want to do is gather my teenage son into my arms, I am painfully aware of how little shelter that provides." In nothing that Gay said in her congressional hearing did she seem aware that Jewish students might have felt the same way after October 7.

In a memorandum[3] to faculty on August 20, 2020, she wrote: "The calls for racial justice heard on our streets also echo on our campus, as we reckon with our individual and institutional shortcomings and with our Faculty's shared responsibility to bring truth to bear on the pernicious effects of structural inequality." Gay continued: "This moment offers a profound opportunity for institutional change that should not and cannot be squandered.... I write today to share my personal commitment to this transformational project and the first steps the FAS will take to advance this important agenda in the coming year."

As the great German sociologist Max Weber rightly argued in his 1917 essay on "Science as a Vocation,"[4] political activism should not be permissible in a lecture hall "because the prophet and the demagogue do not belong on the academic platform." This was also the argument of the University of Chicago's 1967 Kalven Report that universities must "maintain an independence from political fashions, passions, and pressures."

This separation between scholarship and politics has been entirely disregarded at the major American universities in recent years. Instead, our most elite schools have embraced the kind of "institutional change" that Gay has championed. Look where it has led us.

It might be thought extraordinary that the most prestigious universities in the world should have been infected so rapidly with a politics imbued with antisemitism. Yet exactly the same thing has happened before.

A hundred years ago, in the 1920s, by far the best universities in the world were in Germany. By comparison with Heidelberg and Tübingen, Harvard and Yale were gentlemen's clubs, where students paid more attention to football than to physics. More than a quarter of all the Nobel prizes awarded in the sciences between 1901 and 1940 were awarded to Germans; only 11 percent went to Americans. Albert Einstein reached the pinnacle of his profession not in 1933, when he moved to Princeton, but from 1914 to 1917, when he was appointed professor at the University of Berlin, director of the Kaiser Wilhelm Institute for Physics, and a member of the Prussian Academy of Sciences. Even the finest scientists produced by Cambridge felt obliged to do a tour of duty in Germany.

Yet the German professoriate had a fatal weakness. For reasons that may be traced back to the foundation of the Bismarckian Reich[5] or perhaps even further into Prussian history, academically educated Germans were unusually ready to prostrate themselves before a charismatic leader, in the belief that only such a leader could preserve the purity of the German nationalist project. Today's progressives engage in racism in the name of diversity and inclusion. The nationalist academics of interwar Germany were at least overt about their desire for homogeneity and exclusion.

Marianne Weber recalled how, in the wake of the 1918 Revolution,[6] her husband Max had explained his theory of democracy to the former supreme military commander, General Erich Ludendorff:[7]

> Weber: Do you think that I regard the Schweinerei[8] that we now have as democracy?
>
> Ludendorff: What is your idea of a democracy, then?
>
> Weber: In a democracy, the people choose a leader whom they trust. Then the chosen man says, "Now shut your mouths and obey me." The people and the parties are no longer free to interfere in the leader's business.
>
> Ludendorff: I should like such a "democracy."
>
> Weber: Later, the people can sit in judgment. If the leader has made mistakes—to the gallows with him!

Rudy Koshar's study[9] of the university town of Marburg in Hesse illustrates the way this culture led German academia toward the Nazis. The mainly Protestant student fraternities already excluded Jews from membership before World War I. In March 1920, in the turbulent aftermath of the revolution that had overthrown the imperial regime and established the

Weimar Republic, a student paramilitary group was involved in a murderous attack[10] on Communist workers. In the national elections held four years later, the Völkisch-Sozialer Bloc—of which the early Nazi Party (the NSDAP) was a key part—won 17.7 percent of the Marburg vote.

Lawyers and doctors, all credentialed with university degrees, were substantially overrepresented within the NSDAP, as were university students (then a far narrower section of society than today). To middle-aged lawyers, Hitler was the heir to Bismarck. For their sons, he was the Wagnerian hero Rienzi,[11] the demagogue who united the people of Rome.

Even a man who considered himself a liberal, as Max Weber surely did, was susceptible to the allure of charismatic leadership when the fledgling democracy seemed so weak. Three years after Weber's death in 1920, Germany was plunged into disastrous hyperinflation.[12] For many German academics, Hitler's appointment as chancellor in January 1933 was a moment of national salvation.

"Right down to the last, deepest fiber in myself, I belong to the Führer and his wonderful movement," wrote the Nazi lawyer Hans Frank in his diary after a concert he had attended with Hitler on February 10, 1937. "We are in truth God's tool for the annihilation of the bad forces of the earth. We fight in God's name against Jews and their Bolshevism. God protect us!" Such thoughts helped him and many other lawyers to come to terms with the systematic illegality that characterized the regime from the very outset.

German academics acted as Hitler's think tank, putting policy flesh on the bones of his racist ideology. As early as 1920, the jurist Karl Binding and the psychiatrist Alfred Hoche published their *Permission for the Destruction of Life Unworthy of Life*,[13] which sought to extrapolate from the annual cost of maintaining one "idiot" "the massive capital…being subtracted from the national product for entirely unproductive purposes."

There is a clear line of continuity from this kind of analysis to the document[14] found at the Schloss Hartheim asylum[15] in 1945, which calculated that by 1951, the economic benefit[16] of killing 70,273 mental patients—assuming an average daily outlay of 3.50 Reichsmarks and a life expectancy of ten years—would be 885,439,800 Reichsmarks. Many historians were little better, churning out tendentious historical justifications for German territorial claims in Eastern Europe that implied massive population displacement, if not genocide.

A critical factor in the decline and fall of the German universities was precisely that so many senior academics were Jews. For some, Hitler's antisemitism was, therefore—not unlike woke intersectionality in our own time—a career opportunity.

For German academics of Jewish heritage, particularly those who had married gentiles and converted to Christianity, it was disorienting.

The case of Victor Klemperer, a convert to Christianity married to a gentile, is illustrative. A veteran of World War I, Klemperer was appointed professor of Romance Languages and Literature at Dresden University of Technology in 1920. "I am nothing but a German or German European," Klemperer wrote in his diary, one of the most illuminating testaments of the German Jewish nightmare. Throughout the 1930s, he maintained that it was the Nazis who were "un-German." "I...feel shame for Germany," he wrote after Hitler had come to power. "I have truly always felt German."

Yet the atmosphere at German universities grew steadily more toxic even for the most assimilated of Jews.

In April 1933, under the Law for the Restoration of the Professional Civil Service, all Jewish civil servants, including judges, were removed from office,[17] followed a month later by university lecturers. Klemperer recorded his agonized reaction in his diary:[18]

> March 10, 1933... It is astounding how easily everything collapses...wild prohibitions and acts of violence. And with it, on streets and radio, never-ending propaganda. On Saturday...I heard a part of Hitler's speech in Königsberg [the East Prussian university at which Immanuel Kant had spent his life].... I understood only a few words. But the tone! The unctuous bawling, truly bawling.... How long will I retain my professorship?

He managed to hang on to his chair for two years. On May 2, 1935, however, the blow fell:

> On Tuesday morning, without any previous notification—two sheets delivered by post. "On the basis of para 6 of the Law for the Restoration of the Professional Civil Service I have...recommended your dismissal." ...At first, I felt alternately dumb and slightly romantic; now there is only bitterness and wretchedness.

Five months later, to add insult to injury, he was barred from the university library reading room "as a non-Aryan." What followed was a relentless whittling away of his rights as a citizen.

The Nazis' antisemitism led, of course, to one of the greatest brain drains in history. Over two hundred of the country's eight hundred Jewish professors departed, of whom twenty were Nobel laureates. Albert Einstein had already left in 1933 in disgust at Nazi attacks on his "Jewish physics." The exodus quickened after the pogrom known as the Night of Broken Glass in November 1938. The principal beneficiaries of the Jewish brain drain were, of course, the universities of the United States.

Yet for Klemperer, emigration—least of all to Palestine, then a British "mandate" but also the location of the "national home for the Jewish people" promised[19] by the British government in 1917—was out of the question. It would have been an admission that the Nazis were right: that he was, in fact, a Jew, not a German. As he put it: "If specifically Jewish states are now to be set up...that would be letting the Nazis throw us back thousands of years.... The solution to the Jewish question[20] can be found only in deliverance from those who have invented it."

It was this kind of reasoning that persuaded him and many other Jews to remain in Germany until it was no longer possible to get out. Some chose suicide—for example, the Marburg linguist Hermann Jacobsohn, who threw himself under a train. In the end, Klemperer avoided deportation to the death camps only because of the Royal Air Force bombing raid on Dresden in February 1945, which allowed him to shed his yellow star and lie low until the German surrender.

He remained in Dresden after the occupation of eastern Germany. It was not long before he began to notice resemblances between the language of the new Soviet-backed German Democratic Republic and the Third Reich. Like Hannah Arendt and George Orwell, Klemperer understood that the totalitarianism of the Right and the totalitarianism of the Left had fundamentally similar characteristics. In particular, they loved to impose Newspeak on those they subjugated.

Non-Jewish German academia did not just follow Hitler down the path to hell. It led the way. A few examples will suffice.

SS Oberführer Konrad Meyer, a professor of agronomy at the University of Berlin, was one of the experts who helped devise Heinrich Himmler's "General Plan East" (Generalplan Ost[21]), which, in the expectation of victory over the Soviet Union, was supposed to extend German settlement as far as Archangel in the north and Astrakhan in the south. Meyer's version proposed establishing three vast "marcher settlements" with around five million German settlers. The fate of the peoples currently living there would be either annihilation or ethnic cleansing.

In 1940, a graduate student named Victor Scholz submitted a PhD thesis[22] at the University of Breslau with the title "On the Possibilities of Recycling Gold from the Mouths of the Dead." He had carried out his research under the supervision of Professor Herman Euler, dean of the Breslau Medical Faculty.

At Auschwitz, SS Gruppenführer Carl Clauberg, a professor of gynecology at Königsberg, sought to find the most efficient way to sterilize women. Among the techniques he experimented with was the injection, without anesthesia, of caustic substances into the uteruses of prisoners.

Anyone who has a naive belief in the power of higher education to instill ethical values has not studied the history of German universities in the Third Reich. A university degree, far from inoculating Germans against Nazism, made them more likely to embrace it. The fall from grace of the German universities was personified by the readiness of Martin Heidegger, the greatest German philosopher of his generation, to jump on the Nazi bandwagon, a swastika pin in his lapel. He was a member of the Nazi Party[23] from 1933 until 1945.

Later, after it was all over, the historian Friedrich Meinecke tried to explain "the German catastrophe"[24] by arguing that excessive technical specialization had caused some educated Germans (not him, needless to say) to lose sight of the humanistic values of Goethe and Schiller. As a result, they had been unable to resist Hitler's "mass Machiavellianism."

The novelist Thomas Mann—who, unlike Meinecke, chose exile over complicity—was unusual in being able to recognize even at the time that, in "Brother Hitler,"[25] the German educated elite possessed a monstrous younger sibling, whose role was to articulate and authorize their darkest aspirations.

The lesson of German history for American academia should by now be clear. In Germany, to use the legalistic language of 2023, "speech crossed into conduct." The "final solution of the Jewish question" began as speech—to be precise, it began as lectures and monographs and scholarly articles. It began in the songs of student fraternities. With extraordinary speed after 1933, however, it crossed into conduct: first, systematic pseudo-legal discrimination and, ultimately, a program of technocratic genocide.

The Holocaust remains an exceptional historical crime—distinct from other acts of organized lethal violence directed against other minorities—precisely because it was perpetrated by a highly sophisticated nation-state that had within its borders the world's finest universities. That is why American universities cannot regard antisemitism as just another expression of "hate," no different from, say, Islamophobia—a neologism that should not be mentioned in the same breath. That is why Claudine Gay's double standards—with their implication that African Americans are somehow more deserving of protection than Jews—were so indefensible.

That is why rational minds recoiled from her argument that antisemitism on the Harvard campus is tolerable so long as genocide is not being perpetrated.

Well, the backlash against our contemporary treason of the intellectuals has finally arrived.

Donors such as the chief executive of Apollo, Marc Rowan (a Penn graduate), Pershing Square founder Bill Ackman (Harvard), and Stone

Ridge founder Ross Stevens (Penn) have each made clear that their support will no longer be forthcoming for institutions run in this fashion.

On Dec 9, 2023, Penn president Liz Magill stepped down,[26] along with the chair of the Penn board of trustees, Scott Bok. The president of Columbia was another post-Oct 7th casualty. Gay stepped down as Harvard president on January 2.

Yet it will take a lot more than a few high-profile resignations to reform the culture of America's elite universities. It is much too entrenched in multiple departments, all dominated by a tenured faculty, to say nothing of the armies of DEI and Title IX officers who seem, at some colleges, now to outnumber the undergraduates.

In *La trahison des clercs*, Julien Benda accused the intellectuals of his time of dabbling in "the racial passions, class passions, and national passions… owing to which men rise up against other men." Today's academic leaders would never recognize themselves as the heirs of those Benda condemned, insisting that they are on the left, whereas Benda's targets were on the right. And yet, as Victor Klemperer came to understand after 1945, totalitarianism comes in two flavors, though the ingredients are the same.

Only if the once-great American universities can reestablish—throughout their fabric—the separation of *Wissenschaft*[27] from *Politik* can they be sure of avoiding the fate of Marburg and Königsberg.

Excerpts from a Conversation with Sam Harris in October 2017 About Free Speech, Universities, and Cancel Culture

NICHOLAS A. CHRISTAKIS

Free Expression at Universities

In some ways, I'm a little naive in the sense that I believe in institutions. And yet, I'm also skeptical of institutions; I'm worried about institutions. But I also believe in them. And I've devoted my life to academic institutions and to what I take to be the core commitments of modern American universities, which are envied the world over. These commitments center around, for instance, the motto of Yale, which is *Lux et Veritas*. I mean, that's an extraordinary commitment: light and truth.

These institutions are committed to the preservation, production, and dissemination of knowledge. They are guided ostensibly by principles of open expression, reason, and debate and by liberal commitments to the equality of human beings, and their capacity to perfect the world, and the knowability of the world. In my view, they are committed to a belief in the objective nature of reality. I would strongly defend all these principles, and I've devoted my life to them.

In fact, even with respect to the narrower issue of free expression, I have been defending free expression—often for disenfranchised populations—for a very long time. Before I came to Yale four years ago [in 2013], I was at Harvard. My wife, Erika, and I had taken some unpopular stances back then defending the free expression of individuals who were on the side of issues of concern to Black Lives Matter, for instance, and who were protesting for social justice. For instance, there was a high school student who had worn a T-shirt that said, "Jesus was not a homophobe." And we came to his defense. And there were some minority students at Harvard who had some concerns about the final clubs at that institution (those are sort of elite fraternities). They had posted a satirical flyer around campus. Some people were unhappy about that flyer and wanted to squelch the free expression of those students, and we came to their defense.

I am committed to this, maybe naively, but I am indeed committed hook, line, and sinker to this belief that these institutions of higher learning in our society are important, that they are worthy of protection and respect. This is why, when they fail us, I get very sad. I get sad for our society. I get sad for the institutions. I get sad for the students.

…But there's another relevant set of ideas about universities in our society. If you think about these universities, they are supported by tax dollars and the bequests of (primarily) wealthy people. The reason this money is given to these institutions is to further the mission of the preservation, production, and dissemination of knowledge—not to provide faculty with easy lifestyles.

I mean, it's a wonderful thing to be a professor. I see it as a calling. But that's not the purpose of universities. The point is that we are supposed to be that place which preserves Sanskrit, which preserves Shakespeare, which preserves antiquities, which preserves mathematical knowledge and scientific knowledge. And which produces discoveries. We're supposed to be the place that transmits all this to new cohorts of young people and the public at large. That's the role we're supposed to play in society. We are to have a deep commitment to light and truth.

So, I get very upset when fields of inquiry or ideas are proscribed. I think that if our ideas are strong, they should win the battle of ideas. If you're so confident in what you have to say, you should be able to defend it. Your approach should not be to silence your opponents. Your approach should be to win the battle of ideas.

I totally support peaceful protest. I support the right of students to protest. I believe that many of the most important movements—the civil rights movement, the gay marriage movement—many of these movements, which I wholly endorse, have had the lead taken by young people and by people protesting in the streets. This is also part of the American tradition. It deserves respect and cherishing. But you cannot resort to violence or prevent others from speaking.

When we lose sight of these core liberal commitments, I think we wind up spending money and, eventually, spilling blood, which is just heartbreaking.

Objective Reality and Cultural Appropriation

…Earlier, we talked about a commitment to the idea that there is an objective nature to reality. Now, there is a long philosophical debate about this topic. It's a deep and interesting set of ideas about subjectivity. Can we even see the world objectively? Does objective reality even exist? I myself think it does. But you can make an interesting philosophical argument about this.

What about the notion of so-called "social construction," for instance, the idea that the gender of the scientist, say, or racist beliefs in a scientist, might color their objectivity? Of course, they do. We have countless examples of this. We know this from research done by historians and others. We know that it's difficult to be an impersonal observer and that every observer is situated somewhere. I think there is validity to those ideas.

However, I also think that there is an out there, out there, and that it is knowable, and that we can do our best to understand it. When you carry the rejection of objective reality to the extreme that you call it a tool of cis-heteropatriarchy, you really have jumped the shark. You've taken a core idea, which says, "Look, we need to not always believe what we are told," or "We need to understand how a person's position in society might affect what they see," and you have carried it to an absurd extreme.

We know this affects even ostensibly objective phenomena. Historian Emily Martin, for instance, has done some fantastic work (which I teach) on how scientists looking at cell division or menstruation interpret the intrinsic biology by virtue of who they are. But then this idea can be taken by others to such a ridiculous extreme that it becomes absurd.

Similarly, consider the notion of "cultural appropriation." The kernel of the idea there is that some communities of people can be so denigrated that not only are they, let's say, killed and wiped out, but also that their ideas and culture are stolen from them. They can even be effaced. All that's left is a caricature of who they are. There is some truth to this, too. And it's like adding insult to injury: not only might we engage in genocide, but we might take all your ideas, your culture, as well, for ourselves and not even credit you.

The problem is that, again, this idea is carried to a preposterous extreme. Because, of course, the whole history of ideas, and of culture, and of art and music, is one of endless theft. I mean, it's endless modification, and transformation, and exchange of ideas, and thoughts and musical and artistic forms, and so forth. To then start claiming that, like in the Reed College example [in September of 2017], that [this white professor] couldn't teach these things, she couldn't wear a shirt with the expression "Poetry is Lit," because she's appropriating African American slang, it's just a crazy caricature of what is otherwise potentially an interesting philosophical idea to discuss.

I think this is the thing that has made it especially hard for me: that I believe that I have a more than passing understanding of the epistemology here, and I have a more than passing sympathy for some of the concerns that the students have, say, about police brutality, about economic inequality, about racial justice. But I am also deeply concerned with the Maoist abandonment of reason and of discourse, and the kind of dehumanizing and atomizing of people that many of the students and faculty wind up espousing.

I mean, one of the things that has really just depressed me, in the courtyard that day—[*Editor's Note: See explanation of the courtyard experience below*]—was that there was a young African American woman who said to me, very plaintively (and it pulled at my heartstrings), she said: "You cannot understand our predicament because you are middle-aged, and white, and male."

And I said to her that I understood what she was saying but that I nevertheless believe in our common humanity. I believe that all of us— and I still believe this—that all of us, as human beings, can speak to, and understand, each other, united by our common humanity. So, even though I was a different gender, age, and skin color than she was, that I nevertheless could understand her, and that I was *interested* in making the effort to understand her, and I would hope that she could understand me, too.

But the students jeered at this! They jeered in the courtyard that day—which was incredible. There was another student, a minority student, who later wrote a post in the *Yale Daily News*, where he wrote that he had never been more disappointed in his colleagues than when—the titles at the time were that we were the "Masters" of the colleges, but now, we're called "Head of College"—he wrote: "I've never been more disappointed when the Master made the argument about our common humanity and my peers jeered."

My point is that when you abandon the commitment to our common humanity, when you atomize people, when you believe that only certain types of people have authority to use certain types of cultural ideas or tropes, you efface, for me, a fundamental reality of our common humanity and a fundamental tool we can have to interact with each other.

That professor at Reed, the claim that she can't wear a T-shirt that says "Poetry is Lit" is, to me, preposterous, and it violates every basic principle, in my view, that should animate a civilized society.

Yale Halloween Experience

The original email that started the fracas was sent by a dean, a person in the dean's office, a man by the name of Burgwell Howard, who had previously been a dean at Northwestern University. He had sent the same Halloween costume email there, at Northwestern, years before! Then, he decided to resend it five years later at a different university, namely Yale. There had been, to my knowledge, no episodes of students wearing blackface at Yale, or pushing the boundaries in such an extreme way, in the past ten years. Nevertheless, this email was sent out, and it was subsequently very strongly endorsed by the dean of Yale College, Jonathan Holloway, who has since left Yale to be provost at Northwestern.

Actually, in the *New York Times* the previous month, there had been a whole exchange about this Halloween costume guidance issue. In the zeitgeist, people were talking about how this was getting a little out of hand and seemed a bit silly, that universities were providing official guidance on Halloween costumes. I think there were six people who wrote in that article, and five were against Halloween costume guidance, and one was for it. There had also been a number of emails that had come out at Yale, sent by various other administrators, in the run-up to Halloween.

I think the one that Dean Howard sent was maybe the third, and broadest, and most detailed. It had links to acceptable and unacceptable costumes or recommended and non-recommended costumes. This email was coming from a positive intention, I suspect, that is to say that it's not necessary for students to set out to cause needless offense. I think that, in a free society, we have to tolerate offense, but it's not like I'm interested in deliberately offending people.

Anyway, what had happened is that we had been hearing from the students, and Erika in particular had been hearing from her students, that the students felt infantilized by this email. Many of the students were objecting to this. They couldn't believe this. Erika, that day she sent the email, had taught a session of her class, in late October, a class was about child development. It was an animated and intellectually rigorous conversation about what stage of development college students were at and about whether they were capable of choosing their own costumes or negotiating among themselves if they had taken offense, talking to each other, and so forth.

This is more detail than you want, probably. But, earlier in the year, in August 2015, I had sent an email to the students, the four hundred students in Silliman College, our college, about the murders in Charleston—where this man, whose name I'm blocking, thank God, had gone into the African Methodist Episcopal Church, the mother church, in Charleston, and slaughtered nine or ten people at close range, people who had welcomed him into their midst. He was white, and the victims were all black. It was a vile and despicable carnage motivated by racial hatred.

There had been a lot of discourse in the public space that summer. That was the summer when all the Confederate flags began to finally come down. I was very concerned about these events, like many people were. I had organized a series of speakers at Silliman. We had a famous African

American historian from MIT [Craig Steven Wilder] who came and spoke about the history of slavery and American universities. We had some people talking about other aspects of this. Also, I had booked, months earlier, Greg Lukianoff [the head of the Foundation for Individual Rights and Expression], who would come to speak about free speech. I had arranged a series of public speakers.

Anyway, I sent an email in late August or the beginning of September to the students in the college about the aftermath of Charleston. I talked about how, as a public health expert, one of the things that I found most distressing was that Walmart had stopped selling Confederate flags, but it had not stopped selling guns. I argued that, in my view, this had it backward. There was all this focus on *symbolism* but not on practical concerns that, really, we needed to address, such as issues of inequality and issues of violence in our society. And these symbolic things, while important, were potentially distracting us.

We had done this previously. When I had been at Harvard, Erika and I had a similar role there. We would regularly communicate with our students in this fashion, and some would agree, and some wouldn't agree. We had debates there about religious symbols, and public places, and vegetarianism. Could we roast a lamb at Greek Easter in the college courtyard using university money to purchase the lamb? It raised interesting questions for the students to debate.

Anyway, we got all this positive feedback for this, and there had been a lot of students complaining about the Halloween costume guidance email from the dean. That was the history and the background. The *New York Times* article was in the public sphere. Yale students thought it was infantilizing. Previously, we had gotten some praise for engaging the students with ideas. That's what motivated my lovely wife, who has spent her career taking care of battered women, and inner-city children, and homeless substance users, and this had been her life—we're very progressive people, for what that is worth—that's what got her to send this email, which said, "Do you, students, really want or need this level of oversight?"

Just to clarify: my wife's argument was not actually taking a stand one way or the other on whether the guidance was necessary, and one way or the other on the costumes. She was saying, "You students should probably consider whether you wish to surrender this authority to super-ordinates." Fundamentally, this was a left-wing position saying, "You should be deeply skeptical of surrendering power to the administration, and you should talk about that."

…But, you see, the problem is, again, this is where I have some empathy and sympathy for the students, too. This is what was so challenging because, again, you see there's a kernel of truth here. Like we discussed earlier with the notion of cultural appropriation or the claim that scientific objectivity is a tool of oppression—these ridiculously, extreme claims—there's an element of truth, as well, to the students' sense of alienation.

Part of it, again, is developmental. Students who are eighteen to twenty-two years old feel a sense of alienation. We all did, in different ways. Now, if you're a minority student in these institutions, there may be an extra amount of alienation that you feel. I think there are ways that we can discuss that with students. I think there are ways we can reform our institutions. I don't lack sympathy for that, but as Jonathan Haidt has said, I think the fundamental commitment of these institutions is to *lux et veritas*. This has to be done in a way in which we retain a deep and abiding commitment to speaking the truth and having open expression.

…At around 4:00 PM that day [November 5, 2015], the students had assembled in the Silliman courtyard and were chalking remarks on the walkways. Some were very positive, such as "We are one Yale," which I totally endorse. Some were very specific and targeted at us and, I would say, were harassing and threatening (for example, "We know where you live"). And I felt that it was appropriate to model my commitments, and that I had to walk the walk, and not just talk the talk, and that it would have been cowardly to not talk to the students about their strongly held views.

So, I walked out of the Master's Residence and went out into the courtyard around 4:00 PM to witness what the students were writing and just to talk with them. I walked quietly around the courtyard and made a show of reading what they had written to dignify what they were communicating and to model open expression. I just was reading all the prose and these slogans that were written everywhere in colorful chalk.

Unbeknownst to me, the students had just left the larger Cross-Campus courtyard elsewhere at Yale, where they had previously surrounded the dean of the college (Jonathan Holloway), and some fraction of them—I would say 100 to 150 students, or something like that, maybe 150—came to the Silliman courtyard, and they assembled. Eventually, some of the videos that were taken were released. I mean, there were a dozen or two dozen people filming that event, as became clear later. At least six people released videos online, I think, within a few days. I think you can assemble, from different clips that people released, the full hour between 5:00 PM and 6:00 PM (the crowd eventually dispersed at 6:15 or so).

...It was a challenging time, certainly, in my life. It was a challenging time, I think, in some of the students' lives. I was very upset that there was a lot of hostility expressed towards one of the students, in particular. I came to her defense the next day, on November 7, 2015. I sent a tweet out [at 5:49 AM] saying that "no one, especially no students exercising right to speech, should be judged just on [the] basis of [a] short video clip." She was wrongly doxed by another organization—not by Greg, of course. And she got many, many vile threats. I completely repudiate those threats.

This was a young woman who I didn't know well but who, by report, was otherwise a very sensible person. I think she got swept away, as happens with mobs. I think she was not her best that afternoon, though I think she should have known better, nevertheless. Many people were not at their best. I mean, it's very important also to note, as you have suggested by having seen much more of the video footage, that a number of students were very strongly challenged by their emotions that day.

...I think that what I was trying to do in that situation was to try to get the students to realize what they were doing.... I was trying to avoid the circumstance in which the students deindividuated. There is an inflection point during social movements of all kinds, in group dynamics, where you can reach an inflection point where people suddenly feel anonymous and are disinhibited, and when social inhibitions fall away, and people start acting in ways they would not otherwise act. This is well understood.

On Responding to Hate and Mob Behavior

Anyway, on the point of deindividuation, it's very important for the people to feel themselves to be as individuals, and not as part of the crowd, and to feel themselves capable of moral agency. You want the people to be identified by name. "I am so-and-so. I'm not just part of this crowd." You want them to see the person to whom they're speaking as a human being. And so I introduced myself to them and shook their hands as they spoke to me in the crowd.

I have gotten death threats periodically in my life—not many, two, or three, or four times; and I've gotten hate mail the last ten or twenty years about different things. I always respond to it unless I can't cope with the volume of mail I'm getting. A person will send you these very vile things, and then, when you write back to them, they say, "Oh my god. I didn't think you would answer me." Then, they'll say: "I'm actually not a bad person. I'm so sorry I said those mean things to you."

You can literally defang many people who send vile things because they don't recognize that you're a human being on the other end of the line and that you're actually capable of talking to them. Not everybody can be dealt with this way; but some people can be dealt with this way.

There was a wonderful experiment done by a graduate student at NYU, published about six months ago, in which he developed a system for identifying racist speech online, of finding people who were tweeting a lot of very racist things. He developed these little bots; they were actually more like sock puppets. He developed these fake accounts, with either a white person or a black person shown in the little photograph, the avatar, of the Twitter account. The person also either had few followers or many followers, and so had low status or high status. [The 2017 paper was by Kevin Munger, entitled "Tweetment Effects on the Tweeted: Experimentally Reducing Racist Harassment"]

So, he had a group of people who were sending out racist tweets. When they did that—this racist person, let's say, was sending out a very vile, racist tweet, with bad language, to another person—he was able to spot it. The racist person A sends out a tweet to person B. Then, his bot, person C, would respond to person A and say very sweetly, "Hey, man. You probably shouldn't do that. There's a real person on the other end of that."

He found that this simple intervention, especially if you had a white person that did that with high followers—his experiment was to test whether the status of the intervener mattered, and it was always helpful, if I remember the experiment—he was able to show that the simple cultivation in the person expressing hatred of a recognition of the common humanity of the target person attenuated the behavior for months afterward. The racist account reduced or eliminated the racist tweets they were sending.

My point in this example, and in the other stuff we're saying, is that you can actually use these basic liberal principles of our common humanity to redress, and address, wrongs, hatred, and violence in our society. In some ways, you can attempt to tamp down a little bit on certain aspects of mob behavior.

…I hesitate to oppose anonymity because I also think anonymous speech should be allowed. I struggle with the issue of anonymity. I think, on balance, online, it's a corrosive force, but I'm not prepared to say that nobody who's anonymous should be allowed. I've taken this stand publicly: namely, that I think you should have the courage of your convictions. We should, all of us, work together to cultivate a society where we do not demonize people for their beliefs and where we engage them or we ignore them.

If we can create that kind of a culture, I think we could, in parallel, facilitate open expression. If we don't fire people for their opinions…for instance, like the ESPN reporter who recently called Donald Trump a white supremacist, I don't think she should lose her job. I think people should be allowed to have and express their opinions. I don't think we want a society where people are losing their jobs for expressing their beliefs, especially if they do so outside their place of work.

If we do that, this also means that, if we have such a society, then I think we would make it easier for people to have the courage of their convictions and to publicly express their beliefs. These issues all are locked together. This call-out culture and this culture of outrage interconnect with these problems of anonymity and troll-like behavior, and fear of open expression and concealment of beliefs.

On Campus Free Speech

A lot of the cases that are brought up these days as speech that's out of bounds are amazing to me. First of all, there are the well-understood exceptions. Like, one of the distinctions that's often forgotten is the distinction between free expression and harassment. Also, even under the harassment, there's an exception for public figures.

Harassment is when you have *repeated* and, typically, *targeted* speech against a particular individual. For example, famously, the Nazis can march through Skokie, Illinois, the Supreme Court ruled, but they cannot stop in front of my house. The former is free expression; and the latter is harassment. This is well understood. There are boundary cases and difficult cases, of course. But you can't threaten an individual. But, it turns out, you can actually say vile things about public figures like the president or politicians. There's a whole other jurisprudence there, as well, because such figures don't have the same expectation of not being "harassed" or criticized as other people.

There are all these understood differences. There are boundary conditions about inciting violence and imminent danger versus non-imminent danger. For example, there are these left-wing professors who have been calling for white genocide. You can make abstract commitments to white genocide, but it's different than saying, "Let's get guns today and go kill so-and-so."

There are all of these cases. And there's a long history of people thinking deeply about free expression and where the limits are. One of the things that's amazed me is that, in my view and in the view of most people who had experience with First Amendment issues, none of the cases we've been discussing about expression on campus are anywhere near those boundaries.

> ...There are many theories about what's happening. What I would say is that the actors that I think have a duty to address all this are the faculty. Earlier, you quoted from my New York Times piece [Editor's Note: Available in this volume] from a year so ago, and, in that piece, I said that I think it's our obligation, as a faculty, to preserve the

commitment of these institutions to free and open expression. I think it's our duty, as a faculty, to push back against the false claim that speech is violence. I'm not necessarily going to blame the faculty, but I'm going to say that I think the faculty has a duty to oppose these illiberal moves. And I would hope that more and more professors would see this.

...I think all of us together need to work to create a culture of open discourse—certainly within our universities and hopefully in the broader society. At least, that's my view of the kind of society I think we should have and that I would like to live in.

Universities as Dispensers of Parasitic Ideas

GAD SAAD

Much of my academic work has been at the intersection of evolutionary psychology and human behavior in general and consumer behavior in particular (cf., Saad 2007, 2011, 2013, 2017, 2020a, 2021; Saad and Gill 2000; Saad and Stenstrom 2012; Saad and Vongas 2009). While my colleagues in the natural sciences easily accept the fact that human minds are shaped by the dual forces of natural and sexual selection, this has not been the case amongst the great majority of my colleagues from the social sciences. According to the latter group, biology matters in explaining the behaviors of all species, with the exception of one: *Homo sapiens*, which has been referred to as the *human reticence effect* by Ranney and Thanukos (2011). As such, throughout my academic career, I have faced great animus from a wide assortment of academics, including but not limited to postmodernists, radical feminists, cultural relativists, and social constructivists, precisely because foundational tenets of evolutionary biology and evolutionary psychology are contrary to the quasi-religious revealed truths of these various camps. When an evolutionary psychologist posits that there are universal truths arising from a shared biological heritage, this serves as an attack on postmodernism, which purports that there are no objective truths as all knowledge is apparently constrained by subjectivity and epistemological relativism. When the same evolutionary psychologist proposes that there are innate sex differences that arise from sex-specific evolutionary forces, the radical feminists are triggered since they learned in women's studies departments that the two sexes are different solely because of sexist socialization. If an evolutionary-minded philosopher proposes that there are absolute and universal moral precepts, this upsets the cultural relativists, who will retort that imposing a universal moral code on a specific culture is a form of cultural imperialism. After all, the cultural relativist will say, "Who are you to judge whether it is acceptable to cut off the clitorises of five-year-old girls?" If you explain to social constructivists, as I did in Saad (2017), that sex-specific toy preferences are not due to the sexist and patriarchal whims of one's parents, they are offended by my "biological determinism" (a nonsensical concern).

In my doctoral training, I was educated within the behavioral decision theory framework, which explores how humans might depart from axioms of rational choice (see Kahneman, 2011 for a summary of this work). Hence, I am well versed in how people might behave irrationally when juxtaposed against the rationality benchmarks set by classical economists. For example, the transitivity axiom suggests that if an individual prefers product A to product B, and prefers product B to product C, then it must be that she prefers product A to product C (via transitivity). Failure to exhibit this preference implies a form of axiomatic irrationality (Tversky, 1969). But the irrationality that I have witnessed in my thirty-plus-year career as a professor within academia could not be explained using principles from axiomatic rationality. What could explain how otherwise educated and intelligent colleagues could spawn and promulgate the outlandish views covered in the previous paragraph? In seeking to answer this grand question, I searched for a framework that would allow me to explain how so many of my colleagues were espousing truly anti-reason, anti-reality, and anti-science ideas.

As an evolutionary behavioral scientist, I am well familiar with the field of comparative psychology, which explores the origins of human cognition by contrasting a given phenomenon across multiple species (cf. Morwitz, 2014 on how to apply comparative psychology within the field of consumer research). In other words, if one wishes to explore the evolutionary signature of a given behavior, a comparison across multiple species might shed some light. With that in mind, I scoured the vast animal literature and quickly identified neuroparasitology as the relevant framework for my quest to explain the spawning and spread of irrational ideas. The animal kingdom is replete with countless parasite-host interactions across many taxa (Poulin and Morand 2000). Parasites have evolved a wide assortment of parasitic strategies, including which organ they end up parasitizing. Take, for example, the parasitic tapeworm. It lives in the host's intestine. A neuroparasite, on the other hand, chooses its host's brain as its ultimate destination. There are innumerable examples of an animal, which can engage in behaviors that are suboptimal to its survival but that is beneficial to the neuroparasite residing in its brain (cf. McAuliffe 2016, Moore 2002). Terrestrial wood crickets prefer to avoid water, but when their brains are parasitized by a particular hairworm, they merrily jump into water and commit suicide. The hairworm needs to complete its reproductive cycle in water, and as such, it alters the behavior of the hapless wood cricket to suit its interests (Sanchez et al. 2008).

In *The Parasitic Mind: How Infectious Ideas Are Killing Common Sense* (Saad 2020b), I argued that universities are the progenitors of all the parasitic idea pathogens that have sent the West into a death spiral within the abyss of irrational darkness. Specifically, I posited that not only can human minds be potentially parasitized by actual brain worms (e.g., toxoplasma gondii, cf. Flegr 2007), but also, they can be parasitized by ideological rapture. Viewed from this perspective, postmodernism, identity politics, and the DIE cult (diversity, inclusion, and equity), along with its rejection of the meritocratic ethos, cultural relativism, social constructivism, radical feminism, biophobia (fear of using biology to explain human behavior), political correctness, critical race theory, echo chambers bereft of political and intellectual diversity (as occurs in academia), a culture of orgiastic victimhood and never-ending offense, and a progressive bent toward cultural self-loathing and existential self-flagellation all serve as parasitic idea pathogens destroying people's ability to think critically.

Parasitic Ideas—A Canadian Perspective

Since the release of *The Parasitic Mind* in 2020, I regret to say that things have yet to improve within academia in general and Canadian universities in particular. The University of Waterloo, a well-regarded Canadian university best known for its engineering and computer science programs, recently advertised for two open NSERC [Natural Sciences and Engineering Research Council of Canada] Tier 1 Canada Research Chairs at the Cheriton School of Computer Science. Here is the exact call as listed on their website [bold in the original]:[1]

> "Position 1, all areas of artificial intelligence. The call is open only to qualified individuals who self-identify as women, transgender, gender-fluid, non-binary, or Two-spirit.
>
> Position 2, all areas of computer science. The call is open only to qualified individuals who self-identify as a member of a racialized minority."

My undergraduate degree is in mathematics and computer science from McGill University. I studied artificial intelligence with Professor Monty Newborn, who is a world expert on the use of AI in chess (e.g., Deep Blue). I was unaware that my ability to understand artificial intelligence would have been greatly enriched had my professor been non-binary or Two-spirit. The most awe-inspiring course that I ever took in my long career as a university student was Formal Languages, taught by Professor Denis Thérien. I was utterly mystified by the brilliance of Alan Turing and his work on Turing machines. The intellectual depth of his insights borders on a mystical religious experience. Turing is arguably one of the leading minds of the twentieth century, but according to the call from the University of Waterloo, had he applied for one of these two open positions, he would have been turned away. Turing was gay, but he was not a woman, transgender, gender-fluid, non-binary, Two-spirit, or a member of a racialized minority. He "suffered" from being a white male, so apologies, Dr. Turing, but your services are not wanted here. This is happening in the twenty-first century in Canada.

Not to be outdone by the parasitic ideological rapture of the University of Waterloo, another leading Canadian university, the University of British Columbia, put out a call for a Tier 1 Canada Research Chair in Oral Cancer Research in the Faculty of Dentistry. The call made it clear that only certain applicants would be considered for this prestigious professorship:[2]

> ...the selection will be restricted to members of the following federally designated groups: people with disabilities, Indigenous people, racialized people, women, and people from minoritized gender identity groups. Currently, UBC has a gap in representation for people with disabilities. Until such time as this is remedied, the names of those self-identifying as having a disability will be provided separately to the search committee in order for them to follow preferential hiring strategies.

My own university, Concordia University, has been a leading light in the parasitic woke movement. Several years ago, researchers obtained a sizable governmental research grant (New Frontiers in Research Fund) to decolonize light. Heretofore, our understanding of light has apparently been too focused on "white" perspectives, so it is important to incorporate an Indigenous lens. The tagline of the project reads as follows: "Tracing and countering colonialism in contemporary physics." A summary of the project is explained on their official website:[3] "The Decolonizing Light project explores ways and approaches to decolonize science, such as revitalizing and restoring Indigenous knowledges, and capacity building. The project aims to developing a culture of critical reflection and investigation of the relation of science and colonialism." The Decolonize Light project is in line with the recently released five-year strategic plan of Concordia University to decolonize and indigenize the entire curriculum and pedagogy. The plan was explained as follows:[4]

> [it] draws upon the principles embodied in the Two Row Wampum Belt, or Tekani Teiotha'tá:tie Kaswéntah', an ethical framework for how colonial-settler governments are to conduct themselves while living in the lands of the Rotinonhsión:ni—more commonly known as the Haudenosaunee Six Nations Confederacy.
>
> For Donna Kahérakwas Goodleaf, director of decolonizing curriculum and pedagogy in the university's Centre for Teaching and Learning, impressing the principles of the

Two Row Wampum Belt into the strategic plan creates a path where everyone is equal and no worldview is superior.

Hence, a leading research university is arguing in the twenty-first century that the scientific method is only one of many possible epistemologies for understanding the world. My main research areas are evolutionary psychology, consumer psychology, and the psychology of decision-making. I wonder how I will go about indigenizing and decolonizing those fields of study. Hopefully, I will be assigned a patient Indigenous mentor who will guide me through the process.

On May 17, 2024, my university organized a one-day symposium titled "Periods on Campus: Sustainable Research, Activism, and Advocacy for Menstrual Equity in Higher-Ed."[5] The opening paragraph of the call for contributors reads as follows:[6]

> This one-day symposium will bring researchers, student advocates and diverse voices together in a higher educational setting, to discuss the importance of menstrual equity, and the positive impact it can have when considering and including sustainable approaches. We hope to enhance the quality of student life by moving meaningfully towards a culture of menstrual dignity on-campus, showing that higher-ed institutions care beyond the education of students. Menstrual equity is about ensuring that menstruators have access to the menstrual products they need, without facing financial or social barriers. It's about breaking down the barriers that can make menstruation a difficult and even stigmatized experience and recognizing that menstrual equity is a basic human right. Inaccessibility to menstrual products, and lack of supports and services, directly impacts the experience of menstruators on campus. It's time for change!

As someone who has published scientific papers on various behavioral effects of the menstrual cycle (cf. Saad and Stenstrom 2012), I was unaware that menstruation was a contentious human rights issue. Furthermore, I have learned that women should be referred to as "menstruators" since I suppose transmen too can menstruate. It is always preferable to use inclusive and empathetic language as dictated by the ever-changing and capricious whims of the politically correct language police.

As arguably the professor who has been the most outspoken about wokeism and its associated parasitic ideas for several decades, it has not been an easy relationship with my university, given its full commitment to woke nonsense. Between 2008 and 2018, I held the Concordia University Research Chair in Evolutionary Behavioral Sciences and Darwinian Consumption (two five-year mandates). When it came time to reapply for the chair, I was turned down several years in a row even though, using objective bibliometrics, it should have been a formality. For nearly a decade, my university has largely ignored all of my professional accolades, including the release of my latest two books (one of which was an international bestseller), many scientific publications in leading academic journals, and numerous honors and awards (e.g., in 2022, the prime minister of India, Narendra Modi, sent me a personal commendation letter in celebration of India's Republic Day regarding my books and ideas;[7] in 2017, I testified in front of the Canadian Senate and delivered a lecture on Parliament Hill;[8] a *Wall Street Journal* article was recently written about Elon Musk's appreciation and support of my work).[9] I recently had an MSc/PhD course canceled for the first time in nearly three decades as a professor. When the semester began, I had a sizable enrollment, and then, out of the blue, on the same day, nearly all the students dropped the course in unison (even though I had yet to hold a class that semester). I am now being forced to make up the canceled class during the upcoming summer semester. I was recently advised that my undergraduate course on evolutionary consumer psychology would likely be canceled both semesters in the upcoming year. Instead, I am being forced to teach two new courses that are outside of my areas of interest and expertise. Effectively, I am being treated as though I were a sessional lecturer hired on a teaching contract rather than as a chaired professor with thirty-plus years as a distinguished scholar. The bottom line is that while the outside world greatly appreciates my contributions in defending reason, science, and the ethos of meritocracy, my own university views me as a pariah. I cannot sit idly while the edifices of reason are being destroyed one brick at a time. I refuse to apply for research grants that require diversity, inclusion, and equity statements. I refuse to indigenize my pedagogy. I refuse to teach "principles" from gender ideology that negates the reality that humans are a sexually dimorphic and sexually reproducing species comprised of two phenotypes: male and female. I refuse to teach that the scientific method is one of many ways of knowing. I refuse to teach that biology matters for all species on earth except one called *Homo sapiens*. I refuse to be uniquely an "ally" to a specific group of students based on their skin hue, ethnicity, or gender orientation. I am a mentor to all students who wish to learn, irrespective of their irrelevant identity markers. This should be lauded, and yet, it is viewed with derision in today's academic climate. I am the professor of the people, and not the professor of some people more so than others.

Call to Action—How to Inoculate Our Universities from Parasitic Ideas

Universities exist to create and disseminate knowledge. They are meant to enrich the human spirit by expanding the boundaries of what is known. They should not be centers of political and ideological activism. The idea that the pursuit of some knowledge should be forbidden (cf. Kempner, Merz, and Bosk 2011) on ideological grounds is anathema to the mission of a university. The pursuit of truth is a deontological principle and not a consequentialist one. Identity politics and associated movements, such as the DIE cult, need to be eradicated from science. Meritocracy is de rigueur when it comes to adjudicating academic excellence. Students must be inculcated with the intellectual reflex to welcome opposing ideas rather than shrivel into a fetal position because they are triggered and offended by those who think differently from them. Academics should be trained to be intellectual Navy SEALs who boldly venture into unchartered territory rather than being meek invertebrate castrati deeply concerned about adhering to the prevailing orthodoxy. The scientific method is the greatest epistemological tool that human minds have ever created. It liberates us from the confines of our irrelevant identity markers. There is no Indigenous way of doing mathematics, no Albanian way to study light, and no Lebanese-Jewish way to explore the evolutionary roots of human behavior. There is simply mathematics, physics, and evolutionary psychology. The quicker that we return to a commitment to science, reason, evidence-based thinking, meritocracy, and the promotion of individual dignity over tribal collectivist machinations, the faster we will cure our ailing universities from the cancerous parasitic ideas that have been allowed to flourish within our hallowed halls of academia.

In *The Parasitic Mind*, I implored people to activate their inner honey badger. The honey badger (*Mellivora capensis*) is the size of a small to medium-sized dog, and yet it can withstand hundreds of bee stings, confront a group of adult lions, and escape from the death grasp of a constrictor and return to kill it whilst fighting off jackals who wish to steal the dead snake.[10] It has been referred to by the *Guinness Book of Records* as "the most fearless animal in the world."[11] As the old saying goes, "honey badgers don't give a f**k." This is how we must fight against the parasitic lunacy that has engulfed our institutions. Do not be afraid to challenge bad ideas. Do not presume that others will do it on your behalf. Do not cower away because of your careerist concerns. We all have a burden to bear when we speak out, and yet, if we all decide to diffuse the solemn responsibility to speak out onto others, slowly, we will be led to the abyss of infinite lunacy. I am optimistic by nature. I believe that if the silent majority, all of whom despise these idea pathogens, were to speak out in unison, we would return our universities and, more generally, our societies to their former glory.

Spotlight on Scientific Censorship

ANNA KRYLOV AND JAY TANZMAN

"Nothing in life is to be feared; it is only to be understood. Now is the time to understand more, so we may fear less."
Marie Sklodowska-Curie

Censorship—the suppression of facts and ideas—is as old as history itself. Censorship has been invoked to protect people's minds from corruption by bad ideas, to shield religious truths from heresy, to protect the feelings of the faithful from blasphemy, and to ensure the safety of the state in times of war. Suppression of facts and ideas is antithetical to the production of knowledge; yet, from its inception, science has been a target of censorship. Despite the key role science plays in reducing human suffering, providing solutions to pressing problems of humankind, and improving the lives of people worldwide, censorship in science is endemic in even the most advanced democratic societies.

A new and disturbing chapter in the history of scientific censorship suggests that censorship is worsening. Science journals and publishers have pledged to censor scientific articles that are alleged to be "harmful" to a particular group or population, a practice that violates the guidelines of the Committee on Publication Ethics (COPE n.d.). The practice began with scientific journals retracting articles in response to the demands of online mobs but has since been codified into policy by various editorial boards and scientific publishers.

Censorship is objectionable on both philosophical and pragmatic grounds. On the philosophical side, the notion that the public must be protected from dangerous or harmful knowledge is antithetical to liberal Enlightenment values, according to which knowledge is power that the public is capable of using responsibly. On the practical level, by hiding selected facts, censorship distorts our understanding of the world, undermining our ability to solve challenging problems. Moreover, censorship leads to distrust in science. When scientists hide selected facts to promote their political agendas, the public rightfully perceives them as politically motivated agents rather than objective and trustworthy experts.

Despite the long history of scientific censorship and its current prevalence, the mechanisms by which censorship operates, the agents who impose censorship, their motives, and the ultimate costs of censorship have not been systematically investigated. A recent paper in the *Proceedings of the National Academy of Sciences* (*PNAS*) by Cory Clark and thirty-eight coauthors, "Prosocial Motives Underlie Scientific Censorship by Scientists: A Perspective and Research Agenda" (Clark et al. 2023), takes a stab at this issue. The paper lays out important questions regarding the nature and consequences of censorship and calls for systematic research on the subject.

One of the paper's coauthors, psychology professor Steve Stewart-Williams (2023), summarizes the evidence for the current wave of scientific censorship and self-censorship, as well as the rise of censorious attitudes among scientists, which motivated the paper:

- Increasing numbers of scientists report being sanctioned for conducting politically contentious research (German and Stevens 2022).
- Retractions of papers have become more and more common over the last decade, and at least some of these appear to have been driven primarily by concerns other than scientific merit. One group of scholars even retracted their own paper, not because it was scientifically flawed, but because it was being cited by conservatives in ways the authors didn't approve of (AlShebli et al. 2020; Savolainen 2023a).
- Several lines of research suggest that studies reaching politically unpalatable conclusions may have a harder time negotiating the peer-review process than they would if the conclusions were in the

opposite direction (Stewart-Williams et al. 2022). As Clark et al. (2023) note, "When scholars misattribute their rejection of disfavored conclusions to quality concerns that they do not consistently apply, bias and censorship are masquerading as scientific rejection."

- Recent surveys suggest that many academics support censuring or censoring controversial research, with support being strongest among younger scholars (Honeycutt et al. 2023).
- Unsurprisingly, recent polls also suggest that many academics now self-censor on even mildly controversial topics (German and Stevens 2022).
- A large number of academics express a willingness to discriminate against conservatives when it comes to hiring, publications, grants, and promotions. Unsurprisingly, conservative scholars are particularly likely to self-censor (Inbar and Lammers 2012).
- A growing number of journals have explicitly committed to judging scientific papers not just on the quality of the research but also on their (supposed) social or political impact. "In effect," note Clark et al., "editors are granting themselves vast leeway to censor high-quality research that offends their own moral sensibilities" (*NHB* 2022).

Table 1. Taxonomy of scientific censorship

Types of censorship	
Hard	Authorities (e.g., governments, universities, academic journals, professional societies) exerting power to prevent dissemination
Soft	Formal or informal social punishments or threats of them (e.g., ostracism, reputational damage) aimed at pressuring the target
Censors	
Government	Political figures and other governmental institutions
Institutions	Universities, professional societies, journals, publishers, funding agencies, and other organizations
Individuals	Peer scholars, activists, donors, reviewers, or other members of the public
Self	Scholars choosing not to pursue or disseminate their own controversial ideas
Motivations of censors	
Self-protection	Protect one's own reputation
Self-enhancement	Elevate one's own status as virtuous or otherwise valuable
Benevolence	Protect the target of censorship from negative consequences
Prosocial	Protect third parties from the censored content
Punitive	Control narrative and punish the target of censorship
Outcomes of censorship	
Success	Prevents censored content from reaching all or some of the intended audience
Conflict	Creates public controversy, persuading some that the content has been discredited, and others that illegitimate censorship has occurred
Backfire	Censorship attempt brings more attention or legitimacy to the content

Note. Different censorship motives are not necessarily mutually exclusive and, in some cases, may be positively related.

Table 1 from the Clark et al. (2023) paper presents a taxonomy of censorship. The paper reveals an interesting fact: as the table shows, *it is often scientists themselves who are the censors.* This phenomenon is not new—a famous historical example is the case of Galileo Galilei, who was censured by the Roman Catholic Church for his revolutionary theory of heliocentrism. Although Galileo's punishment was decided by the Catholic Church, his persecution was driven primarily by Aristotelian scholars who objected to his theory on philosophical and scientific grounds and appealed to the Catholic Church's authority to punish him.

The paper also explains that the motive for modern scientific censorship is often pro-social. A better understanding of why people engage in censorship and what the downstream costs of censorship are could lead to more effective strategies to combat it. Coauthor Lee Jussim points out,

> University administrators are now well-versed in supposed threats to social justice; far fewer know much about or have deep commitments to academic freedom. Consequently, the immediate and downstream costs of censorship are rarely considered or weighed against the supposed benefits of not causing offense. No wonder we have seen a rising tide of scientific censorship (Krylov and Tanzman 2023a).

Lead author Cory Clark expresses hope that the paper

> will raise the standards of our science leaders and decision-makers who aim to obstruct science based on their personal moral intuitions. At minimum, they should be held to the same standards as the rest of us who strive to create and disseminate science, and make their case with data (Krylov and Tanzman 2023a).

Motivated by the publication of this foundational paper (Clark et al. 2023), we have compiled the following virtual collection of scientific papers, viewpoints, and op-eds to document the modern rise of censorship in science.

The Virtual Collection

Numerous commentaries on Clark et al. (2023) have been published, including op-eds (al-Gharbi and Clark 2023, al-Gharbi and Barbaro 2023, al-Gharbi 2023, Alonso 2023), blog posts (Stewart-Williams 2023, Jussim 2023, Kuntz 2023, Coyne 2023, Jacobs 2023), and a YouTube interview (Revkin et al. 2023).

Not all responses to Clark et al. (2023) were wholly in agreement with the paper. Some op-eds pointed out additional aspects of the censorship phenomenon but supported the main thesis of Clark et al. (2023), that is, that more research and discussion is needed. Darlow and Gray (2024) published the critical comment "Censorship or Inclusion?" in *PNAS*. Their critique, asserted without supporting evidence, was that censorship is needed to protect minorities from suffering harms that are inevitable as long as the groups remain underrepresented as investigators in science. The critique was countered by Clark et al. (2024) as follows:

> Darlow and Gray appear to agree with our main point that scientific censorship is often driven by prosocially motivated scientists. They suggest this censorship is warranted, whereas we call for caution. As described in our paper, many sources of data could enable metascientific investigation of the frequency, costs, and benefits of this form of censorship, but in the absence of such investigation, we are "left to quarrel based on competing values, assumptions, and intuitions," and Gray's letter exemplifies just such an outcome.

Other criticisms of Clark et al. (2023) have included the following:

- Kuntz (2023) found it odd that the paper did not mention the main ideological driver of present-day censorship: Critical Social Justice (CSJ), sometimes referred to as "wokeism." We agree that much of today's pro-socially motivated censorship stems from the CSJ movement (Krylov and Tanzman 2023b).

- Alonso (2023) quotes an ethicist who claims, "There are thousands and thousands of examples of scientific articles published in good scientific journals that lead to real tangible harm." The claim, asserted without evidence, exemplifies, as noted by Clark et al. (2023), that harm is assumed rather than demonstrated by credible research data. At the same time, the purported benefits of censorship have yet to be illustrated.
- Jacobs (2023) questions the concept of self-censorship, characterizing it instead as an act of "prudential judgement."
- Coyne (2023) writes:

    The article lacks tangible examples of how odious this kind of censorship can be. Examples really hit home, especially when you see how hypocritical and sneaky authors and journals can be, even when acting pro-socially.

We agree with Coyne and present, below, a compilation of papers that enumerate such odious examples.

Examples of Censorship

Examples of past and present censorship in the sciences were analyzed by Stevens et al. (2020). By comparing and contrasting past and present examples, the paper provides insight into the evolution of censorship. The authors emphasize the role activists and social media play in censorship, a topic not covered in depth by Clark et al. (2023). Stevens et al. (2020) introduce the concept of the *outrage mob*, namely:

    A group or crowd of people whose goal is to sanction or punish the individual, individuals, or organization they consider responsible for something that offends, insults, or affronts their beliefs, values, or feelings. This group or crowd demonstrates a flagrant disinterest in any further explanation from the target or targets and attempts to carry out punishment often by enlisting authorities with the power to level sanctions on the target or targets.

Stevens et al. (2020) also stress the role of complacency of academic leadership:

> Although outrage mobs often trigger the punishment process, in Western democracies, mobs no longer actually burn witches at stakes. For most punishment to occur in academia, some authority has to agree to implement the mob's punishment. That is, mobs do not get academics fired; it is high level administrators, such as deans, provosts, and university presidents that implement firings. Mobs do not get papers retracted; that is the decision of editors and editorial boards. Thus, the key turning point in whether an academic outrage mob is effective at punishing an academic for their ideas is usually the action of authorities.

Today's ideologically motivated paper retractions evoke parallels with historical and fictional book burnings. Taking these parallels seriously, "The New Book Burners: Academic Tribalism," a chapter in the newly published *The Tribal Mind: The Psychology of Collectivism* by Jussim et al. (2024), takes a deep dive into academic censorship. The authors explain the motives that drive book-burners:

> We use the term "book burning" as did Bradbury: both descriptively and metaphorically to include burning of actual books, but, especially within academia, to calls to retract, remove, and memory-hole published papers. In the present chapter, we focus on factors that have undergirded book burning for thousands of years: a sense of righteous victimization and a desire by the book burners to impose their values and norms on others.

The authors explain that cognitive rigidity and censoriousness are manifestations of tribalism (or political sectarianism) and authoritarianism.

They describe examples of "book burning [of] peer reviewed articles"—retractions of published papers in response to outrage campaigns—such as Gilley's viewpoint essay in *Third World Quarterly* (Gilley 2017), Hudlicky's retrospective essay in *Angewandte Chemie* (Hudlicky n.d.), Gliske's paper on a new theory of gender dysphoria in *eNeuro* (Gliske 2019), and a block-retraction of five papers from *Perspectives on Psychological Science*.

Several studies have investigated the psychology of censorship, for example, the motivations underlying censoriousness. In their paper "Harm Hypervigilance in Public Reactions to Scientific Evidence," Clark, Graso, et al. (2023) demonstrate that people often overestimate harmful, and underestimate helpful, reactions to science and that these tendencies are associated with greater support for scientific censorship. The paper also presents consistent evidence for *motivated confusion*:

> Those more offended by scientific findings reported greater difficulty understanding them. This finding relates to the philosophical concept of "dismissive incomprehension," the tendency to deflect dissonant claims by characterizing them as incomprehensible.

In "Sex and the Academy," Clark and Winegard (2023) present statistics revealing differences in censorious attitudes between men and women. They discuss surveys taken on college campuses showing that women tend to be more sympathetic to censorship than men, a tendency that is often driven by prosocial motives (i.e., preventing "harms").

While the above studies take a broad look at censorship as a social phenomenon, several articles provide concrete examples of academic censorship in various domains. In "Royal Society of Chemistry Provides Guidelines for Censorship to its Editors," Krylov et al. (2022) call attention to censorship guidelines in chemistry publishing that were recently imposed by the Royal Society of Chemistry. The guidelines emphasize that "it is the perception of the recipient that determines offense, regardless of author intent" and define offensive content so broadly (e.g., any content "likely to be upsetting, insulting or objectionable to some or most people") that nearly any paper could be censored. This policy is still in place.

The topic of institutionalized censorship is further discussed in "Critical Social Justice Subverts Scientific Publishing" (Krylov and Tanzman 2023b). This paper, published in the special collection *Perils for Science in Democracies and Authoritarian Countries* (European Review 2023), discusses censorship in STEM publishing in a historical context and explicitly points to CSJ as the main ideological driver of present-day censorship. The paper presents examples of suppression of research findings and raises an alarm about recent proposals to suppress "harmful" research at the funding stage.

The corruption of science and education by CSJ is also discussed in other papers in the collection, for example, in "The Notion of Truth in Sciences and Medicine, Why It Matters and Why We Must Defend It" by Bikfalvi (2023) and "The Universalism of Mathematics and its Detractors: Relativism and Radical Equalitarianism Threaten STEM Disciplines in the US" by Klainerman (2023).

Another alarming example of the institutionalization of censorship was an editorial published in *Nature Human Behavior* (*NHB* 2022), which announced that the journal's editors would censor research they considered harmful. The editorial is a censorship manifesto from its opening line, "Although academic freedom is fundamental, it is not unbounded," to its proposed roadmap of how censorship will be implemented:

> There is a fine balance between academic freedom and the protection of the dignity and rights of individuals and human groups. We commit to using this guidance cautiously and judiciously, consulting with ethics experts and advocacy groups where needed.

This anonymous manifesto has generated considerable attention. Among thoughtful critiques, we highlight "Nature Human Misbehavior: Politicized Science is Neither Science nor Progress" by Jonathan Rauch (2022), author of the seminal books *Kindly Inquisitors: The New Attacks on Free Thought* and *The Constitution of Knowledge*. Rauch takes apart *NHB*'s arguments and provides compelling examples of research that could have easily been censored under *NHB*'s new policy but instead were published and led to important breakthroughs in our understanding of human nature and ultimately advanced important social causes.

Another in-depth critique of the *NHB* manifesto, "The Fall of Nature," has been published by Bo Winegard (2022) in *Quillette*. A more recent, eloquent critique with many excellent examples of published research findings that could easily have been censored under *NHB*'s policy is provided by Alan Sokal (2024a) in "How Ideology Threatens to Corrupt Science," reprinted in this volume.

Examining past censorship can help us better understand present-day censorship. In "Discovering Natural Selection Was Like 'Confessing a Murder,'" Geher (2022) discusses the harms of self-censorship using a famous example: the reluctance of Charles Darwin to publish his theory of evolution. As he wrote to the botanist Joseph Dalton Hooker, "it is as if one were confessing to a murder." Darwin's discomfort was so great that he reported vomiting regularly, thinking about the implications of the theory of evolution on the perceived special place of *Homo Sapiens* among the species.

Geher writes, "Darwin's ultimate decision to publish about evolution shaped our understanding of the world profoundly." He argues that suppressing important scientific findings does humanity a grave disservice. He concludes:

> Not all research findings are going to be popular. And not all scientific ideas are going to be loved by all. But publishing ideas and findings that might cut against the grain, at the end of the day, helps us better understand the nature of the world. And that, as I see it, is kind of the point of science. Right?

Several articles (Savolainen 2023a, 2023b; Gilley 2022; Hill 2018; Marcus 2020, 2021a, 2021b; Krauss 2021; Reynholds 2020; Oransky 2020; Wright 2023; Baily 2023; McMillan 2023; Pomeroy 2024; Jussim 2024a, 2024b) dissect recent ideologically motivated retractions and self-retractions—these are among the odious examples of censorship that Jerry Coyne (2023) alluded to.

Some examples are but comical, such as the removal of the accepted paper "Meta-Analysis: On Average, Undergraduate Students' Intelligence Is Merely Average" from *Frontiers of Psychology* after a Twitter mob complained about the article causing offense (Pomeroy 2024). *Everyone* must be above average; to state otherwise would be hurtful! Notably, the complaints that triggered the decision to remove the paper were based only on the paper's abstract, as the manuscript itself had not yet been made public. As Pomeroy notes, *"At Frontiers in Psychology,* it seems that users on X are now part of the peer review process."

A recent example of a poster and abstract being removed from a conference follows a similar script (Jussim 2024a, 2024b). Unsubstantiated or even false Twitter complaints are sufficient to trigger a retraction in violation of existing conference or publication rules. Jussim writes:

> SPSP took down a duly accepted poster by John Gaski that reviewed international Pew surveys finding [that] majorities or near majorities in many Muslim-majority countries supported at least some violence directed at civilians.... Why did they take it down? Well, the trigger was that it was denounced on Twitter as "Islamophobic" and "bad science." Some even claimed it had no data, which was absurd, if one actually looked at the poster rather than, say, the deepfake presented on Twitter.

Several commentaries discussed the case of the late Tomas Hudlicky, whose peer-reviewed and accepted paper (Hudlicky n.d.) was removed, without a retraction notice, from the journal's website and who was subjected to relentless harassment and ostracism (Sydnes 2021, Krylov et al. 2022, Deichmann 2023). Hudlicky's case was the first incident of brazen ideologically motivated censorship by a major chemistry journal (*Angewandte Chemie,* published by Wiley) against their own policies in response to online mob outrage.

It is noteworthy that the ostracism and relentless harassment of Hudlicky continued long after his paper was removed from the journal's website, reminiscent of the public shaming campaigns against Soviet dissidents (e.g., Sinyavsky and Daniel, Bukovsky, Solzhenitsyn, and Sakharov), in which newspapers and journals published letters that would begin along the lines of "I have not read these malicious anti-Soviet writings, but I am deeply insulted and outraged by them and condemn the author."

Fear of mob outrage often leads well-meaning colleagues to advise self-censorship. A well-known example is Harvard economist Roland Fryer, whose research led to the politically unpopular conclusion that police use of deadly force in the US is not racially biased (Fryer 2019). In Fryer's words, "I had colleagues take me to the side and say, 'Don't publish this. You'll ruin your career'" (Grabien 2024). His friends might have had a point. Shortly after publishing a preprint of his work, Fryer was subjected to a Title IX investigation based on what he describes as false accusations and was severely sanctioned by Harvard (Taylor 2019).

Additional examples of ideologically motivated suppression of scientific facts come from the field of biology. Biologists Coyne and Maroja (2023) [Ed: see this volume] warn of the disastrous consequences of suppression and ideological distortion of knowledge in their field.

Ideologically motivated censorship in the sciences has also been recently discussed by Sokal (2024a, 2024b), who recounts recent examples of censorship and ostracism of scientists who expressed unpopular opinions (Sokal, 2024b). He emphasizes the harm that such censorship causes to science and the public.

> The bottom line is this: It is never justified to distort the facts in the service of a social or political cause, no matter how just. If the cause is truly just, then it can be defended in full acceptance of the facts about the real world; if that cannot be done, then the cause is not just.
>
> And when an organization that proclaims itself scientific distorts the scientific facts in the service of a social cause, it undermines not only its own credibility but that of science generally. How can the public be expected to trust the medical establishment's declarations on other controversial issues, such as vaccines—issues on which the medical consensus is indeed right—when it has so visibly and blatantly misstated the facts about something so simple as sex?

As the examples above reveal, cases of censorship in science are often reported in non-academic outlets. In our experience (in particular, with Clark et al. (2023) and Abbot et al. (2023)), even when such articles reflect rigorous research and analysis and are written by authors with academic credentials, citing them in a viewpoint article in a scientific journal is met with resistance from reviewers and editors. Whether such resistance reflects legitimate concern about the rigor of the scholarship or is itself ideologically motivated censorship is often unclear because editors often mask ideological censorship as methodological criticism. As Clark et al. (2023) point out:

> Contemporary scientific censorship is typically the soft variety, which can be difficult to distinguish from legitimate scientific rejection. Science advances through robust criticism and rejection of ideas that have been scrutinized and contradicted by evidence. Papers rejected for failing to meet conventional standards have not been censored. However, many criteria that influence scientific decision-making, including novelty, interest, "fit," and even quality are often ambiguous and subjective, which enables scholars to exaggerate flaws or make unreasonable demands to justify rejection of unpalatable findings.

Although it is often difficult to prove that a paper purportedly rejected for methodological shortcomings was actually rejected on ideological grounds, in some cases, the real reason for rejection is too obvious to conceal. We illustrate with two examples. The first is "Resistance to Critiques in the Academic Literature: An Example from Physics Education Research" (Reichhardt et al. 2023). The authors presented a rebuttal of a paper recently published in the journal *Physical Review—Physics Education Research*. That paper, titled "Observing Whiteness in Introductory Physics: A Case Study" (Robertson and Hairston 2022), arguably reads more like a hoax than a scholarly paper relevant to physics education. But the rebuttal treated the paper seriously and offered a substantive, professional, and detailed critique (Krauss 2023). The rebuttal, submitted to the same journal as the original paper, was rejected by the editors *explicitly* on ideological grounds, namely, that the rebuttal was "framed from the perspective of a research paradigm that is different from the one of the research being critiqued." To put that in plain English, the authors used scientific methods to debunk a postmodernist paper.

The second example is the story of how a paper with the seemingly mundane title "In Defense of Merit in Science" (Abbot et al. 2023) wound up being published in the *Journal of Controversial Ideas*. As Coyne and Krylov (2023) explain in their *Wall Street Journal* op-ed "The 'Hurtful' Idea of Scientific Merit," the concept of merit was considered by the editors of the *Proceedings of the National Academy of Sciences*, to which the paper was originally submitted, to be "hurtful" and to have been "widely and legitimately attacked as hollow."

A conversation about censorship would not be complete without noting that the rise of academic censorship, retractions, and self-retractions in response to mob outrage campaigns is part of a larger social phenomenon—cancel culture—which, in *The Cancelling of the American Mind*, Lukianoff and Schlott (2023) define as,

> The uptick beginning around 2014, and accelerating in 2017 and after, of campaigns to get people fired, disinvited, deplatformed, or otherwise punished for speech that is—or would be—protected by First Amendment standards and the climate of fear and conformity that has resulted from this uptick.

The book documents how the rise of cancellation campaigns on campuses has contributed to the current climate of fear and self-censorship in social and academic settings. Statistics illustrating the extent of cancel culture are also given in the short essay by Lukianoff, *The New Red Scare Taking Over America's College Campuses* (Lukianoff 2023).

In Closing

Clark et al. (2023) conclude:

> We have more questions than we have answers. Although many members of our research team are concerned about

growing censoriousness in science, there is great diversity of opinion among us about whether and where scholars should "draw the line" on inquiry. We all agree, however, that the scientific community would be better situated to resolve these debates, if—instead of arguing in circles based on conflicting intuitions—we spent our time collecting relevant data. It is possible that there are some instances in which censoring science promotes the greater good, but we cannot know that until we have better science on scientific censorship.

We, too, hope that the subject of censorship will receive the scientific attention it deserves. We hope that future studies will shed light on the harms done by censorship and that a deeper understanding of how censorship operates will help restore freedom of inquiry in science.

We began this chapter with the words of Marie Sklodowska-Curie. We conclude with Plato:

> We can easily forgive a child who is afraid of the dark; the real tragedy of life is when men are afraid of the light.

Note added in proof: A recent three-day conference held at the University of Southern California titled Censorship in the Sciences: Interdisciplinary Perspectives, brought together scholars from across the world to discuss modern academic censorship and its effects on science and society. Recordings of all the conference lectures and panels are available on the Heterodox Academy's YouTube channel.[1]

The Tyranny of Victimhood: How Progressives Turned Regressive

MAARTEN BOUDRY

"I had reduced everything to the simple theory that the oppressed are always right and the oppressors are always wrong: a mistaken theory, but the natural result of being one of the oppressors yourself."

George Orwell

"It depends on the context." With these five simple words, uttered during a hearing in the US Congress in December last year, the former president of Harvard University, Claudine Gay, secured her place in history books. It was her answer to a question just posed to her by Republican Rep. Elise Stefanik: "Does calling for the genocide of Jews violate your university's codes of conduct?" Two academic colleagues sitting next to her, the presidents of MIT and the University of Pennsylvania, had been roasted with the exact same question. All three, with a contemptuous smirk on their lips, refused to give a simple affirmative answer, instead repeating the same evasive and lawyerly phrases they had clearly rehearsed beforehand. Calling for the extermination of the Jews on our campus? Well, sometimes you can, it depends, you see.

In defense of her answer, Claudine Gay boasted that her university is "deeply committed to protecting free expression" in the spirit of the US First Amendment. I wish that were true, but it shows a breathtaking level of hypocrisy. This is the same university where students and staff risk disciplinary sanctions if they address someone using the wrong pronouns or make derogatory comments about another one's weight.[1]

As documented elsewhere in this book, the vaunted "freedom of expression" at Harvard is in such a deplorable state that, in the index of the Foundation for Individual Rights and Expression (FIRE), the university received the lowest score ever recorded, coming in dead last out of 203 elite schools surveyed. If you were to so much as dream about uttering the n-word on campus at an Ivy League university, even in an educational context, you would be immediately hounded off campus and expelled. Let alone that you would call for the "genocide of Black people" and that your university president would defend your free speech in Congress.

Ideological Rot

This hearing in Congress provides the most incontrovertible evidence to date (with three sworn testimonies under oath) of an ideological rot that has been spreading for decades in our most prestigious institutions. Except now the festering sore is exposed for the entire world to see—including to the wealthy donors whose money provides the lifeblood of universities like Harvard.

The moment the sore burst open was on October 7, 2023, or very soon thereafter. Just a quick recap. Just days after Hamas's terrorist attack on Israel, before the country returned a single shot toward Gaza, countless student groups and academics at Western universities had already rushed to condone, downplay, or even glorify Hamas's massacre. The day after the attack, more than thirty student groups at Harvard wrote a joint statement holding Israel's "apartheid regime" "solely responsible" for the massacre.[2] Pro-Palestinian demonstrators aggressively harassed Israeli and Jewish students on several campuses, including Harvard, usually without facing any disciplinary sanctions. No fewer than 120 departments of gender studies around the world felt compelled to condemn Israel in the wake of the largest anti-Semitic massacre since the Holocaust, but none of them had a single word to spare for the female victims of sexual violence by Hamas and other jihadist groups. Not just sexual violence but sickening, sadistic torture that almost defies description: genitals stabbed with knives or riddled with bullets, severed breasts, broken pelvises, families burned alive.³

At my own alma mater (Ghent University), hundreds of academics signed an obscene open letter that explicitly refused to condemn Hamas, shifted all blame to "Zionists," and praised Palestinians for their "tenacity and fierce resistance to racism and settler colonialism," which the signatories found immensely "inspiring." An open letter at the University of Amsterdam (UvA), again signed by hundreds of academics, exulted that 2023 "will no doubt be the year admired, recorded and studied for the way in which Palestinians steadfastly resisted colonialism, occupation and survived genocide." The letter even called for the abolition of Holocaust and Genocide Studies at UvA because the program was deemed "complicit in remaining silent on the unfolding genocide of the Palestinians."⁴

Oppressors and Oppressed

Like the Greek hydra, the ancient monster of antisemitism has multiple heads. There is the Christian "head" that regards Jews as murderers of Christ and inspired bloody pogroms in the Middle Ages based on libelous rumors that Jewish communities poisoned the drinking water and drank the blood of Christian children. There is the racialized version of antisemitism on the far right, which regards the Jewish people as a parasitic life form that poisons the healthy body of society. Yet another "head" is the Islamic tradition of antisemitism, which, much like the Christian variant, ultimately stems from the resentment and frustration about the Jews' refusal to accept the latest divine revelation. And then there is the anti-Semitic tradition on the far left, which portrays Jews as an evil cabal of rootless and money-grabbing bankers and capitalists, stretching out their tentacles like an octopus around the world.

In short, there's no dearth of possible historical roots for animus against the Jewish people. And yet it would be too simple to interpret the demonization of the only Jewish state in the world at our universities—and the support for its genocidal enemies—as just the hoary old monster of antisemitism rearing its ugly head. Antisemitism surely plays some role in campus protests, but the real explanation lies deeper. In a sense, the Jews and/or Zionists are merely *collateral damage* of a different and underlying ideology.

The intellectual roots of this ideology are complex, but the conceptual framework is easy to summarize. First, you divide up the world into two mutually exclusive categories: the oppressors and the oppressed. These two groups are locked in a zero-sum struggle where one group's gain can happen only at the expense of the other's loss. There can be no middle ground: either you are an oppressor or a victim. Next, you apply two completely different moral standards: those who belong to a victim group are innocent by definition and incapable of doing anything wrong. If they *seem* to do some horrible things, that is only because they are responding to the grave injustices to which the oppressors have subjected them. This victim group enjoys protection, compassion, the privilege of low expectations, and preferential treatment by way of compensation for their long history of suffering and oppression. As for the oppressor group, they get none of that. The only good thing the oppressors can do is to humbly acknowledge their collective guilt and strive to be good allies of the victim group, by unconditionally supporting their noble struggle (although that does not absolve them).

The dichotomy that I discuss here is the one between *Western* and *non-Western* and, by extension, white vs. black. In recent years, similar victim vs. oppressor dichotomies have cropped up in other kinds of progressive discourse, and their combined approach has become known as "intersectionality." In this intersectional framework, each dimension of sexual, physical, and cultural identity has its own victim/oppressor binary (male/female, straight/LGBT, cisgender/transgender, slim/fat, and so on), each corresponding to its respective system of oppression (sexism, heteronormativity, transphobia, fatphobia, and so on). The psychologist Jonathan Haidt has used the image of a cake that is being sliced up along different axes. Still, these binary divisions have different intellectual roots and manifest in different ways, so it's not always enlightening to lump them together in the way activists tend to do. In fact, classical Marxism provides another example of such an oppression binary (bourgeois/proletarian), but of course, Marxism long predates the rise of modern intersectionality. In any event, I think the division between Western and non-Western is the dominant and most important one, which, as we will see, takes precedence over the others in case of conflict.

Now, we're missing only one element to understand the charade in the American Congress and the apparent outbreaks of vicious antisemitism in the most progressive bulwarks of our society: Jews belong to the oppressor group. The prime reason is that Israel, the world's only Jewish state, is seen as the last surviving project of European *settler colonialism*, in which white Westerners expel, oppress, or exterminate indigenous populations.[5] Another reason is that Jews (in the US but also in Europe) are regarded as a successful "model minority" against which other minorities are often judged, with many holding positions of influence and prestige.

In this binary framework, the ultimate enemy is not so much Israel, but *Western civilization*, with its long history of imperialism, colonialism, and white supremacy. Islamists and jihadists often proclaim that Israel is merely "little Satan," the squire and helper of "big Satan," which, of course, is the United States, Israel's loyal ally. Something similar applies to the virulent hatred of Israel in progressive circles. The small Jewish nation of less than ten million inhabitants is targeted only insofar as it serves as a *proxy* for the West. Similarly, the astonishing support for Hamas and Israel's other enemies has little to do with any genuine ideological affinity but everything with the old maxim: the enemy of my enemy is my friend. The paragliders on October 7 were lionized as noble warriors, not for what they stand *for*, but for what they were construed as fighting *against*: Western colonialism.

Never mind that there are hundreds of thousands of Sephardic and Black Jews living in Israel, as well as hundreds of thousands of Arab Israelis who practice Islam, that Jewish "colonizers" have no homeland to return to, that Jews are just about the most hated group in the world (comprising just 2.4 percent of the US population, they bear the brunt of 60 percent of all religious hate crimes),[6] that Jews have been persecuted almost to annihilation precisely because they *failed* to pass as "white," that Israel is the only liberal democracy in the Middle East and was founded with a U.N. mandate, that Palestinians and Israelis are ethnically indistinguishable—none of this matters in the slightest. In this infantile binary worldview, Jews and Israelis as a class are "white," hence colonizers, hence racists. As some protestors have delicately put it on their placards: "IDF = KKK."7

How Progressives Became Reactionaries

The demonization of Israel and reflexive sympathy for its genocidal enemies is not the only harmful result of this neat division of the world into perpetrators and victims. By presenting Western civilization as the root of all evil, this ideology ultimately ends up with a wholesale rejection of everything Western civilization stands for—science, progress, freedom of expression, human rights—which are seen as cynical covers for imperialism and oppression. Worst of all, it devolves into the most reactionary form of racist essentialism: the view that democracy, rationality, and secularism are exponents of "whiteness," not suitable for people with dark skin.

How is it possible that progressives, of all people, have embraced such reactionary nonsense?

The backdrop for this progressive betrayal of Enlightenment values is the history of Western colonialism. After World War II, the Western powers were exhausted from the war and increasingly unable to hold on to their overseas possessions. Independence movements were springing up everywhere, trying to throw off the yoke of colonial rule. Like a series of dominoes, the Western colonies began to topple, with one success inspiring another. Some of these decolonial movements found inspiration in Western intellectual traditions—such as socialism and communism, Arab nationalism, or Enlightenment universalism. But there were more radical thinkers who saw an implacable struggle between colonizers and colonized and who wanted to radically break with Western civilization. Or at least, in the words of one of these thinkers, to wield Western thinkers such as Marx, Freud, and Hegel "to incriminate the very civilization producing all of them."8 The most influential expression of this Manichean worldview of the West versus the Rest emerged in Algeria, not coincidentally one of the bloodiest anti-colonial battlegrounds. But it was not developed by an Algerian, but by a Black psychiatrist born in the West Indian French colony of Martinique: Frantz Fanon.

The Wretched of the Earth

Like many other postcolonial intellectuals, Fanon was steeped in the culture of his colonial rulers. Born in 1925 into a middle-class family, the Fanon family mainly tried to assimilate into French culture, otherwise keeping a low profile. An uprising against the occupier was not exactly on their minds. The first words young Frantz learned to write at school were *"Je suis Français."* This strong identification with French civilization permanently changed during World War II when Fanon fought with the French against Nazi Germany on the European front. For the first time in his life, Fanon experienced firsthand how his white comrades viewed him and other non-white soldiers as inferior—second-class citizens at best. The humiliations he suffered while in Europe—such as the Frenchwoman who refused his invitation to dance after the liberation, preferring American dance partners—left scars that would be felt for the rest of his life.

In the course of Fanon's all-too-short career—he died of leukemia at the age of thirty-six—this bitterness and resentment led to a savage condemnation of Western colonialism, culminating in his legendary book *The Wretched of the Earth*. For centuries, the colonial occupier had violently oppressed and humiliated its subjects, robbing them of their humanity. The only way the "wretched of the earth" could regain their humanity, argues Fanon, is by murdering the colonizer. Real freedom could not be granted by the occupier or achieved through peaceful protest; it had to be wrested by force. For the psychiatrist Fanon, violence was not only instrumental in shattering the colonial chains, but also a form of therapy, a refreshing tonic with purifying and cathartic effects on the minds of the oppressed. Like the electroshock therapy he prescribed to his psychiatric patients, outbursts of violence could resuscitate the enslaved mind, which had been dulled to passive resignation and feelings of inferiority. "On the level of the individual, violence dis-intoxicates," Fanon wrote, serving as an antidote to the sickening oppression of colonialism: "For the colonized, life can only arise from the decomposing corpse of the colonist." In Fanon's anti-colonial struggle, there is no room for compromise, because (de)colonization is a zero-sum game with only one possible winner and one loser, or as he put it: "The last shall be first."⁹

This was no idle talk. Fanon, to his credit, also walked the walk. In Algeria, he supported and joined the *Front de Libération Nationale*, even when they started massacring French civilians in terrorist attacks, to which the French predictably responded with brutal retaliations. Fanon believed—against his better judgment as a clinician—that such outbursts of bloody terror would ultimately herald a society of freedom and universal brotherhood.¹⁰ In reality, it led not only to a cycle of retaliation and revenge, but also to brutal internecine fighting among the Algerian guerrillas, with executions of dissidents and new forms of oppression. As conservative Islam slowly gained the upper hand in the Algerian liberation movement and began to oppress women and religious minorities, Fanon turned a blind eye and remained silent. The wretched of the earth had only one true enemy, and that was the Western colonizer.

The lyrical and rousing diatribe of *The Wretched of the Earth*, combined with Fanon's heroic self-sacrifice and tragic death at the age of thirty-six, turned him into a patron saint of the anti-colonization and anti-racism movement. His book became required reading for the Black Panthers, South American guerrillas, the anti-apartheid movement in South Africa, and assorted guerrillas and Islamists from Iran to Palestine. More remarkably, some "progressive" European intellectuals started masochistically cheering for this bloody revenge against their own civilization. In a delirious foreword to *The Wretched of the Earth*, the French philosopher Jean-Paul Sartre—at that time one of the most influential intellectuals on the European continent—attempted to surpass the author of the book in anti-Western bloodlust. Only by killing Europeans, Sartre theorized from the comfort of the Café de Flore in Paris, could the colonized ever become *human* again:

> [When] you slaughter a European, you kill two birds with one stone, you get rid of an oppressor and an oppressed at the same time: what remains is a dead man and a free man. [...] We were human at his expense, he becomes human at our expense.

For those progressive humanists who support the Algerian liberation struggle but condemn its barbaric excesses, Sartre had nothing but contempt: "no gentleness can efface the marks of violence; only violence itself can destroy them." Appeals to humanism are just "an exquisite justification for plundering."[11]

While you can feel some sympathy for the vengeful fantasies of Fanon, in view of his humiliating experiences with racism, the bloodthirsty masochism of the white progressive in Paris is above all pathetic. Sartre's behavior is even more revolting if you bear in mind that he had no qualms at all about covering up and whitewashing the bloody reign of Joseph Stalin, a man who literally colonized, starved, and oppressed half of Europe. But for Sartre, as for many "progressives" after him, the only real enemy was Western bourgeois capitalism, and its victims the only true and noble heroes.

Echoes of Fanon's philosophy of detoxifying violence could be heard soon after the carnage of October 7, before the blood had even dried. If you wonder why on earth left-wing professors would describe a sadistic massacre as "exhilarating" or "awesome," or why students decorated pro-Palestinian posters with heroic Hamas paragliders, you should look no further than Fanon and his extreme version of the victim vs. oppressor binary. In an interview with the left-leaning Israeli newspaper *Haaretz*, even Fanon's sympathetic biographer Adam Schatz had to admit that the pogrom of October 7 can be read as a "classic example of Fanonian struggle."[12]

The Western Hunger for Power and Domination

A second pivotal figure in the rise of anti-Western ideology is the Palestinian American writer and literary critic Edward Said. In his brooding autobiography *Out of Place*, written at the end of his life, Said writes that he always felt torn between his very British first name—borrowed from the Prince of Wales—and his Arabic family name. The offspring of a wealthy Palestinian-Christian family (his father also had a US passport), Said grew up in the twilight of the British Empire, traveling between Cairo and the British Mandate of Palestine. Like Fanon, Said writes that, despite his privileged and affluent background, at crucial moments in his life, he felt that white

Europeans were treating him as a second-class citizen. Although Said was not very politically conscious in his youth and early adulthood, even in the aftermath of the founding of Israel in 1948 and the *Nakba*, he gradually became more committed to the Palestinian cause when he studied at Princeton.

In 1978, Edward Said threw a bombshell into the Western canon, the aftershocks of which would be felt for decades. *Orientalism*—a good contender for the most influential academic book of the twentieth century—is one long indictment of the deep-seated prejudices Western civilization has been harboring for centuries against Eastern cultures, especially Islam. Before Said, "orientalism" was a neutral concept, referring to Western artistic movements drawing inspiration from Eastern styles, or to scientific studies and travel reports about the East. Ever since Said's seminal book, however, "orientalism" has become a dirty word. In almost the entire Western canon, according to Said, the Orient appears as a stereotypical essence (mysterious, exotic, sensual, bloodthirsty), serving as a mirror image to sharpen the self-esteem of Western civilization (rational, enlightened, civilized, masculine): "European culture gained in strength and identity by setting itself off against the Orient as a sort of surrogate and even underground self."

Said's focus on the *discourses* of Western domination and colonialism rather than its actual practice was heavily indebted to the French postmodernist Michel Foucault. For both thinkers, violence is not only inflicted through bayonets, cannons, or torture racks, but also in less visible and more insidious forms, namely through oppressive *discourses*. This means that even apparently peaceful practices and institutions can really be deeply violent at their core. And just as Foucault directed his critical gaze mainly on the innovations of modernity that seemed most innocuous (such as schools, hospitals, and asylums), Said attacked primarily those aspects of Western civilization that seem the *least* offensive: the thirst for knowledge and the attempts to understand other cultures. In Said's eyes, the orientalist does not simply investigate, study, and measure his object of research—he destroys, dominates, and plunders it. Said believes that this pernicious orientalist discourse extends all the way to the American empire of the twentieth and twenty-first centuries. Even today, Western universities, media, and think tanks are deeply complicit in the imperious ambitions of the countries in which they operate.

Orientalism made a splash both at home and abroad and was translated into thirty languages. For the first time, as the Palestinian historian Tarif Khalidi wrote, here was a book "by one of us telling the empire basically to go f— itself." Providing further evidence of deep-seated orientalist prejudices, the Western academic world offered Said distinguished professorships (at Berkeley, Yale, and Johns Hopkins) and showered him with abundant prizes and honors throughout his career.

Fully in line with the moral dichotomy of Frantz Fanon—an avowed source of inspiration—Said's diatribe in *Orientalism* and later works were almost exclusively a one-way street. Only the prejudices of the perpetrator towards the victim commanded Said's critical attention. A comparative study of the ways in which cultures harbor prejudices about each other was of no interest to him. In fact, Said wrote next to nothing about those indigenous cultures and their own discourse of oppression, such as the caste system in India. By treating Eastern cultures and civilization mostly as hapless victims of the all-powerful West, Said deprived them of moral autonomy. The Other is not freed from its chains but condemned to the role of passive plaything and puppet. In his sweeping condemnation of Western civilization, Said also fell prey to that which he claims to combat. For instance, writing about nineteenth-century Europeans, Said produces this sentence without a hint of irony: "Every European, in what he could say about the Orient, was a racist, an imperialist, and almost totally ethnocentric."[13] And mind you, just like Sartre, his ire was above all directed at the liberal and democratic West. Even though Soviet cultural imperialism and persecution of Muslims came much closer to what Said laid at the doorstep of the democratic West, the communist bloc mostly gets a free pass.

Although he never used the word, it is no exaggeration to call Edward Said the godfather of the concept of "Islamophobia," the favorite immunizing tactic to shield the regressive beliefs and practices of Islam from criticism. Throughout his work, Said attributes to the West an irrational fear of Islam, which has been part and parcel of its imperial ambitions for centuries. An irritating effect of this Foucauldian style of "discourse analysis" is that it always deflects attention away from the question: how *accurate* were those representations of Islam? To what extent do Muslims really support violence and terrorism in the name of religion, the death penalty for apostates and adulterers, or the subjugation of women (*spoiler alert*: to an alarming extent)? It's like doing a discourse analysis of how the planet Venus is "represented" in the Western media—as a scorching and inhospitable wasteland, as a cautionary tale about runaway greenhouse warming, as the brightest celestial body in our firmament—without ever considering the question: but are these representations correct? Does our discourse about Venus correspond with the real planet out there? All in all, Said's work was a wonderful gift to the conservative *mullahs* and *ayatollahs* who wanted to cleanse the Arab world of toxic Western ideas and fend off unwelcome criticism of their religious dogmas.

The West Is Always to Blame

The victim/oppressor binary inspired by Fanon, Said, and others continues to loom large over many leftist geopolitical analyses, even apart from Israel. In this one-note symphony, "we" are always and everywhere to blame for whatever happens. If the West intervenes somewhere, then we are war-mongering imperialists in the pocket of the military-industrial complex and guilty of all the misery that ensues. If the West refrains from intervention, we are no less guilty, this time because of negligence and callous indifference to the suffering of non-white people. And if the West backs one of the warring parties from the sidelines, it is attacked on both sides: because we are doing too little to help the "good guys" (which gives the bad guys the upper hand) or because we are adding fuel to the fire and playing into the hands of the bad guys. Heads I win, tails you lose.

In fact, the best predictor of moral outrage among Western progressives about an international conflict is how easy it is to frame the West as the culprit: sympathy for the victims is directly proportional to how close their oppressors are to the West. In other words: no "Teach-Outs" at our universities for the massacres in Sudan, Ethiopia, or Syria, because those are just *subalterns* killing each other. But the Palestinians of Gaza, on the other hand, can be framed as victims of Israel, which in turn is an ally of the United States—now *that* kindles the revolutionary fervor. For years, activists expressed no interest at all in the Houthis in Yemen, an Iran-backed Shiite terrorist group that has unleashed a civil war there. Until January 2024, when these selfsame Houthis started shooting rockets at Western ships in the Red Sea as punishment for its support of Israel. All of a sudden, you could hear chants at university campuses: "Yemen, Yemen make us proud, turn another ship around!" And you could hear deluded academics hailing the Houthis as "the most humane force in modern history," who should be applauded for their "remarkable act of revolutionary solidarity."[14] Who cares that the official slogan of the Houthis is "God Is Great, Death to America, Death to Israel, Curse on the Jews, Victory to Islam," or that the group literally stones and crucifies homosexuals and would happily slit the throat of any of the Western activists that are cheering them? As long as the freedom fighters are aiming their missiles at Western ships, they're on the right side of history.

Although ostensibly standing up for victims of Western imperialism, such analyses have the effect of denying moral autonomy to any geopolitical players except the West. Just as Frantz Fanon remained blind to the theocratic and reactionary tendencies in the Algerian independence movement, many anti-Israel activists today have no interest at all in the noxious ideology of their "freedom fighters." The role accorded to Hamas is that of heroes in an anti-colonial Western drama, regardless of how they regard themselves. In reality, of course, Hamas is a genocidal and theocratic organization that aims to wipe Israel off the map, as everyone can read in their original charter in 1988 and in countless pronouncements from Hamas leaders ever since. The notion that Hamas would lay down its arms in return for the removal of settlements in the West Bank or for *any* two-state solution (whatever the borders) is a delusion on the part of secularized Westerners who fail to take other people's religious beliefs seriously.

The ultimate betrayal of self-professed "progressives"—I can't help resorting to scare quotes again—occurs when the demonization of whiteness is extended to the *realm of ideas*. By labeling ideas and beliefs as "white" just because they happened to have originated in Western Europe,

Enlightenment and modernity become targets of decolonization in their own right. All those lofty principles and ideals of all those dead white males of the European Age of Reason? Nothing but cynical rationalizations for oppression and exploitation, which indigenous non-Western peoples should have no truck with.

Before you know it, you end up with full-blown racist nonsense according to which liberalism and secularism are just "white" notions that unsuited to brown or Black people. By openly abandoning Enlightenment universalism in this way, progressives have become the allies of religious reactionaries and fanatics in non-Western cultures, as the Indian historian Meera Nanda has argued. They have made it all too easy to dismiss liberation struggles *within* non-Western cultures as a covert form of Western imperialism, of a "colonial modernity which privileges a Western conception of reason and modernity."[15] When it comes to Islam, you then get a narrative, as the Pakistani Canadian ex-Muslim Ali Rizvi told me in *Quillette*, in which "the only good Muslim is a conservative Muslim."[16] Or in which, as Yasmine Mohammed has written in *Unveiled*, Western progressives are empowering their own worst enemies.[17] If anything is guilty of "Orientalism" in the pejorative sense of Edward Said, it must be the offensive assumption that misogyny, antisemitism, and hatred of homosexuals just happen to be part of "their" culture, which we should all respect.

Wellspring of Hilarity

Fortunately, there is also some mirth to be found in all this insanity. In recent years, we have witnessed how victim/oppressor narratives, when applied to relatively innocuous issues, have generated an endless stream of silliness that is hard to distinguish from parody. Especially efforts towards "inclusive" language, intended to protect one or another designated victim group, are an endless source of hilarity. Some are discussed in this volume, including the effort by New Zealand to require indigenous knowledge to be granted equal status in school curricula, the Yale controversy over Halloween costumes, and the *Lancet* renaming women "bodies with vaginas." It is easy to make light of inanities like these—which, for lack of a better word, have become known as "wokeness"—but when the same ideology is wielded to condone and justify horrific terrorist attacks, I am significantly less amused. It is all well for privileged students at Western universities to wallow in performative radicalism about "decolonization" and "offensive language" precisely because, at the end of the day, nothing much is at stake. It's just empty posturing that mainly functions to signal moral virtue within certain ideological tribes. But when it comes to the conflict in Gaza, there *is* something at stake. If Israel becomes increasingly shunned and abandoned by its Western allies, the future of the only liberal democracy in the region looks bleak indeed. In the worst-case scenario, the current conflict in Gaza could escalate into an open war with Iran and its proxies or even with other Arab nations.

The most charitable explanation of the one-sided demonization of Israel by Western progressives is that, unlike progressives in Israel itself, they no longer understand what it means to live in a country that has been existentially threatened since its founding, with multiple terrorist groups and nations having vowed to wipe the "Zionist entity" off the map. It is easy to shout anti-Zionist and pro-Hamas slogans if you're thousands of miles away, safely ensconced in your elite campuses. If these same students were living next door to a terrorist group that had just massacred a music festival and half a city and has vowed to repeat the same horror over and over again, I think they would sing a different tune.

It is time for these pampered students, academics, and administrators to "check their privilege," to echo one of their own favorite phrases. In the wake of October 7, it has become abundantly clear what kind of moral monstrosities you end up with if you start dividing the world into evil oppressors and noble victims. If elite universities want to stem the rapidly dwindling trust in academic institutions, they need to consistently apply free speech principles and adopt institutional neutrality across the board, and they need to tackle the ideological rot that they have allowed to spread and even cultivated for decades (mostly in the form of the massive DEI bureaucracy). The last thing they should do is to abet and give cover to this pernicious ideology in US Congress for the eyes of the whole world to see.

The Two Universities Redux

ABIGAIL THOMPSON

Two universities coexist on campuses across America. They share the same physical space but otherwise occupy different worlds. In one are scholars who champion free speech and oppose academia's politicization. They uphold the university's mission, as articulated, for example, by the Princeton Principles: "The core mission of the university, and its distinctive contribution to the American republic, is the pursuit of Knowledge and advancement of knowledge through scholarship and teaching. This mission is sustained by freedom of inquiry, freedom of expression, and equality before laws and campus regulations."[1] We'll refer to the part of the academy holding this perspective as "Knowledge U." Most of the sciences are clustered here.

In the other are those who view the university primarily as an agent of social change. Scholarship is almost incidental to political activism, and advocating for desired ends may require deplatforming speakers with "unacceptable" views. "Acceptable" views cannot be questioned and align closely with an extreme progressive agenda. A recent faculty job advertisement at my institution, UC Davis, illustrates this: "The Chicana/o Studies Department is rooted and aligned with the decolonial project that emerged from the Third World Liberation Front and envisions Chicana/o/x Studies as a space for liberation, resistance, freedom, and decolonial and abolitionist futurities." Society's ills and their causes are accepted facts; the role of educators is to galvanize students to take action to address them. A statement on our Asian American studies department website is explicit: "The way I would describe it is that to major in Asian American Studies is to major in social justice." We'll call this faction "Dogma U."

The divide is profound. The faculty of Dogma U are skeptical of anything termed "objective truth," while members of Knowledge U privately doubt whether what is happening in Dogma U qualifies as education. This split is not confined to the University of California; it is everywhere. The public image of the university, presented to state legislatures and donors, is that of Scholar U. Emphasized are scientific breakthroughs driving the economy, advancements in agriculture, renewable energy, and all of the benefits of real science. Much less is heard about the activities of Dogma U. At present, Dogma U appears to be in ascendance, with the dire consequences discussed below.

A 2018 article titled "The Two Universities" by Mark Bauerlein contrasted Dogma U unfavorably to Knowledge U. This provides a sobering point of comparison between the state of the university in 2018 and now. Bauerlein believed that the problems were confined to irrelevant departments that were attracting fewer and fewer students. He wrote: "The costs and benefits must be kept in mind…. We can't blame campus leaders for softening up on the 'values' side of campus when, relatively speaking, so little is at stake. The intellectual corruptions of higher education are real, but they affect a shrinking part of the institution…."[2]

Bauerlein concluded that there was not much to worry about: "For most students and professors and administrators, the preoccupations of activist students are small irritants in an otherwise well-functioning and well-funded venture."[3] In 2018, I might have agreed. Faculty in Knowledge U thought we could safely ignore the other side of campus. We were wrong. What was festering then has metastasized, and it is deeply troubling.

The reaction on campus to the Hamas massacre of October 7, 2023, revealed the depth of the Dogma U corruption. Far from there being "so little at stake," we awoke on October 8 to find pro-Hamas Israel-hatred entrenched on campus, its adherents tenured. With the use of university resources and institutions, Dogma U pursues its goals on paid university time. The devastating outbreak of antisemitism on campuses across the country represents Dogma U in full flower.

At UC Davis in October, just days after October 7, the student group calling itself Students for Justice in Palestine (SJP) cosponsored a teach-in on the conflict with seven Dogma U departments: Asian American Studies; Chicana and Chicano Studies; Gender, Sexuality and Women's Studies; Middle East/South Asia Studies; Cinema and Digital Media; and Native American Studies. There, a professor stated: "[the] Zionist Enterprise was by far the most powerful and insistent force for antisemitism on the planet," and "[Israel is] A planetary vision converted into a genocidal project," and "I would condemn Hamas when Zionists condemn the entire Israeli state.... Until then, only solidarity for the oppressed."

Since then, anti-Israel demonstrations have run amok, with university property vandalized, classes disrupted, and bystanders harassed and intimidated. A student leader of SJP at UC Davis posted on social media: "I wake up daily and pray for their death and their forever sentencing to hell. Zionists are visious [sic] they are evil they are not people they are not humans, they are devils."[4] Far from being marginalized for such dehumanizing language, SJP is supported by an enthusiastic group of Dogma U faculty, Faculty for Justice in Palestine, and the university provides extra funding for collaborations with this group.

On the one hand, UCD is still running trainings where attendees are instructed that asking someone, "Where are you from?" is a microaggression because it might be interpreted as "You don't belong here." On the other hand, calling Zionists "devils" is defended by campus administrators as crucial for the robust intellectual give-and-take on a vibrant, diverse campus. This induces dizzying cognitive dissonance. It's impossible to maintain a system in which rationalizing the Hamas terror attacks and glorifying the terrorists is A-OK, but saying, "I believe the most qualified person should get the job," is unacceptable.

Bauerlein said, "The preoccupations of activist students are small irritants in an otherwise well-functioning and well-funded venture." The university encampments in the spring of 2024 showed this is not the case. Dogma U created a nightmare for Jews in 2023–24. The effect of anti-Israel propagandizing on the next generation is profound. In a December 2023 Harvard-Harris poll, 51 percent of eighteen- to twenty-four-year-olds think the "long term solution to the Israel-Palestinian dispute" is "for Israel to be ended and given to Hamas and the Palestinians." This is not because they believe in peaceful coexistence: 58 percent think that "Hamas would like to commit genocide against the Jews in Israel." And then we get to frank antisemitism: 67 percent of eighteen- to twenty-four-year-olds think that "Jews as a class are oppressors and should be treated as oppressors."[5]

One might think these appalling results represent a failure of higher education; in fact, they represent Dogma U succeeding as Knowledge U retreats.

While these attitudes are doubtless amplified in social media, on campus they originate from the Dogma U faculty. Does anyone genuinely think a professor hired in response to the job advertisement above keeps personal political views out of the classroom? That's absurd; incorporating extremist views into their teaching is in the job description.

This underlying philosophy was on full display at the January 2024 meeting of the University of California Regents. The regents were deliberating a straightforward proposal to prevent departments from publishing contentious political statements on official department websites, exemplified by a statement on October 17, 2023, from the so-called UC Ethnic Studies Faculty Council on the Israel-Gaza war. The statement and variations were prominently posted on multiple University of California departmental websites, effectively declaring these programs off-limits to Jews. The Ethnic Studies faculty objected to the description by UC President Michael Drake and Regents Chair Rich Leib of the October 7 massacre as an act of terrorism. The faculty wrote: "A community trying to free themselves from decades-long ethnic cleansing and genocide is not the same as one of the world's strongest state militaries…to hold the oppressed accountable for 'terrorism' re-inscribes a colonial narrative…."[6] That is, it is their view that to describe Hamas's heinous actions on October 7 as a terrorist attack is to "re-inscribe A colonial narrative…."

The most heated pushback to the regents' neutrality proposal came from departments for which taking political stances is the central part of their job. The student-regent expressed it this way (boldface added): "The [proposed] policy constrains experts [e.g. in Ethnic Studies] from effectively communicating crucial information," and it "fails to acknowledge that scholars…[in] programs like Ethnic Studies rely on research, theory and factual data…our expressions are not rooted in subjective opinions.... The nature of our statements is inextricably interlinked with our research and expertise as scholars, aligning with the essential responsibility for which we are employed by the UC."[7] Here we see Dogma U masquerading as Knowledge U. The claim is that it's their data, their research, their scholarship, that has led inevitably to the conclusion that Hamas is not a terrorist organization. He went on to liken these departmental political statements to statements by biologists who discover hazardous chemicals in the water system and utilize their department's webpage to inform the public of the danger. This is a dishonest and increasingly common tactic found in Dogma U—use jargon and pseudoscience to present opinions as scientifically validated. The speaker was correct that these political statements are indeed part of "the essential responsibility" for which these faculty are hired by UC. Faculty-activists are the end result of a system that rewards the hiring and promotion of faculty on the basis of political activism, not scholarship.

Dogma U faculty are making their voices heard. Indeed, they are using official university websites to broadcast contentious political opinions, severely infringing the academic freedom of their colleagues and students.[8] They are incorporating their contentious political opinions into their classes. So, whose speech is being suppressed?

The stifling of free speech on campus is selective. Most of the suppression is directed towards what are considered on campus to be "conservative" views.[9] Notable examples include Dorian Abbot, canceled at MIT; Carole Hooven at Harvard; Kyle Duncan at Stanford Law; and, in the spring of 2024, a former IDF soldier driven out of UC Berkeley by an angry mob. Dogma U faculty support for cancellations is enthusiastic. During the Dorian Abbot debacle, a UC colleague posted: "Freedom of speech does not protect you from the loss of your privileged expectations." Who knew that giving a scientific talk in your research area was a "privileged expectation"? However, after some firms indicated that they would not hire from the current crop of college protestors, Dogma U faculty expressed outrage at those cancellations. Some privileged expectations appear to be more sacred than others.

Can anything be done? The first step is for members and advocates of Knowledge U to acknowledge that there is a significant problem, currently being manifested as virulent antisemitism. The next is to understand that the problem is systemic. Too many activists-pretending-to-be-scholars are tenured, and too great a proportion of the administration views their role as supporting them. The core of the university as understood by Knowledge U supporters is surrounded by rot. How did this happen?

A few crucial things occurred in the last decades. The administrative class became professionalized and nationalized. A typical college president or chancellor, even a dean, arrives after a costly national search, lectures us for a few months on the history of our own campus, including some remarks on the football team (go Aggies!), and convenes a committee to write a strategic vision plan. Then they start up a few expensive and ill-conceived initiatives, partnered by their numerous deans, vice-deans, associate and assistant deans, vice-provosts, associate vice-chancellors, vice-presidents, and recently also the innumerable DEI staffers. After five or ten years, they leave campus for their next, higher-paying gig. Their focus is on their careers and on the next five years, not the next fifty—certainly not on the core mission of the university. Simultaneously, the number of administrators has ballooned.

Why haven't the Knowledge U faculty objected? A few have, and a few saw this coming long ago, but almost no one paid attention. The incentive for people in Knowledge U is to keep their heads down to work on things that are fascinating and rewarding. After all, that's why they became scholars in the first place. If this means ignoring the downward spiral, pretending, in public, that what is being produced over in Dogma U is scholarship, that there is only one university, so be it. Anything for a quiet life. This is a losing strategy. First, they came for the humanities, then they came for the social sciences, indeed, but the physics and mathematics departments are now on the list.

Much of the recent downward pressure has come from the enormous diversity, equity, and inclusion (DEI) bureaucracy. DEI nationally has had a setback since October 7, 2023, as Jews, on campus and off, came to the realization, shocking for some, that "inclusion" in DEI was strictly for politically favored groups, and Jews weren't on that list. Indeed, Jews are oppressors, the DEI reasoning goes, so Jews are the enemy. But DEI on my campus is thriving; the number of full-time positions, as well as the budget for DEI, has more than doubled in the last five years. Hopes to expand the DEI umbrella to shelter the Jews are both misguided and doomed to failure. DEI is a keystone of the institutionalized support for antisemitism.

The DEI influence on faculty hiring has been amplified by changes in the faculty hiring process, with much greater influence wielded by administrators. Cluster hires, special hiring initiatives, search waivers for preferred candidates—all are ripe fields for administrative interference. UC also introduced a requirement for all faculty job applicants to submit a diversity statement, which plays a screening role, often a decisive one, in the hiring process. This both invites and results in the filtering of faculty by political viewpoint. This partly explains the significant leftward shift in the political affiliations of faculty and administrators on campuses over the past decade. Conformity, it appears, is our destiny, if not our strength.

Changing this trajectory of decline is a daunting task. We must dismantle DEI bureaucracies, eliminate or reform politicized departments, significantly reduce the size of university administrations, and vigorously defend free speech and academic freedom—all simultaneously. This indeed means defending the academic freedom of colleagues spouting pernicious nonsense—but it requires a concurrent relentless attack on the nonsense itself. We need to point out clearly that departments focused on social justice activism are not legitimate participants in academia, any more than departments of astrology, witchcraft, or homeopathy. Our universities and our society march forward together; if we don't want to live in the wasteland of a Dogma Society, we must stop it now.

SECTION 2
Ideological Corruption of Academic Disciplines

While the general environment for open debate and free inquiry at universities and research institutions has been problematic over the past decade, specific disciplinary attacks across the entire realm of scholarship have been nothing short of shocking. In this section, authors from a wide variety of disciplines, including mathematics, biology, medicine, law, archaeology and paleontology, and even classics describe attacks on scholarship at the very heart of their fields.

Mathematics, among all fields, should be as far removed from the vicissitudes of politics as any field of human inquiry. The language of mathematics is universal, and it is essentially completely divorced from human affairs. Yet, as both Sergiu Klainerman and John Armstrong separately describe, the universality of mathematics is now being questioned and is often represented as either "white," "male," or "Western" at its basis and so requires comprehensive "decolonization" to overcome its intrinsic racism and sexism.

Bruce Pardy describes how it came about in Canada that the legal system, and an activist Supreme Court in particular, have been politicized over recent decades, resulting in, among other things, the legitimization of overt discrimination today and the curtailing of free speech rights in order to somehow make up for past discrimination against women, minorities, and indigenous peoples. The adage "two wrongs don't make a right" comes to mind.

Jerry Coyne and Luana Maroja compellingly explore a similar ideological assault on biology. A wide framework of ideas, discovered and refined over hundreds of years of experimental work, is being systematically dismantled in ways that threaten to undermine how the field will progress in the coming decades. Coyne and Maroja explore six specific examples of the ongoing assault on their fields.

Sally Satel, a psychiatrist with part-time affiliation at the Yale Medical School, describes, within the context of her own near cancellation, how social justice medicine—or *indoctrinology*, as she calls it—poses a danger to open inquiry and scientific rigor in academic medicine.

Elizabeth Weiss, an anthropologist and the subject of two different cancellation experiences, describes how the Native American Graves Protection and Repatriation Act (NAGPRA) has allowed indigenous creation myths to overtly restrict and interfere with scientific research in a way that would be inconceivable were they Judeo-Christian creation myths.

Finally, in one of the more surprising discussions, Solveig Gold and Joshua Katz, both formerly associated with the classics department at Princeton University, describe a shocking development there that leads them to explore how misplaced notions of equity have impacted more broadly on academic scholarship. The development? In the spring of 2021, the decision was made to eliminate its language requirement for undergraduate majors so that students could graduate with a degree in classics without ever taking a single course in Latin or Ancient Greek.

Radical Equalitarianism and Mathematics

SERGIU KLAINERMAN

In my article, "The Universalism of Mathematics and its Detractors" (*European Review* 2023), I have tried to sound the alarm about recent trends of deconstructing mathematics based on what I called radical equality dogma (RED1), precept which claims that "human beings are roughly identical in terms of talents, intelligence (both cognitive and emotional), interests, motivation, powers of concentration, ability to perform various cognitive tasks, and that, consequently, every visible disparity between groups of individuals has its origin in some form of bias and discrimination." These, as explained by I. Kendi in his book *How to Be an Antiracist*, can only be cured by further discrimination. Thus, he writes: "The only remedy to racist discrimination is antiracist discrimination. The only remedy to past discrimination is present discrimination."

This is the exact opposite of the traditional notion of "fair treatment," once viewed as the gold standard of how Americans are supposed to judge and treat each other, according to which "individuals should only be judged by their intrinsic merits and actions, with no regard for their race, ethnicity, religion, nationality, sex or other personal characteristics and no reference to other individuals, or societal or institutional considerations. Unlike fairness, centered on the individual, that is measurable and dealt with by individual-centered measures, equity understood as equality of outcome requires one to compare individuals based on all possible identity groups to which they belong, a task impossible to accomplish without either setting all against all, in a dystopian state of nature society, or by appealing to a strong authoritarian or totalitarian state to impose its will on how resources are to be divided and how people are to be rewarded or punished."

A similar dogma, which I will call RED2, holds that all civilizations, cultures, or countries have contributed, and continue to contribute, equally to all important cultural developments in science, arts, music, literature, mathematics, philosophy, and so on, and that any assertion to the contrary is a manifestation of cultural hegemonism, better known as cultural colonization, the ultimate sin to which the "West" is uniquely guilty of.

According to both versions of RED, mathematics, far from being the universal discipline we used to think it was, is, in fact, heavily dependent on the group that practices it. It may thus be called ethnocentric, racist, or sexist since most of what is taught in schools and universities was produced, allegedly, by European white males. Here are the words of Professor Rochelle Gutierrez, holder of many prestigious prizes in education:

> Algebra and geometry perpetuate privilege because emphasizing terms like Pythagorean theorem and the symbol π gives the impression that math was largely developed by Greeks and other Europeans." Moreover, "On many levels mathematics itself operates as whiteness. Who gets credit for doing and developing mathematics, who is capable in mathematics, and who is seen as part of the mathematical community is generally viewed as white." She also claims in the postmodern style of the times, "Things cannot be known objectively; they must be known subjectively.

In a peer-reviewed article, "A Framework for Understanding Whiteness in Mathematics Education" (*Journ. of UME* 2016), a professor of mathematics education at Vanderbilt University, Luis A. Levya, states:

> The framework developed and presented here illustrates three dimensions of white institutional space, institutional, labor, and identity, that are intended to support mathematics educators in two ways: (a) systematically documenting how whiteness subjugates historically marginalized students

of color and their agency in resisting this oppression, and (b) making visible the ways in which whiteness impacts white students to reproduce racial privilege.

Professor Levya delivered an invited address at the 2023 Joint Mathematics Meeting, the most prestigious yearly gathering of mathematicians in the US, with the title "Undergraduate Mathematics Education as a White, Cis-Heteropatriarchy Space and Opportunities for Structural Disruption to Advance Queer of Color Justice." Laurie Rubel, a math education professor at Brookline College, claims (in the same *UME* journal) that both "meritocracy" and "colour-blindness" are ideological precepts that hold back racial minorities from succeeding in math classes.

More recently, the Australian National University mathematics professors Rowena Ball and Hongzhang Xu wrote (*Mathematics Without Borders 2023*):

> Efforts to decolonize mathematics curriculums, professional societies, and the discipline itself are gathering pace in universities, schools and communities. Questions are being asked louder, more often, and more pointedly, such as: What constitutes mathematical knowledge? What is included in mathematics? Who gets to say? How did it happen that British-European mathematics colonized the minds and curriculums of the whole world? Is the runaway process of development of that mathematics starting to saturate?… Students of under-represented and minority groups and colonized peoples are starting to be more critical about accepting unquestioningly the cultural hegemony of mainstream mathematics." "Mathematics has been gatekept by the West and defined to exclude entire cultures. Almost all mathematics that students have ever come across is European based," she explains." "We would like to enrich the discipline through the inclusion of cross-cultural mathematics." According to her argument, schools should teach less algebra or calculus to leave more time to teaching about

the indigenous "lived mathematics," which, she writes, is less interested in "numbers and arithmetic and accounting" and more on "recognizing and classifying patterns," that is "the science of patterns and periodicities and symmetries."

Leaving aside some of her more incredible claims, such as that of the ability of the Aboriginal Australians to predict eclipses, from a purely anthropological point of view, Professor Ball is not wrong. Yes, it is important to know more, as part of the cultural history of the world, how different societies have developed the rudimentary, or less rudimentary, mathematics needed in their day-to-day "lived experience." But such considerations have little to do with the subject of mathematics as such—a uniquely structured body of knowledge that starts with numbers and arithmetic and, yes, accounting and builds on it, layer by layer, the branches of mathematics we learn in schools and universities: geometry, algebra, calculus, differential equations, topology, group theory, and so on. To understand and make use of this magnificent, uniquely coherent, and "unreasonably effective" structure, it is entirely irrelevant how different layers have developed. The entire body of mathematics can be studied and applied by any country, culture, or group of people anywhere in the world, with no reference whatsoever to where and how its content has developed. This is precisely what the great German mathematician David Hilbert meant when he wrote, "Mathematics knows no races. For mathematics, the whole cultural world is a single country"—a unique cultural stage where individuals all over the world are free to participate, according to their personal interests and talents, to the greatest and most enduring human achievement—what Hilbert called "the highest pinnacle and highest height of the culture of rigorous knowledge."

This is not to say that the history of mathematics is not important; on the contrary, it is a fascinating and inspiring subject, with a lot to add to the cultural history of mankind, but not in the least needed to learn and understand any specific mathematical content. It is fascinating to wonder about the mysterious personality of Pythagoras or learn about the idiosyncratic one of Newton, but neither knowledge is needed to understand either the Pythagoras' theorem or the calculus. More importantly, we don't need to delve into Newton's confusing method of fluxions, with its awkward notations; the modern exposition of the subject in a decent high school manual is greatly superior. Similarly, it would make no sense to learn arithmetic using any other numbering system than the one introduced by Indian mathematicians in the first to the fourth centuries, adopted by the Arabs in the ninth century, and spread throughout Europe starting with the tenth century. The Indian Arab system in use today is vastly superior to the ones previously used by the Egyptians, the Greeks, the Romans, the Chinese, and so on. It would be equally foolish to try to learn algebra using the ninth-century text of al-Khwarizmi or even the more advanced sixteenth-century one of Viète; the modern texts, at least the ones before the advent of ethno-mathematics, have far better notation and content and are thus a lot easier to understand and apply.

A lot of the confusion surrounding the pronouncements made by professors Gutierrez, Ball, and Levya, and many others, stems from a fundamental misunderstanding of the nature of mathematical concepts. Are they invented or discovered? If invented then, like all other cultural artifacts, they reflect the unique idiosyncrasies of the societies or the individuals that have created them. It would then be justified to entertain the idea that the discipline of mathematics might be enriched "through the inclusion of cross-cultural mathematics." But mathematics deals with its unique, abstract form of objective reality. The theorem of Pythagoras is no less valid today than the day it was first discovered, probably in ancient Babylon four thousand years ago. Euclidean geometry is still valid within its precise axiomatic third-century BC framework, even though mathematicians have later discovered that there are other types of geometries equally valid within a broader axiomatic system. Like all other scientists, mathematicians discover objective reality; they do not create it. If mathematics was, in fact, a cultural artifact, like music, literature, or the arts, it would be impossible to explain its extraordinary effectiveness in the physical sciences, weather prediction, engineering, or artificial intelligence.

It is also true that mathematicians, just like chemists or biologists, rely on their creativity when making specific choices in their probing of objective reality. These choices may reflect the cultural environment in which the discoverer operates but do not affect the reality of what is discovered.

America was there before Columbus or whichever non-indigenous explorer first discovered it. The history of its discovery does indeed reflect the European cultural environment in which Columbus lived, but that does not call into question the reality of the continents we refer to as the Americas.

If RED2 is dangerous for how mathematics and science are taught, putting at risk the education of the very people (underrepresented minorities) that it purports to help, RED1 is having a measurable adverse impact on how individual contributions to research and scholarship are being assessed and rewarded by academic institutions and various government and nongovernment funding agencies. The recent, highly concerning peer-reviewed article "Politicizing Science Funding Undermines Public Trust in Science" (I. Efimov et al. 2024; Ed: See appendix of this book) documents how federal funding agencies in the US are "changing the criteria by which they distribute taxpayer money intended for scientific research" based on diversity, equity, and inclusion (DEI) considerations rather than scientific merit. These heavy-handed bureaucratic efforts are all justified in the name of combatting existing biases and discrimination. In so far as these biases are real, such efforts may be justified, but are they?

The recently published book *Matheuses*, authored by three prominent women researchers in France, strongly promoted by the prestigious Centre National de la Recherche Scientifique (CNRS), claims that the present academic system is heavily stacked against women everywhere and calls for a radical transformation of mathematics. Here is a translation from the original French text, "Mathematics needs an internal and collective transformation of practices, based on the refusal to build the discipline on the personal success of a few individuals considered exceptional, and on the systematic rejection of all naturalizing approaches for women and men, but also of questions of taste, talent and merit."

The book starts from the obvious reality that women are far less represented in the mathematical sciences and the occupations based on them and attributes this fact entirely to societal biases (that is self-fulfilling societal expectations that women are naturally less talented or simply less interested in mathematics), as well as to overt discrimination, harassment, and even sexual violence. The authors refuse to consider the possibility that sex differences may have any role in personal preferences or abilities and vehemently reject that any such differences could have biological origins. They go so far as to claim that any dearth of women in informatics is due to "brutal eviction based on sexual violence." According to the authors, "*Les filles sont repoussées de façon humiliante, rejetées parce que jugées sans valeur et dégoutées et découragées*" (Young women are pushed away in a humiliating fashion, rejected because they are considered without value, they feel disgusted and discouraged [by this treatment]). Even the fact that women are in the majority in medicine, biology, pharmacy, or veterinary sciences is presented in a negative light, as if somehow women are pushed by society to these disciplines, which the authors claim to be inferior to the more "noble" mathematically based ones, in which men happen to be overrepresented.

To support its radical claim of ongoing, unrelenting discrimination, the authors deny the existence of natural differences in cognitive aptitudes. Nobody has more "natural talent" in math than anyone else, and intelligence is not an intrinsic quality of a person but the result of personal interactions and training. This RED view of human nature leads the authors to state, "*L'intelligence des femmes s'arrête là où la domination commence*" (Women's intelligence stops where domination starts). The question, according to the authors, "is not to know who is intelligent but to understand who decides the attribution." Thus, in a postmodernist fashion, intelligence is "what is measured by tests of intelligence, or by the school system" in which the child is educated. The school is nothing more than a social "alchemy which transforms social inequalities into school inequalities." Or, in a Marxist doctrinaire fashion, it claims that school "transforms social differences into natural differences, and then uses the pretended natural differences to justify social differences." The book also states, with just a minimum of equivocation, that "the consensus in life and social sciences is that intelligence or the so-called 'mathematical talent' are principally, if not totally, acquired." Really? Is it also the consensus that musical or literary or sports talent are also socially constructed?

The book contains (page 166) these astonishing accusations: "Les savoirs mathématiques légitiment les hiérarchies raciales" (Mathematical knowledge legitimizes racial hierarchies). "La communauté mathématique entretient l'exclusion et l'oppression" (The mathematical community sustains exclusion and oppression). And it ends with this Leninist type slogan:

> "Les filles sont l'avenir des mathématiques féministes, antiracistes et égalitaires dont le pouvoir ne serra plus oppresseur, mais émancipateur" (Women are the future of feminist, antiracist, and egalitarian mathematics, whose power will no longer be oppressive but emancipating).

The book, highly advertised by CNRS, is only the most extreme version of a RED-informed ideology, now widely accepted throughout academia, willing to accept nothing else than full numerical parity in all areas of the sciences and engineering in which men are still overrepresented. Universities, research institutes, and professional societies have put in place various bureaucratic measures by which they hope, in time, to achieve this goal. Here is, for example, how Susan Wexler, the former home secretary of the US National Academy of Sciences, describes the selection process of new members of the academy:

> We assign slots based on the diversity of the lists of nominees that they have forwarded…. Classes presenting a more diverse list get extra slots." The next year, the council reviews how those slots have been filled and adjusts the distribution based on performance. "If [the selectors] used them to pick a bunch of white guys from Harvard, they get penalized." The home secretary is also quoted as saying that the process has been "spectacularly" successful in recent years. "I was one of nine women in my class [of 60 in 1998]. Last year (2020), we had 44 [out of 120], and this year (2021) it's essentially even.

Once more, such efforts would be understandable, even if heavy-handed, if the claims made in the *Matheuse* book were correct. But are they? The best recent refutation comes from an evolutionary biologist. In her excellent essay "Why Do Men Dominate Chess?" (*Quillette* 2024), Carole Hooven elegantly demolishes the RED1 myth that the underperformance of women in chess is due to nothing more than overt societal discrimination. Chess is, of course, in no way equivalent to mathematics; it is just a game, after all. But it is well known that interest and ability for math correlate well with those for chess. The level of participation of women in chess is remarkably like that of their participation in higher-level math or the STEM fields that require it. Moreover, the success of girls in chess competitions mirrors their level of success in math competitions, such as the International Mathematical Olympiad (IMO).

To understand why men dominate chess, Dr. Hooven, Harvard PhD and author of a popular book on the importance of testosterone on human behavior (*The Story of Testosterone 2021*), analyzes various hypotheses. One of them, the greater male variability (GMV) hypothesis, is based on the observation that, though men and women exhibit an equal average level of intelligence (with maybe a slight advantage for women), males exhibit greater variability than females in a range of traits, such as height, intelligence, preferences for leisure activities, risky behavior, and competitive drive. Dr. Hooven points out that the GMV hypothesis is used to explain sex differences in the STEM disciplines. She writes:

> The idea is that even if there's no male-female difference in average math or physics ability, there would still be more men at the very high (and low) end of the ability spectrum. These are the extreme outliers who are most likely to earn prestigious faculty positions, file many patent applications, and win career achievement awards. And there is, in fact, strong evidence supporting the hypothesis; many traits do tend to be more variable in men than in women.

A second explanation is the participation rate hypothesis (PRH). The idea here is that even though schoolboys and girls may have similar aptitudes for chess, the higher number of boys who become interested in playing competitively results in more of them eventually reaching the highest echelons of performance. This phenomenon could also explain the higher achievement rates of high school boys in mathematical competitions, such as IMO.

She then goes on to undermine the explanatory power of PRH by pointing out that in Scrabble, a game completely dominated by women in terms of recreational participation rates, men still outperform women at a competitive level. She points out that the same goes for bridge, a game also heavily dominated by women at the recreational level.

Her article concludes with what seems to be the most realistic explanation, one maybe also consistent with the GMV hypothesis,

> that men are more likely than women to exhibit a "rigid persistence" in an activity, by which the passion controls the individual ("obsessive passion" in the literature). In anecdotal terms, we are talking here about the man who drops everything to become, say, a 16-hour-per-day video-gamer, or a day-trader, or chess addict. Yes, some women take on these kinds of fixations. But men do it more often, and with greater intensity.

Finally, Dr. Hooven states, in stark contradiction to the main premise of the *Matheuse* book:

> Ultimately, sex differences in complex behaviors and skills are always a product of interactions between biology on the one hand (that is, our genes and their relatively fixed effects, such as hormone levels and body size) and our environment on the other (that is, factors such as our family circumstances, social dynamics, and cultural norms). Interactions between the two shape not only our skills and abilities, but also any emerging group differences. But none such complicating factors change the fact that the sex gap in chess [or mathematics] is real and persistent.

These differences have only statistical relevance; individual women can, and often do, exhibit the same obsessive passion for chess or mathematics as men. I know that to be true by my direct professional experience with extremely talented women mathematicians.

Yet differences between human beings in mathematical abilities and interests are "real and persistent." There is no better illustration of this fact than the educational gender equality paradox (GEP), identified by G. Stoet and D. C. Gerry ("The Gender-Equality Paradox in Science, Technology, Engineering, and Mathematics Education." *Psychological Science* 2018). They found that countries with a high level of gender equality have some of the largest STEM gaps in secondary and tertiary education.

> For example, Finland excels in gender equality (World Economic Forum, 2015), its adolescent girls outperform boys in science literacy, and it ranks second in European educational performance (OECD, 2016b). With these high levels of educational performance and overall gender equality, Finland is poised to close the STEM gender gap. Yet, paradoxically, Finland has one of the world's largest gender gaps in college degrees in STEM fields, and Norway and Sweden, also leading in gender-equality rankings, are not far behind. We will show that this pattern extends throughout the world, whereby the graduation gap in STEM increases with increasing levels of gender equality.

Even though statistical disparities are to be expected, one can still ask the question of whether mathematically gifted girls continue to be discriminated against and discouraged from pursuing STEM careers. There is no doubt that societal norms and expectations, and sometimes ugly prejudice and predatory behavior, did in the past prevent the full participation of women in science. But is that still true today? The recent comprehensive meta-analysis study, "Exploring Gender Bias in Six Key Domains of Academic Science," by Stephen J. Ceci and coauthors, explores gender bias in six key domains of academic science: tenure-track hiring, journal peer reviewing, grant funding, letters of recommendation for faculty applicants, salary, and teaching evaluations. While they still find a significant wage gap of about 3.6 percent, though not the 18 percent that is often being reported, they find no bias in all other metrics.

In peer-review success and letters of recommendation, we found no systematic gender bias. Manuscripts by women were accepted for publication at equivalent rates to men's, and women applicants for tenure-track jobs received comparably strong letters of recommendation, despite widespread claims to the contrary.

Moreover, in tenure-track hiring, arguably the most important metric of all, their analyses revealed "that women applicants have a substantial advantage over comparable male applicants."

Though numerical disparities are bound to persist, no matter what we do, we can alleviate their negative impact on society by rewarding people based on the interests and talents they have, rather than those they do not, and treat the interests themselves as worthy of equal respect, rather than insist that all individuals are equally qualified to pursue them. The notion that mathematics and physics are nobler occupations, as the *Matheuse* book claims, nobler than raising and educating children, taking care of the sick, gardening, building, plumbing, or any other occupation without which civilized life would not be possible, is patently false and should be widely rejected.

And yes, we should continue to make all efforts, short of bureaucratic dictates, to make it possible for talented women, indeed all talented individuals, to thrive in STEM and the "hard" sciences.

How Do You Decolonize Mathematics?

JOHN ARMSTRONG

In May 2020, George Floyd was murdered by a white police officer in Minneapolis, sparking international protests. In response to these events, in June 2020, the journal *Nature* declared its support for the Black Lives Matter movement, stating, "We recognize that Nature is one of the white institutions that is responsible for bias in research and scholarship. The enterprise of science has been—and remains—complicit in systemic racism, and it must strive harder to correct these injustices and amplify marginalized voices."

This has led *Nature* to champion a program of decolonization of science. Writing a guest editorial for *Nature*, Nobles, Womack, Wonkam, and Wathuti (2022) explain, "It is so important for science curricula, research and academic spaces to go through decolonization processes. These are not political or ideological acts, but part of science itself—an example of science's self-correcting mechanism in the pursuit of truth."

Decolonizing the curriculum in arts and humanities is popularly understood to mean studying more texts by black and minority authors, but it is not immediately clear what it would mean to decolonize mathematics. This was acknowledged in another editorial in *Nature* (2023): "One common argument is that decolonization is irrelevant to the practice of mathematics: the solution to a quadratic equation doesn't, after all, depend on a mathematician's identity or protected characteristics." Nevertheless, the editorial strongly argued that "we have nothing to fear from the decolonization of mathematics" and endorsed the view that decolonization is not a political program but is simply the pursuit of truth.

This directly contradicted my understanding. Only two months before, I had organized an open letter opposing proposals made by the UK's Quality Assurance Association for Higher Education (QAA) that all university mathematical curricula should present a "decolonised" view of mathematics. This letter was signed by, among others, seven Fellows of the Royal Society. In that letter, we observed that the theory of decoloniality is a postmodernist critique of the "European paradigm of rational knowledge" (Quijano 2007), and we argued that the history of mathematics shows it to be far from European.

Were we, or was *Nature*, misrepresenting the program of decolonization? If the mathematics community was to follow the QAA's advice and present a decolonized view of mathematics, we surely needed a clear understanding of what the term means and the evidence to support the program. To this end, I began a systematic review of a sample of the literature on decolonizing mathematics. This article summarizes what I found.

Eight of the thirty-seven papers took a distinctly theoretical approach to decolonization, and all of these cited Quijano. Quijano had no particular interest in mathematics but took aim at "European" knowledge more broadly:

> The colonizers also imposed a mystified image of their own patterns of producing knowledge and meaning. At first, they placed these patterns far out of reach of the dominated. Later, they taught them in a partial and selective way, in order to co-opt some of the dominated into their own power institutions. Then European culture was made seductive: it gave access to power…. Cultural Europeanisation was transformed into an aspiration (Quijano 2007).

For Quijano, knowledge is subjective, and what constitutes knowledge is determined by who holds power. This view is far from unique to Quijano; it is a central tenet of postmodern thought.

Another highly cited author within the sample was D'Ambrosio, who introduced the concept of *ethnomathematics*. D'Ambrosio states that "we should not forget that colonialism grew together in a symbiotic relationship with modern science, in particular with mathematics." D'Ambrosio argues that "[belief] in the universality of mathematics" is becoming hard to sustain in the face of anthropological research. Ethnomathematics studies indigenous approaches to mathematics but also widens the scope of what should be considered as mathematics. Often, in approaches derived from ethnomathematics, it is unclear whether an indigenous practice is itself mathematical or whether it can simply be modeled mathematically. For example, Eglash, Bennett, Drazan, Lachney, and Babbitt (2017) describe a lesson on Anishinaabe boatbuilding using wooden strips that explains how these can be modeled using Bezier curves. However, the Anishinaabe themselves did not historically do this.

Particularly highly cited in the sample was Bishop, cited by twelve out of thirty-seven authors. In his paper "Western Mathematics: The Secret Weapon of Cultural Imperialism," he argues that "to decontextualise, in order to be able to generalise is at the heart of western mathematics and science; but if your culture encourages you to believe, instead, that everything belongs and exists in its relationship with everything else, then removing it from its context makes it literally meaningless." Bishop asserts that western mathematics can be distinguished from other forms of mathematics by its values of "rationality," abstraction, "power and control" and "progress." He argues that by promoting these four values, mathematics promotes Western "cultural hegemony."

One obvious problem with these views is that mathematics did not originate in Europe. Our contemporary number system, which makes crucial use of the concept of zero, had its origins in India and became more widely known through the writings of the Persian mathematician al-Khwarizmi and the Arab mathematician al-Kindi. It was then popularized in the West by Fibonacci in his book *Liber Abaci*. The word algorithm comes from the name al-Khwarizmi. The word algebra derives from the title of one of his books. One could see Fibonacci's work as an astonishingly successful precursor of ethnomathematics, but with the crucial caveat that Fibonacci did not deny the universality of mathematics.

Bishop himself acknowledges that the term "western mathematics" is "in a sense…inappropriate," due to the historical and contemporary development of mainstream mathematics outside the West but then argues that the term is "thoroughly appropriate" since it was "western culture which played such a powerful role in achieving the goals of imperialism." Aikenhead (2017) provides an alternative resolution of the issue:

> The Eurocentric impulse to appropriate from other cultures can account for how European mathematicians throughout the centuries seem to have imported ways of mathematizing from earlier cultures but then reconstructed those ideas to fit the European mathematical philosophy or ideology of the time.

Mathematics is frequently described as Western or European throughout the decolonization literature, which often appears to manufacture a myth of mathematics as European before proceeding to demolish it.

Another challenge for postmodern accounts of mathematics is that the same mathematical results have often been discovered independently internationally, strongly suggesting that culturally independent truths do exist. The Maya civilization was using the concept of zero long before Columbus set sail to the Americas. The Keralan mathematician Madhava of Sangrama (or perhaps one of his followers) discovered the following remarkable formula for π

$$\frac{\pi}{4} = \frac{1}{1} - \frac{1}{3} + \frac{1}{5} - \frac{1}{7} + \frac{1}{9} - \frac{1}{11} + \cdots$$

centuries before it was rediscovered first by Gregory (1671) and again by Leibniz (1673).

Similarly, the claim for deep synergies between mathematics, imperialism, and capitalism appear hard to justify in the light of mathematical history. Mathematics has been used in societies of all political stripes since the dawn of civilization.

Despite its international origins, many of the texts were concerned by the possibility that teaching mathematics might impose a Western worldview. Twelve of the texts in the sample were focused on teaching indigenous students in countries such as Canada, New Zealand, and Australia with minority indigenous populations, and eleven of these emphasized the importance of centering indigenous beliefs. In practical terms, these authors promoted including aspects of indigenous culture in education. Garcia-Olp, Nelson, and Saiz (2022) suggest having students compute the volume of tipi. Eglash, Bennett, Drazan, Lachney, and Babbitt (2017) describe "a unit called 'wigwametry' using circular model wiigiwaam construction to allow Native students to investigate the value of the mathematical constant Pi" as well as lessons on Anishinaabe boatbuilding. Aikenhead (2017) discusses the risk of tokenism and stereotyping in such approaches:

> Sterenberg (2013a) repeated an insulting word problem "Imagine a band of 250 Aboriginal People. Each tipi can hold approximately eight people. Calculate how many tipis would be needed to house the entire band" (p. 21). This is insulting because Indigenous people would not divide themselves in the hypothetical way stated in the word problem. Relational and spiritual factors would dominate. And the required hypothetical state of mind itself, captured in the word imagine, is a value embraced strongly in the culture of school mathematics but would generally be foreign to an Indigenous culture in the context of people choosing a tipi to enter, because it conflicts with how to live in a good way.

In addition to questioning the privileging of mathematics over indigenous beliefs, many of these texts recommended using approaches to teaching mathematics derived from critical mathematics education or critical pedagogy more broadly. The term "critical mathematics education" was introduced by Frankenstein (1983), who suggests teaching division by getting students to show that "each of the richest 160,000 taxpayers got nine times as much money as the maximum AFDC grant for a family of four." This is necessary to avoid supporting "hegemonic ideologies," as in conventional mathematics education, "even trivial math applications like totaling grocery bills carry the ideological message that paying for food is natural."

While Frankenstein pioneered this approach within mathematics, it derives from Paulo Freire's highly influential approach of critical pedagogy. Kincheloe and Steinberg (1997) explain that "critical pedagogy is the term used to describe what emerges when critical theory encounters education." According to Frankenstein: "Freire's epistemology is in direct opposition to the positivist paradigm currently dominant in educational theory. Positivists view knowledge as neutral, value-free, and objective, existing totally outside of human consciousness."

The final significant strand of literature on decolonizing mathematics found in our sample comes from South Africa. It has some similarities with the literature from countries with indigenous minorities but differs in that there is an occasional hint of skepticism. Mudaly (2018) argues that South African policy needs to provide more examples of how to implement decolonization in practice. He found that "practicing teachers, including some from African and Indian communities, could not identify aspects of a colonised curriculum, so they did not know how they could easily decolonise it. Many of their responses were superficial and they provided examples and solutions that appeared to be coerced and contrived.... Whenever they attempted to illustrate the use of indigenous knowledge in mathematics, the only example they chose was the cylindrical hut with a conical thatched roof." Schubring (2021) remarks that "Walsh's (Mignolo & Walsh, 2018) programmatic claim that decoloniality should 'displace Western rationality as the only framework and possibility of existence, analysis and thought' is quite strong."

In summary, the dominant theme of the literature on the decolonization of mathematics was to question the epistemic privilege of mathematical knowledge. This chimes with the analysis of *Nature* (2023), which states that debates on decolonization "reprise aspects of an older, more academically focused debate on whether—or to what extent—scientific knowledge is socially constructed."

However, there was another important recurring theme that I have not touched upon. This is the emotional rather than intellectual motif found throughout the literature that decolonization is a moral and spiritual necessity. Kearns, Tompkins, and Lunney Borden (2018) state, "We are reminded that decolonizing work has a spiritual dimension to it, as Mi'kmaw Chief PJ Prosper so eloquently stated, 'moving from the head to the heart.'" Aikenhead (2017) observes, "I have come to realize that for my contribution to go above tokenistic culturally relevant content, it is not only the content that must be decolonised but also myself." Fernandes, Giraldo, and Matos (2022) say they "understand decoloniality as a choice: a decision to share a political agenda engaged in struggle, resistance and insurgency against the various traces and effects of coloniality we are traversed by."

The emotional appeal of decolonization provides an explanation for why a police shooting in Minneapolis might lead to calls for mathematics curriculum change in the UK. The Black Lives Matter movement has focused heavily on symbolic gestures and on admissions and accusations of guilt for complicity in the horrors of colonialism. Symbolic gestures may be used to provide some emotional support even if they cannot be analyzed rationally.

It is worth commenting on the absences in the literature. I found no papers that directly addressed the question of why we should decolonize the curriculum by teaching the "problematic" history of the development of mathematics. There were no studies that showed that students wished to be taught in this way, and no studies that showed this would benefit minority students. Yet this is precisely the approach to the program of decolonization promoted by the QAA, which suggests that we teach that "some early ideas in statistics were motivated by their proposers' support for eugenics, some astronomical data were collected on plantations by enslaved people, and, historically, some mathematicians have recorded racist or fascist views or connections to groups such as the Nazis."

There may be merit in teaching the history of mathematics to students, but we should not teach a skewed or ideologically biased version of that history. Mathematics dates back thousands of years and has flourished throughout the twentieth and twenty-first centuries. It misrepresents mathematics, and mathematics teaching, to say, quoting a mathematician at Durham University: "Theorems or techniques have names associated to them and most of the time, those names are of nineteenth-century French or German men." The names of Arabic digits, "algebra," and "algorithms" pay homage to their non-Western origins. Central theorems, such as the Fundamental Theorems of Algebra and Calculus, do not have names associated with them at all. The major fields of topology, probability, and statistics only acquired their mathematical foundations in the twentieth century. The practice of modern mathematics is resolutely international. While it is true that some mathematicians are Nazis, one should also consider that many of the great names of twentieth-century mathematics and science were Jews persecuted by the Nazis.

If we bias our teaching of the history of mathematics to emphasize the "problematic," we risk indoctrinating students. This is inevitable if we do not acknowledge the existence of debate. Yet decolonization is consistently presented as a moral necessity rather than an area of debate.

Many, if not most, mathematicians do not accept the principles of postmodernism underlying decolonization. Instead, they see mathematical and scientific knowledge as privileged over other beliefs, irrespective of whether they are European. The scientific rejection of the indigenous European belief of heliocentrism is regarded by many mathematicians as one of the triumphs of scientific and mathematical history. The use of statistics has played a major role in the contemporary rejection of indigenous European beliefs about racial superiority. For a mathematician who believes in this epistemic privilege of science to adopt a postmodern approach to race sends a clear signal that they do not see issues of race as important.

This is why the media sees decolonizing mathematics as comic. The public knows that the concept of racist mathematics is nonsensical, and they, rightly, view those who claim to be able to decolonize a linear-algebra class as buffoons.

Decolonizing mathematics is the reductio ad absurdum of the decolonization program. But if decolonizing mathematics is absurd, perhaps so is decolonizing science, art, or history. Arts and humanities departments have embraced decolonization with gusto; this is just one symptom of their wider embrace of postmodern theory. For the time being, most university mathematics students (in the UK at least) are receiving a genuine education. However, within the wider university, indoctrination in postmodern thought is increasingly the norm. As mathematicians and scientists, we have a duty to all the students at our institution, and we should not allow the spread of indoctrination and science denial to go unchallenged.

(See Armstrong, J., & Jackman, I. (2023). The decolonisation of mathematics for full details of all references.)

How the Law Went Woke

BRUCE PARDY

According to Professor Eric Kaufmann, Canada was the world's first woke nation. "Wokeness is increasingly becoming its identity," Kaufmann wrote in the *Telegraph* in 2021,[1] "As activists destroy the past to arrive at Year Zero, the world's first postnational state is rising from the ashes." How did it get this way? Kaufmann cites a weak attachment to history, the highest levels of immigration in the West, and an official culture of politically correct ultra-sensitivity to minorities.

No doubt these factors have played a significant role. But "Canada's guiding project of equity and diversity," as Kaufmann puts it, can be laid in good part at the feet of the Supreme Court of Canada. In the great white north, over a period of decades, the court has turned the law from a bulwark against radical change into a sword that compelled it. The institution that many Canadians thought would push back against woke insanity instead has pushed it forward.

Here are just three of the many ways in which the court has turned the law in Canada into an instrument for cultural revolution.

(1) Justice Is Not Blind: Turning Equality into Equity

In June 2021, an Ontario high school student tried to sign up for a summer program. He was rejected because he was white. The "SummerUp" program, sponsored by the Ontario government, was open only to black students. His father filed a complaint with the Ontario Human Rights Tribunal alleging racial discrimination. In November 2022—seventeen months later—the tribunal dismissed the complaint.[2] White people, wrote the tribunal, cannot claim discrimination. "[A]n allegation of racial discrimination or discrimination on the grounds of colour is not one that can be or has been successfully claimed by persons who are white and non-racialized."[3]

It's not just the Human Rights Tribunal. For almost thirty-five years, the Supreme Court has eviscerated legal equality. At the heart of Western legal culture lies the principle that the law applies equally to all. Justice should be blind, the idea goes, meaning that it should not care if you are rich or poor, man or woman, gay or straight, black or white. Everyone is subject to a common set of standards. Over time, the Supreme Court has flipped this idea on its head. Equal application of the law has been supplanted by the opposite proposition: different rules should apply to different groups to produce equal or comparable outcomes.

As of 1985, British Columbia required lawyers to be Canadian citizens. Mark David Andrews, a British citizen and permanent resident of Canada, who was otherwise qualified to be a lawyer, challenged the requirement. In 1989, his case became the Supreme Court's first decision[4] under s. 15(1) of the Canadian Charter of Rights and Freedoms. Section 15(1) of the charter guarantees the right to equality:

> Every individual is equal before and under the law and has the right to the equal protection and equal benefit of the law without discrimination and, in particular, without discrimination based on race, national or ethnic origin, colour, religion, sex, age or mental or physical disability.

The court could have concluded that section 15 meant that the same rules and standards should apply to everyone. The law, they could have said, should not be arbitrary and should not take irrelevant factors into account. A law that criminalizes loitering if you are black but not if you are white, for example, is unequal because it treats different groups differently in a manner not relevant to the prohibition that it creates. Had the court approached equality in this way, it could have assessed whether the BC requirement treated people equally: whether being a citizen was relevant or irrelevant to being a lawyer. If it was irrelevant, the requirement would be unconstitutional.

But the court went down a different road. In a passage that still influences equality jurisprudence to this day, Mr. Justice McIntyre wrote:[5]

> ...a law which treats all identically and which provides equality of treatment between "A" and "B" might well cause inequality for "C", depending on differences in personal characteristics and situations. To approach the ideal of full equality before and under the law...the main consideration must be the impact of the law on the individual or the group concerned...the admittedly unattainable ideal should be that a law expressed to bind all should not because of irrelevant personal differences have a more burdensome or less beneficial impact on one than another.

This passage describes not equality of treatment, but equality of burden or outcome, also known as "substantive equality" or "equity." *Andrews* launched the Supreme Court on its journey towards equity, not equal treatment, as a constitutional right. Equity requires *unequal* treatment for members of different identity groups. Since *Andrews*, the Supreme Court has decided numerous cases under s. 15 and has never wavered from its commitment that equity, not equality, is what the constitution requires.[6]

Section 15(1) of the charter has an exception that was not relevant in *Andrews*. Section 15(2) states:

> (2) Subsection (1) does not preclude any law, program or activity that has as its object the amelioration of conditions of disadvantaged individuals or groups including those that are disadvantaged because of race, national or ethnic origin, colour, religion, sex, age or mental or physical disability.

Section 15(2) allows for programs of "affirmative action" that discriminate against members of some groups to promote the fortunes of others. Affirmative action gives preferential treatment to some and excludes others. Governments might offer employment opportunities, for instance, for which only certain racial groups are eligible, or from which others are excluded.

We know that section 15(2) is a specific and narrow exception because of what it says. The opening words of 15(2) are "Subsection (1) does not preclude…" Section 15(2) provides for specific programs that seek to achieve substantive equality. It conflicts with the principle of equal treatment in 15(1). If it did not, there would be no need for the words "Subsection (1) does not preclude…." Indeed, there would be no need for 15(2) at all. If 15(1) and 15(2) both protect substantive equality, then 15(2) is redundant.

Yet the Supreme Court has made 15(2) the general rule instead of the exception. "Section 15(1) and (2) work together to promote the vision of substantive equality that underlies s. 15 as a whole,"[7] the court wrote in a 2008 decision allowing Aboriginal fishing but not non-Aboriginal fishing. In 2020, it went further, deciding that section 15 does not merely allow discrimination against certain groups but sometimes requires it.[8] The court declared unconstitutional the pension program of an RCMP job-sharing scheme allowing employees to work part-time. The voluntary program was open to men and women. Since part-time employees work fewer hours, they would earn lower pension entitlements. More women than men chose to enroll, and as a result, more women than men ended up with lower pensions. The rules of the program, which were the same for men and women, were unconstitutional.

Like the charter, human rights codes tend to have an exception that allows for affirmative action. Also like the charter, those exceptions are becoming the general rule. A program that discriminates against the "correct" groups will tend to fit within the exception as a matter of course. Statutes that purport to prohibit discrimination authorize it instead.

Thanks to forty years of jurisprudence from the Supreme Court, preferential measures, distinguishing between people by their color, lineage, gender, and sexuality, are becoming the order of the day.

(2) Free Speech: The Right to Speak the Proper Thoughts

Elon Musk has called it insane.[9] Critics cited in the *Telegraph* described it as "draconian."[10] According to global law firm Norton Rose Fulbright, it is "the most significant expansion of Canada's hate speech laws."[11] "It" is Bill

C-63, the Liberal government's online harms act. It will crush Canadians' online freedom of speech.

Once passed, the act will haul people before the Canadian Human Rights Tribunal for social media posts that "detest or vilify" a person or group on a protected ground. You might not be told who made the complaint. New "digital safety" agencies will make rules and impose penalties on online platforms. Courts can impose house arrest for those who "on reasonable grounds" may utter so-called hate speech in the future. Minor offenses motivated by "hatred" can be punished by life imprisonment.

It may be a government bill, but the Supreme Court made it possible. In the United States, "hate speech" is constitutionally protected under its First Amendment. But in Canada, the Supreme Court has said that governments can censor speech that "detests or vilifies" a person or group.[12] The Liberals have cleverly used the Supreme Court's own language to draft C-63.

But it's not just Bill C-63. For a country with a constitutional guarantee of free speech, Canada already has a remarkable array of legislated censorship, which exists because of the court's narrow interpretation of the scope of free expression.

In 2017, Morgane Oger ran as an NDP candidate in the BC provincial election in the False Creek riding in Vancouver. Born a male, Oger regarded herself as a woman. William Whatcott, an outspoken religious activist, believed Oger to be a man, and campaigned against her even though he was not running himself. Oger narrowly lost the election. She brought a human rights complaint under section 7 of the B.C. Human Rights Code. Much like Bill C-63 proposes to do, section 7 prohibited public statements that "indicate discrimination…or is likely to expose a person or a group or class of persons to hatred or contempt" on a prohibited ground.

The tribunal found[13] that Whatcott had discriminated against Oger and exposed her to hatred or contempt in violation of section 7 of the Code. You can speak in favor of voting for transgendered people, the tribunal wrote, but you can't speak against voting for them. You cannot say what you think unless you think the right thing.

Of course, freedom of expression is never absolute. "Your money or your life" is speech but can also be an assault. Announcing that your colleague cheats on his taxes is defamatory unless you can prove it is true, while relentlessly hounding the object of your desire with unwanted propositions may constitute harassment. But exceptions make the rule: freedom of expression should not depend upon social approval of its content. Free speech should not have to make sense. You should be free to speak your thoughts not because they are rational or correct but because they are yours. In Canadian law, thanks to the Supreme Court, commitment to that idea has been vanquished.

(3) Deference: Empowering the Woke Managerial State

In November 2022, the Ontario Divisional Court ruled that the Ontario College of Psychologists can reeducate Jordan Peterson.[14] In November 2022, the college ordered Peterson, a prominent public intellectual and emeritus University of Toronto psychology professor, to undergo remedial education because of statements he made on social media. Peterson argued that the order infringed his freedom of expression. But the Divisional Court deferred to the college, concluding that the order was a reasonable exercise of the regulator's authority. In the era of judicial deference to the administrative state, constitutional rights do not mean what many Canadians think they mean.

The Supreme Court of Canada is again the culprit. In most circumstances, regulatory bodies are not required to apply the law correctly, the court has said, but only "reasonably." In other words, in certain contexts a regulator can be legally wrong if it is not unreasonably wrong. "'Administrative justice' will not always look like 'judicial justice,'" the court wrote in 2019 in its leading administrative law case, *Canada (Minister of Citizenship and Immigration) v. Vavilov*,[15] "a court…does not ask what decision it would have made in place of that of the administrative decision maker…or seek to determine the 'correct' solution to the problem."

But the story gets worse. In cases involving the Charter of Rights and Freedoms, regulatory bodies can infringe charter rights if they do so "proportionately." In its 2012 decision in *Doré v. Barreau du Québec*,[16] the Supreme Court said that regulators must balance charter protections against the statutory objectives the regulator is attempting to achieve. Where a court is reviewing a decision that may infringe a charter right, as in the Peterson case, regulators are entitled to deference if the decision "falls within a range of possible, acceptable outcomes."

Our government was not supposed to work this way. In principle, the administrative branch is powerless except for specific statutory mandates. Courts supervise government agencies, including professional regulatory bodies, to ensure they remain within those legislated powers. "I know of no duty of the Court which it is more important to observe, and no powers of the Court which it is more important to enforce, than its power of keeping public bodies within their rights," wrote Lindley M. R. in an 1899 UK case,[17] "The moment public bodies exceed their rights they do so to the injury and oppression of private individuals."

But the Supreme Court has slowly shifted the legal ground beneath our feet. The country has moved away from the rule of law back towards rule by executive fiat. Judicial deference grants control not to a monarch but to a professional managerial class. Broad discretion in the hands of administrative bodies has become the foundation of our modern system of government.

Deference from the courts makes regulatory witch hunts possible, when carefully crafted in the guise of "unprofessional conduct," "incivility," or "undermining public trust in the profession." Professional regulators can turn their focus from ensuring competence and ethical practice to political compliance. In Peterson's case, complaints did not come from clients nor relate to his psychology practice, which he ceased in 2017. Instead, they largely targeted Peterson's tweets on political topics such as Justin Trudeau and his chief of staff, an Ottawa city councillor, the climate change agenda, and transgender activism.

In principle, remedies and punishments are to be imposed only after a matter has been adjudicated and an accused found guilty. Yet the college insists that the order is not disciplinary. In fact, the college's committee did not conduct a disciplinary hearing and never concluded that Peterson committed professional misconduct. But courts have deferred to regulators in this respect as well. And the practice is growing. The Law Society of

Ontario proposes to give its Proceedings Authorization Committee similar powers to require lawyers to submit to reeducation, even if the committee concludes that there are no grounds for a disciplinary hearing.

Peterson is not alone. Across the country, regulators are becoming ever more insistent on ideological concurrence. Law societies are instituting politically laden "cultural competence" requirements. Medical regulators have sanctioned doctors for expressing medical opinions contrary to approved COVID-19 narratives. Nurses and teachers cannot safely voice doubts about transgenderism or "anti-racist" agendas inside their institutions. The courts' deference empowers the tyranny of the administrative state.

Conclusion: Progressive Hegemony at the Supreme Court

The Supreme Court of Canada is sometimes described as less politicized than its American counterpart. That is not because the Canadian court is apolitical, but because one political persuasion is dominant. The judges, of course, are not always of one mind on the legal issues before them. During some periods, dissenting opinions have been common. But unlike the US court, the Canadian court tends not to engage in ideological struggles. It does not need to. The struggle is already won.

In Canada, conservative governments have been ineffective at staffing the court with judges aligned with their philosophical persuasions. But also, conservative governments in Canada have not always been very conservative, and rarely so with their judicial appointments. Moreover, judges are lawyers first, and the legal profession itself has become predominantly woke. Provincial law societies, which regulate the profession, are pushing to incorporate comportment with DEI and other social justice causes as licensing requirements. Most Canadian lawyers are trained in Canadian law schools, which have come to champion the social justice agenda, some more explicitly than others. The woke progressive pipeline to the courts is secure. But that is a story for another day.

The Ideological Subversion of Biology

JERRY A. COYNE AND LUANA S. MAROJA

We're familiar with the culture wars that pit progressive leftists against centrists and those on the Right. In the past, skirmishes about politics and sociocultural issues were restricted largely to the humanities. But—apart from the "sociobiology wars" of the '70s (Schnettler 2020) and our perennial battles against creationism—we biologists kept out of these struggles. After all, scientific truth would surely be immune to attack or distortion by political ideology.

We were wrong. Scientists both inside and outside the academy were quick to begin politically purging their fields. Campaigns were launched to strip scientific jargon of words deemed offensive, to ensure that results that could "harm" people seen as oppressed were removed from research manuscripts, and to tilt the funding of science away from research and toward social reform. The American government even refused to make genetic data—collected with taxpayer dollars—publicly available if analysis of that data could be considered "stigmatizing." In other words, science—and here we are speaking of all STEM fields (science, technology, engineering, and mathematics)—has become heavily tainted with politics as "progressive social justice" elbows aside our real job: finding truth.

These changes have been a disaster. By diluting our ability to investigate, withholding research support, controlling the political tone of manuscripts, and demonizing research areas and researchers, ideologues have cut off whole lines of inquiry. This will decrease human well-being, for, as all scientists understand—and as the connection between heat-resistant bacteria and PCR tests demonstrates (Yellowstone Volcano Observatory

2020)—we never know what benefits can come from research driven by pure curiosity. But nourishing curiosity has a value all its own. After all, it doesn't make us healthier or wealthier to study black holes or the Big Bang, but it certainly enriches our lives. Thus, the erosion of academic freedom in science by progressive ideology hurts us both intellectually and materially.

Although biology has clashed with ideology in the past (e.g., the Soviet Lysenko affair, creationism, and the anti-vax movement), the present situation affects *all* scientific fields. What's equally unfortunate is that scientists themselves have become complicit in their own muzzling.

Here we give six examples of how our field—evolutionary biology—has been impeded or misrepresented by ideology. Each example involves a misstatement spread by ideologues, followed by a brief explanation of why each statement is wrong. Finally, we give what we see as the ideology behind each misstatement and then assess its damage to scientific research, teaching, and the popular understanding of science. Our ultimate concern is biology research, but research goes hand in hand with teaching and the public acceptance of biological facts. If certain areas of research are stigmatized by the media, public understanding will suffer, and a loss of interest in teaching and in research will follow. The misrepresentation or stigmatization by the media ultimately deprives us of opportunities to understand the world.

While here we concentrate on our own field of evolutionary biology, we add that related ideological conflicts are also common in sciences such as chemistry (Krylov 2021), physics (Robertson and Hairston 2022), math (Pais 2013), and even computer science (Ali 2022). In these other areas, however, the clashes involve less denial of scientific facts and more effort toward purifying language, devaluing measures of merit, changing the demographics of scientists, altering how science is taught, and "decolonizing" science ("Decolonizing Science Toolkit" 2022). Evolutionary biology has been especially susceptible to attacks on scientific truth because it deals with the most fraught topic of all: the origin and nature of *Homo sapiens*. We begin with a misconception about our species that's become quite common.

JERRY A. COYNE AND LUANA S. MAROJA

(1) Sex in Humans Is Not a Discrete and Binary Distribution of Males and Females but a Spectrum

This statement, one of the most common political distortions of biology (e.g., Ainsworth 2018), is wrong because nearly every human on earth falls into one of two distinct categories. Your biological sex is determined by whether your body is designed to make large, immobile gametes (eggs, characterizing females) or small and mobile gametes (sperm, characterizing males). Even in plants we see this dichotomy, with pollen producing tiny sperm and ovules carrying the large egg. The volume difference between eggs and sperm is huge, about sixty thousand times in humans and up to millions of times in other organisms such as birds. It is the bearers of these two reproductive systems that biologists recognize as "the sexes."

Because no other types of gametes exist in animals or vascular plants, and we see no intermediate gametes, there is no third sex. Although many species of animals and flowering plants have hermaphrodites (Jarne and Auld 2006), these simply combine male and female functions (and gametes) within single individuals and don't constitute a "third sex." Further, developmental issues can sometimes produce people who are intersex, including hermaphrodites. Developmental variants are very rare, constituting only about one in 5,600 people (0.018 percent; Sax 2002), and also don't represent "other sexes." (We know of only two cases of true human hermaphrodites that were fertile, but one individual was fertile only as a male [Parvin 1982] and the other only as a female [Schoenhaus et al. 2008].)

Only in protists, fungi, and algae do we find more than two distinct mating classes of individuals, with individuals able to mate with members of any class but their own. Since gametes of all classes are the same size ("isogamous"), biologists call them "mating types" rather than sexes.

Sex is thus a binary (Elliott 2020)—not just in humans but in all animals and plants. And it's a binary because natural selection has *favored* the evolution of a binary. In 1958, the famous evolutionist Ronald Fisher asked: "No practical biologist interested in sexual reproduction would be led to work out the detailed consequences experienced by organisms having three or more sexes; yet what else should he do if he wishes to understand why the sexes are, in fact, always two?" (Fisher 1958, ix).

Although two discrete gamete types are not needed to obtain the well-established advantage of sexual reproduction (McDonald et al. 2016), the evolution of the sexual binary happened multiple times (Dacks and Kasinsky 1999, Lehtonen 2021) as shown by biological observation and mathematical models (Deng 2007). Beginning with an ancestral species having gametes of equal size ("isogamy"), disruptive natural selection promotes the splitting of the population into two groups of individuals producing either small and mobile ones or large and immobile ones ("anisogamy"). Two sexes have thus evolved, and henceforth, the species will resist the invasion of individuals having other types of gametes—that is, other new sexes.

The stability of the two-sex condition is evident when we study sex determination during development, which varies widely across species (Bachtrog et al. 2014). Sexes can be determined by different chromosomes and their genes (e.g., XX vs. XY in humans, ZW vs. ZZ in birds—individuals with like chromosomes being female in mammals and male in birds); different development temperatures (crocodiles and turtles); whether you have a full or half set of chromosomes (bees); whether you encounter a female (some marine worms); and a host of other social, genetic, and environmental factors. Natural selection has independently produced diverse developmental pathways to generate the sexes, but there are just two destinations: males and females, an objectively recognized dichotomy—not an arbitrary spectrum of sexes.

But despite the facts, the dichotomy of sex—especially in humans—has recently come under ideologically based attacks. Even in apparently objective discussions of sex and gender, individuals are often said to have been *assigned* their sex at birth (e.g., "AFAB": assigned female at birth), as if this were an arbitrary decision by doctors—a "social construct"—rather than *an observation of biological reality*. Even the Society for the Study of Evolution was swayed by ideology to publicly declare that biological sex should be viewed as a continuum (Society for the Study of Evolution 2018). Teachers have been hounded out of their jobs (Bennett 2022) and deprived of their classes (Hooven 2023) simply for declaring that human sex is binary. As we'll see, this controversy comes from a deliberate conflation of a biological reality, the sexes, with a social construct, gender.

Denying the dichotomy of sex prevents us from understanding one of biology's most fascinating generalizations: the difference between males and females in behavior and appearance. The color, ornamentation, and weapons of males compared to their absence in females, a difference seen in species such as deer, birds, fish, and seals, result from *sexual selection*: the process, first suggested by Darwin (1871), in which males compete with each other for access to females. This involves either direct antagonism between males or males appealing to female preferences through their color, ornaments, and behavior. These differences ultimately come from females investing more in reproduction than males, starting with expensive eggs.

Ultimately, this puts the burden of parental care on females. Tied up in offspring production and rearing, females become the sex less available for mating, even when the ratio of males to females is one-to-one. Sexual selection also explains behavior: why, in most species—including our own—males are more promiscuous than females. For a male, fertilization involves merely expending a little sperm. For a female, eggs are few and expensive, pregnancy is long, and offspring are hard to tend to and feed. Antlers, plumes, peacocks' tails, elaborate male mating dances, and bird songs: these and a host of other traits make sense only as the evolutionary results of having different-size gametes.

Why do so many people resist the sex binary? Because it's in their ideological interest to conflate biological sex with *gender*—one's social identity or "sex role." Unlike biological sex, gender does form a sort of a continuum so that gender, but not sex, can be more of a continuum (Wright 2023). Still, even gender distributions are *bimodal*: most people identify with male and female genders.

And why do people distort the truth? We suspect that some of those whose gender doesn't correspond to one of the two biological sexes, and their allies, want to redefine sex so that, like gender, it forms more of a continuum. While jettisoning the sex binary is meant well, it also distorts a scientific fact—and all the evolutionary consequences that flow from that fact.

(2) Virtually All Behavioral and Psychological Differences Between Human Males and Females Are Due to Socialization

Evolution and genetics are often claimed to play *no role* in these differences. This well-known "blank slate" ideology (Pinker 2002) asserts that all humans, including males and females, are born with the same behavioral propensities, and whatever behavioral or psychological differences we see derive purely from socialization, including economic or environmental influences.

To a biologist, this kind of blank-slateism—which may stem partly from the Marxist faith in the infinite malleability of humans—is wrong. Multiple studies clearly show that there are lots of *average* differences between male and female behaviors (Archer 2019, Stewart-Williams and Halsey 2021), including sexual interests, parental care, aggression, promiscuity, risk-taking, interest in people versus things, empathy, fearfulness, spatial abilities, violence, and traits connected to social relations. It's important to realize that we're talking about *averages* here: there's a lot of overlap between the distributions of male and female behaviors, so individuals can show characteristics seen more often in the other sex. Some women, for example, are more aggressive than the average man. And we must add that socialization *is* a likely contributor to many behavioral differences between men and women.

But we cannot assert that these average differences result *solely* from socialization. It's likely that the average differences in the behaviors listed above have an evolved and genetic basis. Over millions of years, natural selection certainly caused some behaviors of males and females to diverge. We know from multiple sources, including evaluating the general likelihood of an adaptive explanation; looking for behavioral parallels in other species; determining whether a sex difference in behavior is ubiquitous among different human cultures, including hunter-gatherers; testing whether the behavior is influenced by reproductive hormones such as testosterone; and seeing if the behavior appears at the expected time of development. Risk-taking and male-male aggression, for example, are strongest during the peak reproductive years of young adulthood—just as we expect if these are behaviors that evolved to help men secure mates.

But to many, even suggesting a biological basis for sex differences in behavior is taboo, perceived as a kind of misogyny. A recent example is Chelsea Conaboy's declaration in the *New York Times* that "maternal instinct is a myth that men created" (Conaboy 2022). She argues that well-known differences between men and women in attentiveness and behavior toward their children are due to socialization. The obvious retort from biology is that while some human societies do push maternal care onto women, the greater attentiveness of mothers than fathers to their children—attentiveness triggered by cues such as hormones, lactation, infant crying, and the sight of babies—is seen not only in every human society but, more important, also in thousands of other animal species, including our primate relatives (Blaffer Hrdy 2000). These other species lack the social pressures that, to blank slaters, explain sex differences. It would be an odd coincidence indeed if misogyny and the patriarchy just happen to create a situation in humans identical to that seen in our evolutionary cousins and other distant relatives.

The false idea that human males and females are born biologically identical in behavior and psychology is a form of "biological egalitarianism." This is the view that all groups *must* be the same in important aspects of their biology because if they weren't, one might be tempted to slide from nonidentity into "inequality" and from there into bigotry, misogyny, and other discriminatory behaviors. But as we'll see, there's no logical connection between what we see in nature and how we should regard the dignity, rights, and liberties of different individuals or groups. The first is a matter of reality, the second a matter of ethics—how we socially construct morality.

(3) Evolutionary Psychology, the Study of the Evolutionary Roots of Human Behavior, Is a Bogus Field Based on False Assumptions

The biologist P. Z. Myers joined several other critics of this field (once called sociobiology) when he asserted that "the fundamental premises of evo psych [evolutionary psychology] are false" (Myers 2013). Even social psychologists, who almost universally accept evolution itself, are surprisingly unenthusiastic about the idea that evolution explains important aspects of human psychology, social attitudes, and preferences (Buss and von Hippel 2018).

But this view is misguided (al-Shawaf 2019), for the fundamental premise of evolutionary psychology is simply: *our brains and how they work—which yield our behaviors, preferences, and thoughts—sometimes reflect natural selection that acted on our ancestors.* Nobody denies this for our bodies—palimpsests of once-adaptive traits that are no longer useful ("dead" and nonfunctional genes, wisdom teeth, tailbones, and transitory coats of hair in embryos)—but opponents of evolutionary psychology deny it for our behaviors. There is no scientific reason for such duality. Why should our bodies reflect millions of years of evolution while our behaviors, thoughts, and psychology, molded by the very same forces, are somehow immune to our past? This could only be true if human behaviors lacked genetic variation. Yet research has shown that our behaviors are among the most genetically *variable* human traits (Turkheimer 2000)!

Thus, the "sociobiology wars" of the '70s, launched by E. O. Wilson's eponymous book, continue under a new name, but the subject remains human exceptionalism—somehow, humans are nearly free of the evolutionary forces that molded behavior in other species. It's true that the early days of evolutionary psychology included some "soft" research that proposed dubious and untestable adaptive hypotheses, but now the field has reached an explanatory maturity that has to be taken seriously (al-Shawaf 2020).

Indeed, evolutionary psychology explains, to our best knowledge, several human behaviors. These include why we favor kin over non-kin—and closer kin over more distant kin—why we mistreat stepchildren more frequently than biological children, why males are more aggressive than females, the difference in promiscuity and sexual proclivity between men and women, why men show more sexual jealousy than women, why certain facial expressions convey emotions, why we have fears of snakes and spiders and disgust with bodily fluids, and why we hunger for sugars and fats. Indeed, some of our behaviors, like the propensity to eat things that are no longer healthy, constitute features useful in our ancestors but now useless or even harmful.

By walling off a huge area of research and teaching that involves human nature, the ideological vilification of evolutionary psychology prevents us from understanding our species. As two evolutionary psychologists noted, "Not a single degree-granting institution in the United States, to our knowledge, requires even a single course in evolutionary biology as part of a degree in psychology—an astonishing educational gap that disconnects psychology from the rest of the life sciences" (Buss and von Hippel 2018, p. 156). Without such knowledge, we're left with "social constructs" as the sole source of our behaviors, an explanation that utterly fails to explain the observed data. It goes without saying that when dealing with human behaviors, it's best to have the fullest possible explanations, both social and biological.

The dismissal of evolutionary psychology is motivated by a blank-slate ideology that sees humans as almost infinitely malleable. We've already mentioned that Marxism has almost certainly influenced this attitude. More reasons are outlined in Steven Pinker's book *The Blank Slate: The Modern Denial of Human Nature*, including a disdain for biological determinism; a belief that things that can be learned, such as language, cannot *at the same time* involve capacities that have evolved; the false view that biology is destiny—that what is inherited cannot be changed—and a flat denial that biology plays a large role in human behavior, including similarities *and* differences between individuals or groups. As we'll see, studying genetic differences between individuals or groups is especially taboo, for that work is said to promote bigotry and eugenics.

(4) We Should Avoid Studying Genetic Differences in Behavior Between Individuals

The default assumption of many people is that the genetic differences between people in traits such as educational achievement, IQ, and similar traits shouldn't be studied. In some cases, the very existence of genetic differences is denied despite strong evidence from various lines of research, such as twin studies. Such work is thought to produce a ranking of people, a promotion of bigotry, and an unfair sorting of individuals onto different educational tracks. And yet, even within each ethnic group, variation in virtually every trait, physical or behavioral, has an appreciable genetic component. This goes for traits such as height, blood pressure, the tendency to smoke or drink, neuroticism, and cognitive abilities and educational attainment. For the last two traits, more than half the variation among individuals is based on variation in their genes (Plomin et al. 1994, De Zeeuw et al. 2015). However, these measures reflect variation *within* a population and say nothing about the basis of differences *between* populations or ethnic groups.

This kind of study has become more useful since science developed techniques to sequence the DNA of an individual's entire genome. With that information, and sequencing many individuals, you can correlate each variable DNA position (i.e., single nucleotides) with various traits, determining which bits of the DNA are correlated with variation in a trait. This kind of study (genome-wide association studies, or GWAS [Uffelmann et al. 2021]) has, for example, turned up nearly four thousand areas of the genome associated with educational attainment (Okbay et al. 2022). Fascinatingly, many of these genes are active mainly in the brain (Lee et al. 2018). Using GWAS studies, it's now possible to make predictions about a person's appearance, behavior, academic achievement, and health simply by analyzing the DNA of an individual and calculating their individual "polygenic scores" based on large samples of their population. Adding to ethical challenges, this can even be done on fetal DNA, and companies are now advertising embryo selection based on these scores (Hemptinne and Posthuma 2023).

GWAS analysis offers many possibilities for helpful intervention, especially by monitoring individuals for health conditions they're genetically liable to develop. The usefulness of GWAS scores for educational achievement, however, is far more controversial (Coyne 2021). Although genetic differences play a role in many aspects of what we consider "intelligence," it's easier to equalize people's prospects via social and educational reforms than by using polygenic scores.

Yet understanding genetic variation underlying educational outcomes might one day be useful. For instance, if we discover genetic variants that respond well to educational or social interventions, it might be possible to target these individuals early on. These genetic studies could help identify environmental effects, too: if two people with identical polygenic scores wind up with different outcomes, how did their environments differ? This is why doing such research is worthwhile.

Like our other examples, the resistance to these studies also rests on a blank-slate view of human nature that rejects genetic influences on behavior. Genetic studies of anything beyond physical traits and disease are, it's claimed, linked with eugenics and similar acts of bigotry in the past.

In fact, the fear and avoidance of behavior-genetic research is so strong that even the National Institutes of Health defines races solely as social constructs (National Human Genome Research Institute 2023) and has limited researchers' access to public, taxpayer-funded databases containing information about the genetic constitution, health, education, occupation, and income of anonymous individuals. This restriction apparently applies even to studies that don't involve differences between races, and so it appears to be the US government's attempt to stifle research on behavioral genetics in general (Ritchie 2022)—especially behaviors related to academic and social success.

(5) "Race and Ethnicity Are Social Constructs, Without Scientific or Biological Meaning."

This is the elephant in the room: the claim that there is no empirical value in studying differences between races, ethnic groups, or populations. Such work is the biggest taboo in biology, claimed to be inherently racist and harmful. But the assertion heading this paragraph, a direct quote from the editors of the *Journal of the American Medical Association* (Flanagin et al. 2021, p. 621), is wrong.

Before we handle this hot potato, we emphasize that we prefer the words *ethnicity* or even *geographic populations* to *race*, because the last term, due to its historical association with racism, has become too polarizing. Further, old racial designations such as *white*, *black*, and *Asian* come with the erroneous view that races are easily distinguished by a few traits, are geographically delimited, and have substantial genetic differences. In fact, the human species today comprises geographically continuous groups that have only small to modest differences in the frequencies of genetic variants, and there are groups within groups: potentially an unlimited number of "races." Still, human populations do show genetic differences from place to place, and those small differences, summed over thousands of genes, add up to substantial and often diagnostic differences between populations.

Even the old and outmoded view of race is not devoid of biological meaning. A group of researchers compared a broad sample of genes in over 3,600 individuals who self-identified as either African American, white, East Asian, or Hispanic. DNA analysis showed that these groups fell into genetic clusters, and there was a *99.84 percent match* between which cluster someone fell into and their self-designated racial classification (Tang et al. 2005). This surely shows that even the old concept of race is not "without biological meaning." But that's not surprising because, given restricted movement in the past, human populations evolved largely in geographic isolation from one another—apart from "Hispanic," a recently admixed population never considered a race. As any evolutionary biologist knows, geographically isolated populations become genetically differentiated over time, and this is why we can use genes to make good guesses about where populations come from.

More recent whole-genome work confirms a high concordance between self-identified race and genetic groupings. One study of twenty-three ethnic groups found that they fell into seven broad "race/ethnicity" clusters, each associated with a different area of the world (Banda et al. 2015). On a finer scale, genetic analysis of Europeans showed that a map of their genetic constitutions coincides almost perfectly with the map of Europe itself (Gilbert et al. 2022). In fact, the DNA of most Europeans can narrow down their birthplace to within roughly five hundred miles (Novembre et al. 2008).

Of what use are such ethnicity clusters? Let's begin with something many people are familiar with: the ability to deduce one's personal ancestry. If there were no differences between populations, this task would be impossible, and "ancestry companies" wouldn't exist. But even physical traits can sometimes predict ancestry: AI programs can, for instance, predict self-reported race quite accurately from just chest X-ray scans (Wawira Gichoya et al. 2022).

On a broader scale, genetic analysis of worldwide populations has allowed us to not only trace the history of human expansions out of Africa, but to assign dates to when *H. sapiens* colonized different areas of the world. This has been made easier with recent techniques for sequencing ancient DNA (Reich 2019). On top of that, we have fossil DNA from groups such as Denisovans and Neanderthals, which, in conjunction with modern data, tells us these now-extinct groups bred with the ancestors of "modern" *Homo sapiens* (most of us have some Neanderthal DNA in our genomes).

Further, there's medical value in genetic studies of populations. A fair number of genetic diseases are somewhat associated with ethnicity (González Burchard et al. 2003): maladies such as Tay-Sachs disease, sickle cell anemia, cystic fibrosis, and hereditary hemochromatosis. These associations make both diagnosis and prenatal counseling more efficient, for one can use ethnicity to focus on possible medical issues. The incidence of many ailments has both genetic and environmental causes, requiring considering diet and lifestyle. Yet genetic analysis could help with even these complex ailments. GWAS analysis based on ethnic-specific studies, for instance, might give estimates of the risk of various illnesses by testing infants or even fetuses. If you know you're at risk, then monitoring your lifestyle can reduce the chance of getting seriously ill when you're older.

Fortunately, GWAS data for different ethnic groups are beginning to be collected, and medical researchers already recognize that studies of different ethnicities are essential to understanding disease and reducing health disparities. This is because genetic results from one group may not generalize to results from other groups. A recent GWAS analysis of dementia (Sherva et al. 2023), for instance, discovered that some regions of the genome increase the risk in African Americans but not white Americans; another study showed novel SNPs associated with cardiovascular disease in sub-Saharan Africa populations (Singh et al. 2023); and a third showed a novel gene for Parkinson's disease in African populations (Rizig et al. 2023). This implies that some genes able to predict health conditions will differ between these groups and that possible diagnoses, interventions, or cures might differ as well.

Finally, there are forensic reasons for associating genetics with ethnicity. These involve predicting what a perpetrator or victim might look like (e.g., facial features and eye and skin color) from a sample of tissue or semen or, when using ancient DNA, predicting how people might have looked. We know now, for instance, that some Neanderthals had pale skin and red hair (Ledford 2007) and that dark skin and blue eyes might have been common in European *Homo sapiens* a few thousand years ago (Katz 2018).

But the central question about genetics in the culture wars involves *behavioral* characteristics of different populations, with differences in intelligence being the most taboo subject. In light of the checkered history of this work, it behooves any researcher to tread lightly, for virtually any outcome save the worldwide identity of populations could be used to buttress bias and bigotry. Indeed, even writing about this subject has led to sanctions on many scientists, who have "found themselves denounced, defamed, protested, petitioned, punched, kicked, stalked, spat on, censored, fired from their jobs and stripped of their honorary titles" (Carl and Menie 2019). A well-known example is Bo Winegard, an untenured professor in Ohio who was apparently fired (Flaherty 2020) for merely suggesting the *possibility* that there were differences in cognition among ethnic groups (Winegard et al. 2020). This is why most biologists stay far away from this topic.

The taboo is not whether there are *observable differences* in IQ and life outcomes between groups, for these are well known (Roth et al. 2001) and easily measured using standardized tests (National Center for Education Statistics n.d.). Rather, the issue is what causes these disparities: genetic differences, societal issues such as poverty, past and present racism, cultural differences, poor access to educational opportunities, the interaction between genes and social environments, or a combination of the above. A few methods have been applied to this question, including adoption studies (Weinberg et al. 1992), analysis of ethnically mixed populations (Lasker et al. 2019), and GWAS. The genomic analyses have concentrated on educational attainment—highly correlated with IQ and some measures of success in life—but have focused almost exclusively on white descendants of Europeans. And the predictive power of these ethnically white GWAS scores nearly vanishes when you apply them to other ethnic groups (Okbay et al. 2022). The reason for this decay in predictability involves the genetic differences between groups, including differences between the subset of genes that affect educational attainment, the existence of different variants of the same genes involved in both groups, or differences between groups in how genes and their variants interact with each other and the environment. The upshot is that it is not easy to translate findings from one ethnic group to another; each group needs to be studied separately.

Two other issues make it hard to analyze behavioral and cognitive differences between groups. First, these traits are usually affected by variation at hundreds of genes spread throughout the genome. Second, those genes are *physically connected to other genes* on chromosomes. Taken together, this means that many genes for external appearance (color, facial structure, hair texture)—the very genes that give phenotypic information about ethnicity—are physically linked to other genes, including those for educational attainment. Because genes lying close to each other on the chromosomes are inherited together, we have no way to completely separate genes affecting appearance from those affecting educational attainment. If differences in achievement between groups come at least partly from society treating people differently when they *look* different (e.g., via bigotry and racism), then the societal effect caused by "appearance genes" is conflated with the direct effect of "academic achievement genes."

But despite the difficulty of disentangling the effects of genes and environments, there are still societal benefits to understanding genetic effects *within* different groups. For example, GWAS studies—conducted separately for each ethnicity—could illuminate whether genetic variants associated with educational outcomes differ among different groups or respond differentially to environmental interventions. Imagine, for instance, a gene whose variants were associated with thyroid function. Further imagine that gene variants that *reduced* thyroid function, causing iodine deficiency symptoms, were associated with lower educational attainment than variants with higher thyroid function and that the low-iodine variants were more common in whites than in Asians. (This is not completely fanciful: iodine deficiency can reduce IQ by a full fifteen points [Qian et al. 2005], and genes might affect how well one absorbs iodine [GB HealthWatch n.d.].) A simple intervention might involve iodine supplementation in whites having "low expression" DNA variants but not in those with "high expression" variants (too much iodine is toxic). This example is not far-fetched because we know that different groups have many unique gene forms (i.e., "private alleles" [Rosenberg 2011]), forms that might have important effects on behavior in addition to their own unique interactions with the environment.

It should be clear from this example that the reason for studying genetic differences between ethnic groups is to boost the success of *individuals* whose DNA is known, not to rank different groups for one trait or another. But to do this boosting, we must first understand the nature of genetic differences among groups. Many objections to this kind of work vanish when you realize that while the focus is on population-specific DNA segments associated with achievement, the ultimate goal is to help each person do their best.

In our view, then, research on cognition or educational attainment *within and between groups* should not be demonized, banned, or automatically denied publication, and the data should be publicly available. It goes without saying that scientists should be cautious and vigilant against its misuse or misrepresentation. But it's hard to argue with the idea that the more we understand—including genetics—the more success we'll have with social policies. Indeed, there are good arguments suggesting that stifling research on IQ, or equating this research with racism, will cause more harm than good (Carl 2018). After all, political equality should be a moral imperative, not a conclusion based on empirical data. Ultimately, the value of a human being does not and should not depend on their IQ or years of schooling.

The great evolutionary biologist Ernst Mayr stated it well:

> Equality in spite of evident non-identity is a somewhat sophisticated concept and requires a moral stature of which many individuals seem to be incapable. They rather deny human variability and equate equality with identity. Or they claim that the human species is exceptional in the organic

world in that only morphological characters are controlled by genes and all other traits of the mind or character are due to "conditioning" or other non-genetic factors.... An ideology based on such obviously wrong premises can only lead to disaster. Its championship of human equality is based on a claim of identity. As soon as it is proved that the latter does not exist, the support of equality is likewise lost. (Mayr 1963, pp. 648–649)

(6) Indigenous "Ways of Knowing" Are Equivalent to Modern Science and Should Be Respected and Taught as Such

Because indigenous peoples such as New Zealand's Māori and the New World's Native Americans were the victims of colonialism, their traditional knowledge is often lauded as an alternative version of modern science—a "way of knowing" developed independently from what's called "colonialist science" but seen by many as having equal value. In fact, the New Zealand government requires indigenous ways of knowing to be given equal status to modern science in the classroom—and to other subjects in all secondary school education (Jones 2022). South Africa is also experiencing a decolonization of biology. An article in the prestigious journal *Nature* calls for decolonizing pharmacology in that country, concentrating on local herbal remedies to "anchor the curriculum in local experience" (Nordling 2018). While this adds a homegrown flavor to learning, dropping an anchor in local experience can divert the student from an education in modern pharmacology.

Matauranga Māori, the indigenous way of knowing in New Zealand, is a mélange of empirical knowledge derived from trial and error (including the navigational ability of their Polynesian ancestors and Māori ways of procuring and growing food) but also includes nonscientific areas such as theology, traditional lore, ideology, morality, and legend. Yet all these are considered worthy of teaching as coequal to the methods and results of modern science. Māori scholars, for example, have advanced the improbable claim that Polynesians—the ancestors of the Māori—were the first to discover Antarctica in the seventh century. This claim is surely false (Matthews 2022), probably based on a faulty translation of an oral legend. In fact, Antarctica was first seen by the Russians in 1820. Nevertheless, New Zealand's Royal Society, the nation's most prestigious scientific organization, gave a $660,000 grant to the Māori to explore this bogus narrative (Royal Society Te Apārangi 2021). There's also been a revival of the traditional herbal and spiritual remedies of Matauranga Māori (Paewai 2022), which incorporate chanting as a means of healing (Robin 2022). While local remedies may be helpful, they are almost never tested using the gold standard for medicine: randomized controlled trials.

Indigenous ways of knowing usually include practical knowledge, including observations about the local environment and practices developed over time, including, in the case of Matauranga Māori, ancient methods of navigating and the best way to catch eels. But practical knowledge is not the systematic, objective investigation of nature—free from assumptions about gods and spirits—that constitutes modern science. Conflating indigenous knowledge with modern science will confuse students about what constitutes knowledge and the nature of science itself. Modern science did arise in Western Europe in the seventeenth century, a time when women were denied education. This situation severely restricted people's opportunities but provides no reason to discredit science itself—the best way of generating knowledge about the universe—as "Western" or colonialist. ("Western" has become a misnomer, insulting many people in other countries who now practice science.)

A related issue pitting indigenous culture against modern science is forensic anthropology, the study of ancient societies using human remains and artifacts. In North America, human remains, depending on where they're found, can be claimed by Native Americans as their own, withheld from scientific study because they're seen as members of modern indigenous groups. Indeed, federal law mandates the return of bones and other artifacts to indigenous groups (National Park Service 2023, Weiss 2023). The remains must be reburied, even if there's no clear genealogical connection to modern Native American populations. In the case of Kennewick Man, the indigenous "scientific" claims included a Native American leader rejecting the truth that his ancestors arrived via the Bering Strait from Asia on these grounds: "From our oral histories, we know that our people have been part of this land from the beginning of time," says Mr. Minthorn. "We do not believe that our people migrated here from another continent" (Knickerbocker 2001).

The Biden administration has widened these regulations to refer not just to human remains, but to *any* object that a Native American tribe can claim as "sacred," including relics like potshards and arrowheads. Neither the government nor scientists can challenge Native American claims, which can result (after paying tribal consultants) in either giving the objects back to Native Americans or displaying them according to tribal regulations (Prabhakar and Mallory 2024). Objects can even be covered in museums if their sacred "powers" are deemed too strong (Weiss 2023)! These designations can be based not on history but simply on oral tradition, mythology, or religious sentiment. In this way, and in seeming violation of the First Amendment, Native American spirituality impedes the study of American history.

One victim of this undue respect for "other ways of knowing" is physical anthropologist Elizabeth Weiss. For simply studying 500–3,000-year-old bones from California, Weiss was demoted by her university and banned from accessing her department's bone collection (Flaherty 2022). She's not even allowed to study X-rays of the remains or show *a photograph of the boxes in which they are kept*. Many other universities, such as Berkeley, are sending back or reburying artifacts and old bones (Elassar 2022). The result: valuable human history and anthropology remain off limits because remains and artifacts are considered sacred. Clearly, the best solution would defer burial until *after* scientific study or DNA collection, but policy simply prevents us from learning about our past.

The promotion of "other ways of knowing" comes from a desire to valorize oppressed groups by holding up their culture as having the same epistemic authority as science, a view that philosopher Molly McGrath called "the authority of the sacred victim" (McGrath 2021). In its secular form, this authority derives from postmodern views that science is just one of many "ways of knowing" and that the hegemony of science reflects power rather than accomplishment. This is encapsulated by the motto, espoused by some on both the Right and the Left for decades, that "science is always political" (National Academies of Sciences 2017).

Like biblical creationism, much indigenous knowledge has a substantial spiritual or theological component that comes not from evidence but from authority or revelation. To add any of this knowledge to modern science, you must first separate the empirical wheat from the spiritual chaff. This is what the non-denominational Pastor Mike Aus meant when, after giving up his faith, he described "religious knowledge" this way: "There are not different ways of knowing. There is knowing and not knowing, and those are the only two options in this world" (Aus 2015).

Nearly all the ideologically driven distortions of biology come from one mindset: radical egalitarianism. Thus, the sexes, ethnic groups, and even individuals are *genetically nearly identical* in behavior and psychology (though not in appearance), and behavioral differences are due to socialization and environment. Socialization has become the default explanation for why there are more men than women in math and physics, why males are more aggressive and females more empathic, why there are differences in achievement between social classes and ethnicities, and why some groups are differentially represented in science and academia. While social influences can affect these outcomes, the pervasive evidence for genetic influence on humans makes it unwise to reject *a priori* the influence of hereditary factors. Nevertheless, because the biological data contradict the fashionable blank-slate ideology, its advocates are forced to render their program immune to data, which they do by *twisting the facts of biology to conform to their beliefs*.

Biological egalitarianism damages science in two ways. One is through *deterrence*: the chilling of research that prevents scientists from studying or teaching certain problems (Maroja 2022). This isn't accomplished by direct prohibition of research but by instilling fears into teachers or researchers that discourage them from working on and even discussing such topics. A few public examples are all it takes to deter many others, such as the pillorying of those who teach that there are only two sexes in humans (e.g., Hooven 2023 and Bennett 2022). Further, those who study group differences and their genetics can be dismissed by labeling them as sexists, misogynists, racists, or eugenicists. This has been strikingly effective, for what liberal—and most biologists are liberals—wants to be tarred with those labels? Likewise, those who refuse to accept the equivalence of modern science and indigenous ways of knowing are deemed racist *and* colonialist (McAllister 2022). No wonder that teachers, researchers, and professors censor themselves on these issues.

The other damage involves *direct action*: imposing requirements or punishments on scientists whose research strays from biological egalitarianism. Punishments have ranged from taking classes away from professors (Jaschik 2022), making their lives miserable so they leave academia (Wright 2018), demanding fealty to falsehoods (Morey 2022), direct firing (Winegard 2020), demanding the infusion of mythology into science (Coyne 2022b), rejecting scientific papers because their findings don't respect the "dignity and rights of all humans" ("Science Must Respect the Dignity and Rights of All Humans" 2022), withholding publicly funded data from researchers (Lee 2022), and diverting research funds to ideologically derived projects (the National Institutes of Health once adopted this plan but soon abandoned it; Kaiser 2021).

Beyond this, there are the many attacks on scientific merit as an outmoded way of judging science or hiring scientists (Abbot et al. 2023). We increasingly see calls, coming mostly from the Left, to replace evaluations of merit with more "holistic" schemes that take group identity into account. This has led many universities to require prospective faculty to submit diversity statements as part of their job applications (Foundation for Individual Rights and Expression 2022), as well as eliminating the obligation for prospective students to submit scores on standardized tests such as the MCATs, SATs, and GREs, and even firing professors whose science classes are too difficult (Murphy Marcos 2022).

Science has always been subject to ideological influence and control, beginning with the Catholic Church's censoring of Galileo, whose heliocentric solar system contradicted accepted theology. And those influences have come from both the Right and the Left, including debates about evolution, vaccine efficiency, global warming, fluoridated water, and so on. But what's happening now is different. First, recent attacks on science are more general than before, involving not just single issues but spreading into every field.

The biology wars, for example, now involve much more than evolution and have spread to biological sex, differences between groups, the scientific language we're allowed to use (Cheng et al. 2023), the treatment of biological artifacts, and, indeed, whether there are valid ways of learning about the natural world apart from modern science. And, of course, famous biologists of the past, such as Gregor Mendel (McLemore 2021), Charles Darwin (Fuentes 2021), and Aldous Huxley (Somerville 2021), are denigrated as racist or sexist.

Further, the attacks on science come not just from the public, religious believers, or political authorities, as in the past, but involve scientists themselves—scientists who deem certain research taboo, restrict the availability of publicly funded data, and demand that research papers should be censored if they might be offensive. In the case of the Lysenko affair, Soviet fiat dictated the distortion of genetics and agricultural science (Lysenko 1948), but today, our own colleagues force nature into the Procrustean bed of ideology. Although scientific nonconformity may not be the life-or-death issue it was in Stalin's Russia (Pringle 2008), jobs and research are at risk.

Why is this happening? We suspect political climate changes, including the rise of identity politics, have caused scientists on the Left to use their own fields to signal ideological virtue and membership in a political "tribe." Further, science departments have also been infected by the postmodernism pervasive in humanities departments (Anderson 2021). When combined with the self-censorship of researchers and teachers who fear professional damage, this poses a grave threat.

How can we restore science to its primary mission: understanding nature and the universe? Since ideological pressure comes largely from scientists themselves, including those who dispense grant money and judge research papers, we can't count on scientific arguments to solve the problem. Indeed, radical egalitarianism is itself a form of faith, resistant to facts and rational argument (McWhorter 2021). It is also a pledge of group allegiance. Steven Pinker explained how resistance to evolution did not involve rejecting scientific evidence but rather served as a badge of adherence to a religious ideology that happened to reject evolution on principle. His explanation also holds for the quasi-religious progressive ideology that is damaging biology:

Professing a belief in evolution is not a gift of scientific literacy, but an affirmation of loyalty to a liberal secular subculture as opposed to a conservative religious one. In 2010, the National Science Foundation [NSF] dropped the following item from their test of scientific literacy: "Human beings, as we know them today, developed from earlier species of animals." The reason for that change was not, as scientists howled, because the NSF had given in to creationist pressure to bowdlerize evolution from the scientific canon. It was that the correlation between performance on that item and on every other item on the test (such as "An electron is smaller than an atom" and "Antibiotics kill viruses") was so low that it was taking up space in the test that could go to more diagnostic items. The item, in other words, was effectively a test of religiosity rather than scientific literacy. When the item was prefaced with "According to the theory of evolution," so that scientific understanding was divorced from cultural allegiance, religious and nonreligious test-takers responded the same (Pinker 2018, p. 356).

So, if facts won't turn the tide, what can we do?

An obvious palliative is a form of liberal egalitarianism and morality independent of biological differences. As Pinker noted in *The Blank Slate*: "Equality is not the empirical claim that all groups of humans are interchangeable; it is the moral principle that individuals should not be judged or constrained by the average properties of their group" (Pinker 2002, 340).

We can also keep stressing that the job of scientists is to find truth, not decide how that truth should be used by society. This is not to claim that all research is equally valuable or interesting, nor to argue that science hasn't been misused in harmful ways. But, given the observation that a lot of pure research has led to discoveries that could never have been predicted, we should avoid placing entire areas of work off limits. If some people distort or misuse scientific research for ideological ends, scientists themselves should take the lead in correcting the record.

But perhaps the ultimate solution involves philosophy—emphasizing that *there is no value in looking to nature to determine which of our behaviors are good, moral, or normal*. Doing so always involves two well-known fallacies. The first is the *naturalistic fallacy*—the famous dictum that *is* does not equal *ought*, also phrased as "what is natural is what we should do." The second is the related *appeal to nature* fallacy, which argues that what is natural must be what is *good*.

Both fallacies lead to the same errors. First, if we condition our politics and ethics on what we know about nature, then our politics and ethics become malleable to changes in what we discover about nature later. For example, the observation that female bonobos rub each other's genitals as a bonding behavior has been used to justify why human homosexuality is neither offensive nor immoral. Bonobo behavior is, after all, "natural." Similar same-sex behaviors have been reported in many species and have been used to the same end (Roughgarden 2013). But what if *no* such behavior had been seen in any nonhuman species? Or what if the bonobo observation was shown to be wrong? Would this make homosexual behavior immoral or even criminal? Of course not, because enlightened views of homosexuality rest not on parallels with nature but on ethics, which tells us that there's nothing immoral about consensual sex between adults.

Second, many behaviors that are "natural" because they're found in other species would be considered repugnant in our own species. These include infanticide, robbery, and extra-pair copulation. As one of us wrote, "If the gay cause is somehow boosted by parallels from nature, then so are the causes of child-killers, thieves and adulterers" (Coyne 2022a). But we don't really derive our morality or ideology from nature. Instead, we pick and choose those behaviors in other species that happen to jibe with a morality we already have. (People do exactly the same thing when they pretend to derive morality from religious texts such as the Bible.)

All the biological misconceptions we've discussed involve forcing preconceived beliefs onto nature. This inverts an old fallacy into a new one, which we call the *reverse appeal to nature*. Instead of assuming that what is natural must be good, this fallacy holds that "what is good must be natural." It demands that you see the natural world through lenses prescribed by your ideology. If you are a gender activist, you must see more than two biological sexes. If you're a strict egalitarian, all groups must be behaviorally identical and their ways of knowing equally valid. And if you're an antihereditarian—a blank slater who sees genetic differences as promoting eugenics and racism—then you must find that genes can have only trivial and inconsequential effects on the behavior of groups and individuals. This kind of bias violates the most important rule of science, famously expressed by Richard Feynman: "The first principle is that you must not fool yourself—and you are the easiest person to fool."

The greatest danger is not to the layperson's understanding of science but to science itself. The guiding principle of science—and of academic freedom, on which science depends—is freedom of inquiry. Those who place whole fields of investigation off limits or distort scientific truth for political reasons violate this freedom and deprive us of the intellectual and practical benefits that could come from pure, untrammeled research.

We aren't under the illusion that calling attention to these points and emphasizing the fallacy of the reverse appeal to nature will push ideology out of science. Progressive ideology is growing stronger and intruding further into all areas of science. And because it's "progressive," and because most scientists are liberals, few of us dare oppose these restrictions on our freedom. Unless there is a change in the Zeitgeist, and unless scientists finally find the courage to speak up against the toxic effects of ideology on their field, in a few decades, science will be very different from what it is now. Indeed, it's doubtful that we'd recognize it as science at all.

Social Justice, MD — Medicine under Threat

SALLY SATEL

On January 8, 2021, I gave a department-wide lecture to the Yale Department of Psychiatry. I had once been a resident there and then an assistant professor for five years. I left New Haven in 1993 to pursue a health policy fellowship in Washington, DC, and eventually joined a think tank, but I remained a lecturer in the department. My lecture was about the year I spent, from 2018 to 2019, assisting with treatment efforts in an embattled small town in southeastern Ohio that was reeling from the opioid crisis.

I discussed the "deaths of despair" phenomenon, first described by Princeton economists Angus Deaton and Anne Case, and showed photos of haunted industrial landscapes and the lonely downtown area. I presented national data on the characteristics of individuals who abused prescription pills and on the frequency with which addiction develops. (It's far less common than you've been led to believe.) I spoke about the culture of prescribing in rural mining towns and the myriad factors that caused the crisis. I drew attention to the dimensions of personal agency in addiction. In closing, I highlighted the heroic efforts of Irontonians to boost the economy and the morale of their beloved town.

Exactly four weeks later, I received an email from the chairman of the department, a fine man and brilliant researcher whom I have known since we were interns together in the 1980s. He admitted that he had not anticipated "the extent of the hurt and offense that folks would take" to my presence. He appended an anonymous complaint that he had received from an unspecified number of "Concerned Yale Psychiatry Residents."

The residents told the chairman that my talk, coming only two days after the January 6th attack on the Capitol, "was further traumatizing to us." They wrote that "the language Dr. Satel used in her presentation was dehumanizing, demeaning, and classist toward individuals living in rural Ohio and for rural populations in general.... We find her canon to be beyond a 'difference of opinion' worth debate." My earlier writing on health disparities, which some troublemaker must have alerted them to (after all, the topic of race was nowhere in my talk as almost everyone in Ironton is white) was deemed a "racist canon." They expressed "shock and disappointment" at the chairman's failure to "take a public stand against" me and questioned *his* commitment to the department's anti-racist agenda. "Will you continue to invite Grand Rounds Speakers with racist and classist mindsets, like Dr. Satel?" the residents asked. Although they requested that the chairman "revoke" my lectureship at Yale, he did not do so.

I waited a little over three years before asking to resume my interaction with residents and fellows. The plan was for me to lead a journal club. We would discuss an article chosen for me by the faculty member who was willing to sponsor the visit, about the opioid crisis that appeared in the *Lancet*. "Well, maybe next year," I was told by department leadership. By then, presumably, some of the residents who were discomfited by my talk on January 8, 2021, would have moved on to jobs outside the department. Apparently, my mere presence in the department would re-re-traumatize them.

Think of my near-cancellation as a biopsy of medicine's body politic—and the revealed malignancy therein. Evidence as follows:

One week after the murder of George Floyd in 2020, the Association of American Medical Colleges announced that the nation's 155 medical schools "must employ anti-racist and unconscious bias training and engage in interracial dialogues." Public health schools and agencies launched campaigns to declare racism a public health crisis.

In 2020, the AAMC informed medical schools that they "must employ anti-racist and unconscious bias training and engage in interracial dialogues," as part of its "Statement on Police Brutality and Racism in America and Their Impact on Health." In the spring of 2021, the American Medical Association's Strategic Plan for Advancing Health Equity and Justice in

Medicine advocated "mandatory anti-racism [training]" as part of its vision that all physicians "confront inequities and dismantle white supremacy, racism, and other forms of exclusion and structured oppression."

After the Floyd murder, epidemiologist Jennifer Nuzzo at the Johns Hopkins Bloomberg School of Public Health informed would-be marchers "the public health risks of not protesting to demand an end to systemic racism greatly exceed the harms of the virus." Days later, 1,200 health professionals cheered her on in an open letter.

The problem, of course, is that the job of epidemiologists is to inform the public about risks. Yet, there is no way they can quantify the risk of not marching. But even if they could, it is not their job to tell others what risks are worth taking or what their moral prerogatives should be. These experts allowed their own moral commitments, not objective metrics of risk, to shape their advice.

In 2021, the AMA advocated "mandatory anti-racism training" as part of its directive that all physicians "confront inequities and dismantle white supremacy, racism, and other forms of exclusion and structured oppression." That same year, the AMA produced a fifty-four-page document called *Advancing Health Equity: A Guide to Language, Narrative, and Concepts*. The guide condemns several "dominant narratives" in medicine. One is the "narrative of individualism" and its misbegotten corollary, the notion that health is a personal responsibility. A more "equitable narrative," the guide instructs, would "expose the political roots underlying apparently 'natural' economic arrangements, such as property rights, market conditions, gentrification, oligopolies and low wage rates." The dominant narratives, says the AMA, "create harm, undermining public health and the advancement of health equity; they must be named, disrupted, and corrected."

Fast-forward to June 2024. The *New York Times*, when pro-Palestinian medical students and doctors, some wearing keffiyehs on the job at the University of San Francisco Medical Center, UCSF, were calling for the institution to take a stand against the war between Israel and Hamas and call for a cease-fire. Their chants of "intifada, intifada, long live intifada!" could be heard in hospital rooms. Remarkably, UCSF did not ask activists to stop, but one didn't need to be Hippocrates to know that patient trust in doctors and medical institutions depends on health-care professionals' strict duty to uncouple their political ideology from their clinical work.

These examples are among the most visible evidence of a deeply worrisome experiment underway across American medical schools: the morphing of the role of physicians from healers to activists and of medicine from promoting health to advancing social justice. This new vision is corrupting collegiality and skewing the medical student curriculum away from essential coursework to make room for lectures on intersectionality, implicit bias, and, for first-year medical students at UCLA, a mandatory course on "structural racism" in which the guest speaker led students in chants of "Free, Free Palestine" and demanded that they bow down to "mama earth" while praying for "black," "brown," and "houseless people" who die because of the "crapatalist lie" of "private property."

Here I will focus on some ways in which social justice medicine—or indoctrinology, as I like to call it—poses a danger to open inquiry and scientific rigor in many academic medical centers. I will discuss, in overview, the impact of the social justice agenda on free discourse, on exploration of certain research topics, and on patient care.

The implementation of the social justice agenda has constrained collegial discourse, challenged the maintenance of standards, and suppressed honest analysis of certain problems. In her article in the *Free Press* in 2021, "What Happens When Doctors Can't Tell the Truth?" journalist Katie Herzog interviewed doctors who've been reported to their departments for criticizing residents for being late. (It was seen by their trainees as an act of racism.) I, too, have heard from doctors who stopped giving trainees honest feedback for fear of retaliation. I've spoken to those who have witnessed residents refuse to treat patients based on race or their perceived conservative politics.

In some quarters, there is reflexive attribution of group differences in health to systemic racism. "It's axiomatic at this point," said a colleague who had participated in a group discussion of stress and rising suicide in black youth. The tacit rule was that only fear of police aggression or racial discrimination were allowable explanations, not the psychological torture of bullying by classmates or the quotidian terror of neighborhood gun violence.

The pressure to attribute health differentials to outside forces naturally takes the focus off how patients can be in better control of their own health. It also narrows the scope of understanding the roles of myriad other factors.

Compounding matters, efforts at "addressing and undoing racism and bias" in medicine are so poorly defined as to be useless at best and harmful at worst. Jeffrey Flier, former dean of Harvard Medical School and a graduate of Mt. Sinai School of Medicine in New York City, signed up for educational modules offered by the school in the wake of George Floyd. As Flier described in the *Free Press* in 2024, the sessions comprised sustained accusations of white supremacy in medicine, dismissal of the scientific method in medical research, with a scolding of Flier thrown in for good measure when he gently questioned the value of the vague term "anti-racism" in discussions of improving minority health. "[I]t is *exactly because* the issues of racism and bias in medicine today are so important that precise definitions and rigorous critical discussion are so crucial in medical education," he stated. "Yet, much like a devotee accepting holy writ, we were to forgo questions and simply embrace the doctrine, even without knowing what it means."

Interest in open discussion is nil. When Dr. Flier submitted a paper to *Academic Medicine* outlining some of his concerns about the anti-racism instruction at Mount Sinai—the paper was rejected in two days, he wrote, "without peer review or editorial explanation, which I found surprising."

Few doctors, save for Flier and Huddle and an organization called DoNoHarm.org, complain openly. I understand why. I hope I would have the fortitude to resist openly were I still a full-time faculty member at the Yale School of Medicine—if I were a junior person, I doubt I would. (It's no coincidence that the outspoken are almost all retired or now working in non-medical institutions.) "Most in academic medicine who are troubled by the intrusion of social justice are keeping their heads down and keeping their mouths shut," said my colleague Thomas Huddle, an internist and professor who retired in 2021 from the University of Alabama at Birmingham's medical school, one of few physicians willing to go on the record.

I know of no systematic polling on this issue. But reports of retreat from teaching come up frequently in my discussions with senior clinicians, as senior professors forgo discretionary teaching opportunities lest they offend a student who will complain to a department head or medical school dean.

Another consequence of the social justice agenda in medicine is the suppression of certain topics, both social and scientific. Consider the case of Dr. Norman Wang. In 2020, Norman C. Wang—a cardiologist at the University of Pittsburgh School of Medicine who expressed skepticism about mandatory affirmative action after conducting a careful review of the data—was stripped by his department of his directorship of the electrophysiology fellowship and barred from having contact with medical students, residents, or fellows because his views were "inherently unsafe."

His peer-reviewed paper—"Diversity, Inclusion, and Equity: Evolution of Race and Ethnicity Considerations for the Cardiology Workforce in the United States of America from 1969 to 2019," which appeared in the *Journal of the American Heart Association (JAHA)* in March 2020—was retracted by the journal without Wang's consent. The American Heart Association, which publishes *JAHA*, tweeted that his article "does NOT represent [the American Heart Association's] values." The cardiologist has sued the university and the American Health Association. (In September 2022, a judge dismissed the university from the case.)

Wang made his case on strong empirical grounds. According to an analysis by economist Mark J. Perry at DoNoHarm.org, between 2013 and 2016—the last year for which the AAMC makes data available—medical schools admitted only 8 percent of white college seniors with below-average undergraduate GPAs and below-average MCAT scores. Asian college seniors with those qualifications were offered slots at a rate of 6 percent. At the same time, schools accepted 56 percent of black college seniors with below-average undergraduate GPAs and below-average MCATs and 31 percent of Hispanic students with those scores. This means that a black student in that range was over seven times as likely to gain admission as a white college senior with the same grades and more than nine times as likely to be admitted as a similarly situated Asian senior.

Performance differentials already exist, and compromising standards will almost surely exacerbate the problem. A 2022 study of clinical performance scores by authors from Emory University, Massachusetts General Hospital, and the University of California, San Francisco, among other institutions, analyzed faculty evaluations of internal medicine residents in such areas as medical knowledge and professionalism. On every assessment, black and Hispanic residents were rated lower than white and Asian residents. In the journal *Academic Medicine*, the authors speculated that these differences were attributable to "bias in faculty assessment, effects of a non-inclusive learning environment, or structural inequities in assessment." They failed to rule out another possibility—namely, suboptimal performance based on suboptimal preparedness of students admitted under affirmative action policy.

Another fraught issue is race and genetics. In some quarters, medical researchers are warned not to explore gene contributions to understanding health disparities. Some worry that putting "genes" and "race" in the same sentence will encourage eugenics and the fiction that races are discrete entities defined by biological traits. But studies involving genes and race are simply about population genetics—i.e., the fact that people sharing a geographical ancestry are more likely to have particular gene variants (alleles) in their genome than do people with a different heritage.

Researchers and physicians agree that pharmacogenomics—the elucidation of the relationship between treatment and individuals' unique genomic fingerprint to create personalized therapies—will make the controversy obsolete. But until this gold standard is used widely, group-based genetic analysis will have some value. Even with the caveats in mind, genetic heritage can be relevant to medicine with regard to appropriately dosing certain drugs, more accurately predicting responses to those drugs, using clinical decision-making via algorithms (an especially controversial matter that scientists are currently debating in good faith), and determining heightened risk for certain conditions, such as cardiovascular and renal disease.

A team writing in *Health Affairs* blog in 2020 warned researchers who planned to publish on health disparities to "never offer genetic interpretations of race because such suppositions are not grounded in science." They also proposed that medical journals "reject articles on racial health inequities that fail to rigorously examine racism." The article review process, they say, requires "editors who are well versed in critical race theory." But why? Genetic inquiry across groups is emphatically not "racial science" or scientific racism. The objectivity of research is not a form of complicity in structures of power; it is the very condition for the discovery of treatments that are genuinely universal.

Concerned by the disavowal of such studies, experts spoke up. "For some applications, race may continue to be the best variable to capture the influence on health," wrote John P. A. Ioannidis, Neil R. Powe, and Clyde Yancy in the *Journal of the American Medical Association* in 2021. "Quick dismissal," they cautioned, "may worsen outcomes, especially for the most disadvantaged populations."

One especially clear example concerns kidney function and a gene called APOL-1, or Apolipoprotein L1. Variants (alleles) of this gene elevate the risk of developing kidney disease in people of African ancestry fourfold, necessitating, on average, dialysis or kidney transplant at a younger age relative to other individuals who develop renal failure. A 2011 paper appeared in the *Journal of the American Society of Nephrology*, titled "Apolipoprotein L-1 and the Genetic Basis for Racial Disparity in Chronic Kidney Disease." The explanation is natural selection wherein high carrier frequency of *APOL1* renal disease risk alleles confer protection against infection with subspecies of the protozoan parasite *Trypanosoma brucei* that causes sleeping sickness endemic to sub-Saharan Africa.

In the *New England Journal of Medicine*, five genetics experts, who identified themselves as black, understand phenomena like the APOL-I gene. Ideally, they said, in a 2021 article, race will be replaced with "genetic ancestry" as a variable in medical research and practice but that until more ancestry data are available, "ignoring race" and extrapolating research findings from European-ancestry populations for "the treatment of non-European populations…is neither equitable nor safe." The authors also expressed disappointment that some "curricula promote ideologies that downplay the medical achievements of genetic studies."

The social justice imperative also manifests in treatment decisions. Early in the COVID-19 pandemic, some argued that if hospitals needed to ration ventilators, they should prioritize black patients rather than rely exclusively on standard criteria, such as clinical need or prognosis.

In late 2020, when it came time to distribute the coronavirus vaccine, an assortment of authorities—including legal scholars, public health experts, and state officials—argued for giving high priority to black citizens in the name of "historical injustice." About that historical injustice, there can be no doubt, but the Advisory Committee on Immunization Practices of the Centers for Disease Control and Prevention (CDC) concluded that race should supersede age as a prioritization category because the oldest cohort in America is whiter than the general population.

In reality, the risk was massively greater for older people than younger people, a differential that dwarfed the black-white difference. The CDC was well aware of this. One CDC official even said that a race-based allocation plan would result in up to 6.5 percent more deaths, many of whom would be black senior citizens—the highest risk group. No matter, the official went on to say, "Racial and ethnic minority groups are underrepresented among adults [older than] 65."

Apparently, America's elderly are too white. Elevating "health equity" was about to take precedence at the CDC. The agency's loyalty was to an ideal—not, foremost, to saving the most American lives from infection. The massive public outcry that ensued almost surely helped derail the racial-preference plan; the committee revised its recommendation, thereby preventing thousands of deaths.

That clinical considerations might flow from the politics of group identity, and not from the best interests of individual patients, the North Star of clinical ethics, is a grave error.

"Do no harm" is a covenant that doctors make with their patients, not with political systems and hierarchies.

The fundamental problem with social justice in medicine is that there are no limiting principles to it—the boundary between social issues and medical ones thoroughly dissolves. How will the AMA's new call to "focus attention on *inequitable systems, hierarchies, social structure, power relations, and institutional practices*" (emphasis in original) affect the formation of trainees' professional identities?

We are already seeing the effects on curricula as training in basic medical science, early clinical skills, epidemiology, and bioethics is being displaced to accommodate the anti-racist curriculum. If health is completely at the mercy of social forces, as Social Justice, MD insists, will the importance of self-care be given adequate attention? If certain topics are off limits and every possible hypothesis is not open to investigation, how will knowledge mature and innovation progress? Will a health equity agenda so distort the priorities of medicine that patients will be harmed? How will the adoption of a zealous social justice agenda affect public trust? And will the hyperbole about racism in medicine be self-fulfilling, exacerbating distrust where it may already exist?

The pragmatic imperatives of clinical practice may be the best buffer against ideology. The surgical suite, the emergency department, and the examining room are the definitive, consequential spheres of clinical intervention.

The famous nineteenth-century German statesman and physician Rudolf Virchow called physicians the natural "attorney for the poor." In the clinic and at the bedside, good doctors argue as eloquently as any lawyer for the disadvantaged through their specialized knowledge and compassion. In medical journals, they spread knowledge through dispassionate, truth-seeking methods that speak to all. And in the realm of medicine, they do their best work aiding those who are most vulnerable and in need, regardless of group affiliation. The best way to be an anti-racist doctor is to be a good doctor.

See QR Document for references.

Burying Science Under Indigenous Religion

ELIZABETH WEISS

The study of skeletal remains is under attack by repatriation activists who wish to replace scientific research and dissemination of this information with animistic religious tales. In anthropology, repatriation is defined as the act of returning artifacts and human remains of past peoples that are held at universities and museums to related modern tribes or communities.

When the US's Native American Graves Protection and Repatriation Act (NAGPRA) was passed in 1990, anthropologists viewed NAGPRA as a human rights law that enabled tribes to reclaim ancestral artifacts and human remains; but Native Americans most frequently cited religious reasons to repatriate these same materials. And, in a review of newspaper articles that covered repatriations between 1990 and 2010, Native Americans consistently mentioned the importance of religion in regard to the reason they engaged in repatriations.

NAGPRA was meant to apply only to sacred objects, funerary objects, objects of cultural patrimony, and human remains that could be definitively linked to a modern tribe. There was an acceptance that other artifacts—even when affiliated with a modern tribe—could be retained for research and display purposes.

Human remains and artifacts unaffiliated with any federally recognized modern-day tribe were to be continuously curated and available for research and display. Affiliation was defined as "a relationship of shared group identity which can be reasonably traced historically or prehistorically between a present-day Indian tribe or Native Hawaiian organization and an identifiable earlier group." And it was to be determined through a "preponderance of the evidence based upon geographical, kinship, biological, archaeological, anthropological, linguistic, folkloric, oral traditional, historical, or other relevant information or expert opinion."

With these limitations in place, NAGPRA was seen as a compromise between Native American tribes and anthropologists interested in continuing to study America's past through the use of skeletal remains and artifacts. By 2020, over 90 percent of human remains and qualifying artifacts that were affiliated with federally recognized tribes had been repatriated to modern-day tribes. Some of the remaining materials had not been repatriated due to a variety of circumstances, such as the tribe not being ready to rebury the remains. Unaffiliated human remains and artifacts, and artifacts that did not fit into the above-mentioned NAGPRA categories, remained available for research and display across the nation's museums and universities.

Shortly after the passing of NAGPRA, some scholars—myself included—foresaw that NAGPRA's compromise would inevitably fail, in large part due to the repatriation activists who pushed for a continued greater emphasis on their religious beliefs and a desire to take everything back. For instance, the late Clement W. Meighan (an anthropologist who taught at the University of California, Los Angeles, for nearly forty years), writing in *American Antiquity* in 1992, already wrote that suggestions that "we balance our own values with the professional ethic of cultural relativism by 'compromise and mutual respect' is not realistic" because archaeologists will not compromise their scholarly beliefs and "many Indian activists are not going to compromise away their beliefs (however unsupported by evidence) that every Indian bone of the past 12,000 years belongs to one of their ancestors." Well, Meighan was right—about the Indians, at least!

NAGPRA requires religious beliefs to be taken seriously when it mandates that NAGPRA committees must contain two "traditional Indian religious leaders." Furthermore, the allowance of "oral tradition" as evidence opened the door for Native American creation myths to be used as evidence in repatriation decisions. I and my coauthor, James W. Springer, in our 2020 book *Repatriation and Erasing the Past*, highlighted the many ways in which NAGPRA has breached the constitutional separation of church and state.

Scholars who were skeptical of NAGPRA's ability to maintain a compromise have been proven correct. Administrative changes to NAGPRA have all been in favor of supporting Native American religious beliefs. For instance, they added a category of evidence called "Native American traditional knowledge." This "knowledge" is allowed to be "safeguarded," as it may be considered "confidential to individual Native Americans, Indian Tribes, or the Native Hawaiian Community." It's not to be shared with outsiders and—therefore—completely untestable. Native American traditional knowledge includes "religious" and "spiritual" components. Ironically, Native American traditional knowledge—unlike other traditions—does not need to be passed down through time. They can make up the tradition as they go along. And Native American traditional knowledge is considered "expert opinion"—not to be questioned.

Changes that revolve around consultation on objects held in museums have led to widespread closure of exhibits across the nation—museums will reopen their exhibits after tribal consultations, likely with far fewer objects to display and featuring Native American narratives instead of factual scientific evidence. A good example of such an exhibit is the American Museum of Natural History's Northwest Coast Hall. After $19 million spent and five years of work, the exhibit has fewer artifacts, promotes religious beliefs, such as warnings that some objects may contain powers that can be harmful, and descriptions of "historical events" that involve supernatural encounters with "Thunderbirds."

Even prior to these changes, virtue-signaling museum and university administrators have played key roles in overstepping the law when they go beyond compliance to repatriate remains and artifacts. For example, the McClung Museum at the University of Tennessee, Knoxville, decided to return 164 items that could not be linked to modern tribes. McClung Museum administrators also decided to cover up photos of archaeological excavations and line drawings of burials. In the article, "Is NAGPRA Irretrievably Broken?," historian Ron McCoy writes about the Rochester Museum and Science Center's 2010 decision to repatriate replica medicine masks, which were made between 1935 and 1941 for public display, to the Tonawanda Band of Seneca Indians. The tribal leaders claimed that these were "sacred objects." And, as reported in "Caring for sacred and culturally sensitive objects" written by Miriam Clavir and John Moses for the Canadian Conservation Institute, Navajo consultants have been able to prevent the exhibition of a replica basket because of its "conceptual authenticity."

In California, even with recent repatriation laws, such as AB389, that "prohibit[s] the use of any Native American human remains or cultural items for purposes of teaching or research at the California State University," some universities have gone even further. For example, California State University, Bakersfield, has placed a moratorium on "research, teaching, display, imaging, and circulation of human remains and cultural items (including archival material, notes, movies, and data) that are potentially subject to NAGPRA and CalNAGPRA." CalNAGPRA is the Californian version of NAGPRA. And Cal Poly Pomona administrators have put a "[s]uspension on the collection and use of Native American Ancestral Remains and Cultural Items" that "includes but is not limited to images, renderings, and reproductions of ancestral remains and cultural items that are or have been in a university's collection." These university regulations have effectively buried California archaeology and physical anthropology and, with that, the science that flows from it, such as teaching new forensic students and understanding past peoples' lives. Using the data as comparative samples in a variety of studies has ceased as well. This will affect our ability to understand questions ranging from the peopling of the Americas to the spread of diseases.

Repatriation and the ideology behind it are not just problems in the US; Canada, Australia, and New Zealand face similar attacks on the scientific study of the human past.

In Canada, the adoption of the United Nations Declaration of the Rights of Indigenous Peoples (UNDRIP) has resulted in the repatriation of a twelve-thousand-year-old British Columbian site; tellingly, "elder" Garry Oker of the Dane-zaa celebrated the repatriation by noting: "We can tell a narrative that's truly Indigenous." Whether this narrative is actually factual is another story. In the Canadian Museums Association's publication "More than Giving Back: Repatriation Toolkit," the authors wrote that "photographs, archival documents, and songs, plants, seeds, and language recordings" are all items that should be considered for repatriation. Furthermore, they stated that museums should follow rituals that include restricting which sex can handle the materials, ritual feeding of masks, and periodic smudging of objects. They caution that fire alarms should be deactivated, and a fire extinguisher should be at hand! In a parallel manner to the American Museum of Natural History's concerns about "objects of power," such as a bird bone whistle that may summon supernatural beings or warnings that pregnant or menstruating women should stay away from certain objects of power, Canadian repatriation literature also warns of harm that can come from "power." For instance, in "Caring for sacred and culturally sensitive objects," the authors wrote: "Can the power considered inherent in some objects affect someone outside that belief system? There is no simple answer to this question." Further, they compare radioactivity with sacred power:

> [S]ome sacred and sensitive objects carry inherent power. Yet, what is inherent power? How does it manifest itself and what does this mean for people who come in contact with the object? An analogy for power might be found in the comparison between a sample of granite and a sample of uranium in a natural history collection. Both have a physical integrity that can be described by standard procedures, but the uranium has a powerful, invisible and intrinsic attribute. Those who know how to use it can do so for helpful, or for harmful, purposes. Those who ignore it can be damaged by it. Thus, there are strict protocols for handling it and using it.

Australia, unlike the US, has no federal legislation regarding repatriation, although some territorial legislation prohibits the private ownership of indigenous human remains. There is the Australian Government Policy on Indigenous Repatriation that was established in 2011; it covers the repatriation of human remains, associated notes, and data held overseas and in Australia, but it holds no legal weight. It does pay for repatriation activities.

Yet, Australia is perhaps the country most active in repatriation. Not only are they active in the repatriation of human remains, but also in the

"Return of Cultural Heritage Project" guidelines, museums are advised to return photos, manuscripts, audio and video recordings, and even data from both Australian and overseas institutions. And in "A Repatriation Handbook: A guide to repatriating Australian Aboriginal and Torres Strait Islander Ancestral Remains" (from the National Museum of Australia and the Return, Reconcile and Renew Project), museums and universities are told that there is no excuse to avoid unconditional repatriation of human remains and any associated notes and data. Furthermore, policy from the Australian Museum and Gallery Association states not only that human remains should not be displayed, but also that "images and replicas" of indigenous human remains should not be exhibited. Even if the human remains were obtained legally through trade or sales, they still qualify for repatriation.

In 1985, aboriginal leader Jim Berg repatriated human remains from the University of Melbourne, which set the scene for many repatriations to come. And, in 1991, Kow Swamp Skull—a thirteen-thousand-year-old (although some suggest it is nearly twenty-two thousand years old) artificially modified skull that has been studied to understand the origins of humans—was given to the Yorta Yorta community and buried. Like America's Paleoindians, Kow Swamp Skull's relation to modern communities is tenuous, and its utility to scientific research was tremendous.

The oldest remains to be repatriated are the forty-two-thousand-year-old Mungo Man. Mungo Man was excavated by Australia National University geology professor Jim Bowler in 1974, after rains exposed the remains. Mungo Man was repatriated in 2017 and reburied in 2018, and as told in the *Smithsonian Magazine* article, Bowler, who had once studied to become a Jesuit priest, wrote the letter of support for the repatriation. Bowler lamented the discovery and stated that he asked Mungo Man for guidance. Furthermore, Bowler believes that the spiritual world of the aborigines is real.

Often, academics and scholars view repatriation as a way to atone for colonial "sins." But, for the indigenous community, like for Bowler, repatriation is tied to supernatural beliefs. Throughout the above-mentioned handbook, concerns about ghosts, dangerous spirits, and trauma from seeing remains are mentioned. And the Return, Reconcile, and Renew Project states on its website:

> There are clear spiritual and religious reasons why human remains must not be disturbed. Frequently, it is considered essential that the dead remain in their traditional country in order for their spirit to rest and for the connection between the living, the dead, and the land to be in equilibrium. Disturbing this equilibrium is considered dangerous, leading to sickness and ill-health, both for people and the environment.

Plus, prior to gaining entry onto the website, one must click on an "I agree" button that includes the statement: "Support the process of returning Old People to country so their spirits may rest."

Although Australian aborigines claim that it is important for their health, safety, and well-being to return the human remains to a final resting place—whether that be a cave that is sealed off with concrete or a burial ground that later becomes the site of one of their casinos (as in the burial site for over eighty human skeletons by the Larrakia people in 2002)—this process is not to be rushed, according to the literature. And aboriginal communities may take years to return the remains to their final resting place, even after repatriation has occurred. From repatriation to the "funeral," the remains are held in "Keeping Places" that are run by the aboriginal communities. These places ironically are often not safe, and although theft, vandalism, and arson have occurred, the race of the perpetrators is never mentioned in the above-mentioned handbooks and guides.

New Zealand, too, has been seeking repatriation of materials from abroad and at home. Since 1975, New Zealand has utilized its Protected Objects Act to prevent artifacts relating to the Māori culture, history, or society from being taken out of the country and has actively engaged in trying to reclaim those materials from overseas institutions. Since 2003, the Karanga Aotearoa Repatriation Programme, a government-mandated authority that is housed in the Museum of New Zealand Te Papa Tongarewa, has provided funding and in-country authority for the repatriation of human remains and artifacts. Now, the Karanga Aotearoa Repatriation Programme also provides support, including funding, for the repatriation of materials and human remains from overseas institutions. When human remains are repatriated, the Museum of New Zealand Te Papa Tongarewa initially holds them in their Wahi Tapu, a term that means sacred space. Since the Programme began, over eight hundred human remains have been returned to New Zealand.

The most famous repatriation claims in New Zealand revolve around the tattooed and preserved mummified indigenous heads, commonly called mokomokai. Facial tattoos were a tradition among the Māori until the mid-nineteenth century; the tattoos revealed family relations and status. When a high-status tattooed individual died, their head would be preserved, kept in a carved box, and brought out during sacred ceremonies. Heads of enemies were used as war trophies, mocked, and sometimes mutilated.

Although the first repatriation of a mokomokai occurred in 1985, when the Vienna Museum of Ethnology (also known as the Museum Für Volkerkunde) returned one mokomokai upon the condition that the rest of the collection would remain in this Austrian museum, recently such repatriations have increased drastically and without conditions. For instance, more than 90 percent of repatriations from Germany, which have also been pushed forward as a result of new German museum guidelines, have occurred since 2017.

The mokomokai were traded by the Māori to Europeans in the 1800s; the Māori often received guns in exchange, which they used for intertribal warfare. The trade for these heads ceased as a result of a decrease in demand from Europeans, which was in large part due to the Māori's deceptive practices that included tattooing deceased slaves or even killing slaves and then tattooing them. The most famous collector of mokomokai was Major General Horatio Gordon Robley. Robley, a British army officer and artist, was able to procure a collection of thirty-five mokomokai in the 1860s while he served in New Zealand. Robley had an intellectual interest in the heads as a result of his interest in tattooing and art; he wrote an illustrated book, which contained photos and drawings on the topic, that was published in 1896. Robley offered to sell the mokomokai to the New Zealand Government;

they passed on the offer, and the heads were instead sold to the American Museum of Natural History in New York.

Much of the narrative on repatriation from the Karanga Aotearoa Repatriation Programme revolves around the "righting the wrongs of the past" and "addressing grievance that multiple generations have carried" since Europeans' arrival with Captain Cook in 1769. There is also an anti-science narrative from the Programme; for instance, the Programme's website states: "Repatriation is important as it shows respect for the dead, for cultural beliefs, and for the hurt that has been caused to source communities as a result of the development of science and museum collections." Yet, there is also a religious narrative from the indigenous communities. For instance, in a 2008 request to the British Museum to return seven mokomokai (rejected by the museum), testimony arguing for the burial of these remains came from an indigenous scholar who wrote:

> By not committing the tinana [person] to "mortuary disposal" keeps the physical in a state of limbo, which arrests and disables the wairua (spirit) to return to its resting place, to the origin of our spirit. For Māori that place is Hawaiki [ancient homeland] Nui, Hawaiki Roa, Hawaiki Pāmamao. These actions condemn that group of people to a fate worse than death, a state we might call kahupō, or spiritual blindness.

One may wonder what is behind scholars' support of the repatriation movement. In *Repatriation and Erasing the Past*, James W. Springer and I define repatriation as "any ideology, political movement, or law that attempts to control anthropological research by giving control over that research to contemporary American Indian communities." Here, I expand it to include contemporary indigenous communities (like the Māori in New Zealand, the Australian aborigines, and Canada's First Nations) around the world. Repatriation ideology stems from postmodernist ideology, which divides the world into victims and victimizers. It also does not accept the concept of objective truth or knowledge; rather, it places importance on who tells the story instead of whether the story is accurate. So, repatriation ideology regards indigenous people as victims and Europeans as victimizers.

Hence, in understanding the past of people who developed no written language, the victim narrative is given priority, and these narratives must be treated as unquestionable and equal—or perhaps even superior—to scientific explanations.

Science is distinctly different from the narratives provided by indigenous communities. Paleoanthropologist Geoffrey A. Clark wrote in his 1998 article "NAGPRA, the Conflict between Science and Religion, and the Political Consequences" that:

> Science can be defined as a collection of methods for evaluating the credibility of knowledge claims about the experiential world. Science does not pretend to certainty; it only seeks better and better approximations of it. Scientific conclusions are continuously subjected to critical scrutiny. Science is, therefore, self-correcting.

None of this holds for the religious narratives provided by indigenous repatriation activists. Repatriation ideology places anthropological collections and the research on them into the hands of contemporary indigenous peoples engaging in repatriation who, unfortunately, cannot see beyond their religious beliefs and, thus, can only provide narratives filled with creation myths, ghosts, and malevolent spirits.

Anthropologists have been at the forefront of fighting against anti-evolutionary religious beliefs when they come from Christian creationists, but the same is not true for their support of science and evolution in the face of indigenous animistic creationists. This marriage between religion and anthropology can only be understood when repatriation ideology is viewed through a postmodern lens. As Geoffrey Clark wrote:

> The worldview of Western science is under serious and sustained assault, and there is a danger that "science-like" views of reality will perish in the face of a multipronged attack in which mysticism, religious fundamentalism, creationism, and the believers in the paranormal combine with postmodernist academics to attack the critical realism and

mitigated objectivity which are the central epistemological biases of the scientific worldview.

Unfortunately, repatriation ideology has spread like a cancer from anthropology and is now entering other sciences in the US through the former Biden administration's mandate to incorporate "indigenous knowledge" in nearly all government-funded research, from the Environmental Protection Act to the National Institutes of Health. A preview of incorporating such knowledge can be found in the report on the 2019 Traditional Medicine Summit: Maintaining and Protecting Culture Through Healing, which includes no information about medicine and multiple references to spirituality, God, and the Creator. At the beginning of the conference, the organizers "acknowledged the sacred space devoted to offerings for spirits, guardians, ancestors." Speakers preached that "God makes everything possible" and that it is "vital" for medical researchers to "appreciate the importance of prayers." In New Zealand, repatriation ideology has filtered into science classes, with topics such as Māori knowledge of animals; but in Māori "knowledge," the supernatural and the real are not separated. For example, Michael R. Matthews from the School of Education at the University of New South Wales points out that birds, which are called "manu"—a word really meaning any flying animal—are seen as "messengers from spirit worlds, carrying warnings or reassurances…from deceased loved ones."

Thus, anthropology led the sciences back into the "demon-haunted world" Carl Sagan warned about. And for what purpose? In 1992, Clement W. Meighan asked:

> When scholarly classes in United States archaeology and ethnology are no longer taught in academic departments (they are diminishing rapidly), when the existing collections have been selectively destroyed or concealed, and when all new field archaeology in the United States is a political exercise rather than a scientific investigation, will the world be a better place?

We now know the answer to his question, and it is a resounding "no"!

An Apology for Philology

SOLVEIG LUCIA GOLD AND JOSHUA T. KATZ

In a bid to increase enrollment, especially of students of color, the Princeton Department of Classics in the spring of 2021 made the now-infamous decision to eliminate its language requirement for undergraduate majors so that students could graduate with a degree in classics without ever taking a single course in Latin or Ancient Greek. The decision enraged many alumni and outside commentators, who marveled at the department's soft bigotry.[1] As the linguist John McWhorter put it, "[t]he Princeton Classics department's new position is tantamount to saying that Latin and Greek are too hard to require Black students to learn."[2] The decision also left people scratching their heads: what does it mean to do classics without the languages?

It's a good question.[3] While classics is an interdisciplinary subject, knowledge of Greek and/or Latin has traditionally been the one thing all classicists share, whether their focus is papyrology or archaeology. And more specifically, all classicists have traditionally been trained in the subfield that binds the whole field together: philology.

What is classical philology?[4] The best point of reference for an American audience may be this: it looks a lot like textualist and originalist approaches to constitutional interpretation. Classical philology, as we were trained to do it, is the attempt to understand a text as it was produced in its historical context, with as few anachronistic preconceptions as possible. The classical philologist unpacks each word or phrase by comparing it to other uses of the word or phrase in a given author's corpus—and in the broader corpus of texts to which that author would have been exposed.

Philology was long the core of classics because it, in some sense, provided a unifying purpose for the different subfields: literature, philosophy, history, linguistics, papyrology, epigraphy, numismatics, textual criticism, art history, and archaeology—all are important tools for the philologist who works to uncover a text's context, and philology is an important tool for these subfields in turn. So central was philology to the discipline that from 1869 until 2014, the Society for Classical Studies, the main learned society of classicists in North America, was called the American Philological Association.

But as both this name change and Princeton's elimination of its language requirement reflect, philology is no longer the core of classics.[5] It has been replaced by a form of activist teaching and research—recently bolstered by a $1 million grant from the Mellon Foundation to Sasha-Mae Eccleston and Dan-el Padilla Peralta, professors of classics at Brown and Princeton, respectively, for their initiative "Racing the Classics"[6]—that prizes such presentist concerns as identity (race, gender, sexual orientation, class, indigeneity, and (dis)ability), politics, and the environment. This new approach aims, at turns, to use classics to solve contemporary problems,[7] to project contemporary problems onto the analysis of classical material, and to look at how the reception of classical material has caused or perpetuated many of these problems in the first place.

At the heart of this approach is a methodology—best summarized by Mathura Umachandran and Marchella Ward in the opening chapter of their coedited 2024 volume *Critical Ancient World Studies: The Case for Forgetting Classics*[8]—that

> critiques the field's Eurocentrism and refuses to inherit silently a field crafted so as to constitute a mythical pre-history for an imagined 'West'.... [I]t rejects the assumption of an axiomatic relationship between so-called classics and cultural value.... [I]t denies positivist accounts of history, and all modes of investigation that aim at establishing a perspective that is neutral or transparent, and commits instead to showcasing the contingency of history and historiography in a way that is alert to the injustices and epistemologies of power that have shaped the way that certain kinds of knowledge have been constructed as "objective"

> within the discipline known as classics.... [I]t requires of those who participate in it a commitment to decolonising the gaze of and at antiquity.[9]

In short, there's a whole lot of discussion *about* classics, but very little, well, classics.[10]

And when there is analysis of classical texts, it increasingly looks like this:

> Through a combination of Audre Lorde's Black queer lens and Paul Preciado's trans scholarship on the dildo, I further argue that by imagining Simulus as Black, queer, and/or trans, the power imbalance between Simulus and Scybale is greatly reduced.[11]

This sentence from a scholarly article, chosen more or less at random (and written, incidentally, by a white cisgendered woman, Francesca Bellei), comes from the Spring 2024 issue—a special one titled "Race and Racism: Beyond the Spectacular"—of *TAPA*, the journal formerly known as the *Transactions of the American Philological Association*. Despite its peculiarities, the article in question exhibits a far greater command of traditional philology than much of what makes it into *TAPA* these days[12]—or, for that matter, into the *American Journal of Philology*, which has published two special issues in recent years titled "Diversifying Classical Philology,"[13] with articles that, in the words of guest editor Emily Greenwood, "exemplify a practice of intramural counter-philology—of turning the tools of philological criticism onto the discipline to examine its sources of cultural knowledge, ways of knowing, and its lacunae."[14]

Philology, you see, has become a dirty word—for three primary reasons, best articulated in the writings of Eccleston, Padilla Peralta, Greenwood, and Patrice Rankine.[15] Although their arguments are often jargon-heavy and difficult to parse, we have worked to reconstruct the charges against philology below. It is high time for someone to challenge what has quickly become gospel in our field—and, for that matter, in society at large, where both objectivity and "worship of the written word" have been branded "white supremacy."[16] It is high time, in short, for an apology for philology.

(1) Philology's Racist Past

So, what are the three charges against philology? First, that philology was historically associated with slaveholders, imperialism, and race science.[17] In particular, until recently, classicists revered the legacy of Basil L. Gildersleeve, the founder of the *American Journal of Philology*,[18] who was, however, a slaveholder and fought for the Confederacy.[19]

This charge is undeniable but not all that remarkable: every longstanding field of inquiry has been shaped by people with reprehensible views, many of them involved in reprehensible practices. Those views and practices do not necessarily implicate the inquiry: Schrödinger was a pedophile, but that has no bearing on his cat. Moreover, even if classical philology had been *explicitly* engineered to be a means of supporting, say, slavery (it was not), the philological method would not necessarily be compromised.[20]

To prove that classical philology cannot be disentangled from its slaveholding past, the accusers would have to demonstrate that the method itself is compromised. And they do indeed attempt to do this—their second charge against philology.

(2) Philology's Intrinsic Whiteness

To quote Rankine from 2019, "philology professes to retreat from all contemporary inquiry, fixing its gaze on the past."[21] But this "is and always has been a pretense, and a pernicious lie"[22] because, as Rankine puts it in the introduction to the Spring 2024 issue of *TAPA*, "the race-neutral, colorblind position was simply an unconscious strategy of concealment."[23] That is, "[r]epressing the Oedipal secret of your identity (which is hidden in plain sight), it might never occur to you that your ability to read and write about Euripides…unimpeded by anything but the text, manuscripts, and your outstanding philological training owes to your racial identity, your whiteness."[24]

In other words, white scholars are so accustomed to a culture of whiteness that we do not realize how whiteness colors everything we do, including how we read ancient texts. We may think we have developed an impartial, near-scientific method of textual interpretation, but really philology is a means of blindly projecting our contemporary white biases onto the past under the guise of objectivity.

We—by which we mean Solveig and Joshua, not all "white" scholars—find this charge bizarre because philology, or at least *good* philology, forces us to acknowledge precisely those contemporary biases that Rankine seems to believe it actively conceals. A good philologist knows that when she reads just about any word in Greek or Latin, she will project onto the word a whole host of meanings and references that have accumulated around the word (and corresponding words in other languages) in the past two thousand years. Her task—*the* task of a philologist—is to strip away those meanings and get to the meaning of the word as it was originally written. She does this by reading widely and deeply on the use of the word in the ancient world. She never takes for granted that she knows what a word in a text means simply because she can translate it into her native tongue.

An example in English: we've called this chapter "An Apology for Philology." A naive reader would almost certainly assume that the title is introducing a regretful acknowledgment of philology's failures. However, a careful reader—a good philologist—would consider (1) that the chapter does *not* seem to harp on philology's failures, (2) that the chapter is about classics, (3) that Solveig's main area of expertise is Plato, and (4) that catchy titles often play with non-obvious meanings of words. And in view of all this, the good philologist would, and should, conclude that "apology" here refers not to the usual meaning in twenty-first-century English but rather to the earlier (now secondary) meaning that comes from Greek ἀπολογία, "defense," as in Plato's *Apology*.

What the naysayers are criticizing is not philology; it is *bad* philology. Now, it would seem that their next move is to allege that there is no such thing as good philology—that it is simply impossible to strip away all our contemporary biases when evaluating an ancient source, impossible to "un-race" ourselves.[25] As Greenwood writes, "our way of life is not embodied in ancient Greek or Latin, instead we embody these languages and impart our values to them, no matter how scrupulous or objective we think we are being."[26] To this we say: sure, it probably *is* impossible to strip away *all* our biases. But that does not mean we should throw in the towel, give up trying to understand the ancients on their own terms, and resign ourselves to identity-driven studies. Impartiality, however unattainable, remains a worthy goal.

Besides, we have yet to see evidence that the impossibility of fully "un-racing" ourselves actually does have a significant impact on philological scholarship. Where are all these white supremacist (mis)readings of (e.g.) Euripides? The naysayers never give an example. Or rather, Rankine gives one example in each of his articles under discussion—neither from philological scholarship.

First: "At the heart of classical philology's claim to impartiality and critical distance, the notion that study of the languages *per se* is the pathway to truth and understanding, is race."[27] In a footnote to this sentence, Rankine states that "[a]ny number of examples could support this point" and then cites five pages from E. R. Dodds's seminal 1951 book *The Greeks and the Irrational* that, Rankine claims, are "steeped in the paradigms of anthropology (at the time) and the field's articulation of African primitivism."[28] "His method," Rankine concludes tersely, "is philological."

Dodds was indeed a major philologist, but this book was an explicit effort to make use "in several places of recent anthropological and psychological observations and theories,"[29] and the pages Rankine cites are all examples of this anthropological approach, *not* philology. For instance, Dodds mentions an uptick in superstition among the Tanala tribes in Madagascar as a possible point of comparison for a similar phenomenon in Archaic Greece.[30] There is no careful analysis of text. Moreover, Dodds was well aware that his use of anthropology was historically contingent: "[I]n these relatively new studies the accepted truths of to-day are apt to become the discarded errors of to-morrow."[31] So much for a supposed "claim to impartiality."

Then there is Rankine's other article, in which his example comes from nineteenth-century language instruction. He wildly misrepresents his source in a staggering act of intellectual sloppiness or dishonesty, and it is worth quoting a large portion of the relevant paragraph:

> Once it is clear that motives…are impure, the uneasy connection between race and the Classics is exposed as

iron-clad, rather than incidental.... An example: Denise McCoskey has studied Latin language instruction and the subject of slavery in 19th-century grammar books. In an unpublished paper, she works to "determine the kinds of 'classical values' students were absorbing not by reading, say, Tacitus or Vergil, but by learning noun declensions and completing practice exercises." This may not seem like an ideologically loaded exercise; and yet the seemingly innocuous use of the English "servant" for *servus* in the American context belies real efforts at erasure, the rubbing out of the enslavement of Africans that began in Virginia in 1619. This enslavement and its erasure, by representing the abject status of the "slave" with the far less degraded status of the "servant," impacts the subsequent status of blacks in America as second-class citizens.... Ostensibly innocent, the rendering of *servus* as "servant" obfuscates the relationship between the Roman world that an American student enters through the Latin grammar book and her own contemporary prism.... Equally pernicious, the student does not even learn about the cruelty that was Roman slavery, cruelty now excused by notions of historical relativism and revisionism because "slaves" or "servants," after all, must deserve and desire their status.[32]

Rankine's argument, as we understand it, is that nineteenth-century grammar books translated *servus*, "slave," as "servant," which allowed students to read about the ancient world without having to confront the reality of slavery in their own time.[33] He also seems to suggest that this obfuscation continues in language instruction to this day. And, as the article goes on, Rankine returns to the (mis)translation of *servus* as evidence of our inability ever to view the past objectively: "If a move as seemingly innocent as 'servant' for *servus* belies the neutrality that it seems to present, imagine how cloudy is the view of modern concerns that quicken the study of the past in the first place. Our Oedipal blindness is total."[34] The trouble is that

Rankine's source for the mistranslation, the then-unpublished paper by Denise McCoskey, "belies" his argument.

McCoskey's (now-published) paper[35] identifies a number of problems in contemporary language instruction: for instance, the *Cambridge Latin Course*'s use of the sentence *servī erant laetī*, "the slaves were happy."[36] There is no mention, however, of anyone nowadays teaching *servus* as "servant." More important, though, is that there is scarcely any mention of *servus* being translated as "servant" in the nineteenth century either. McCoskey gives only two examples, in passing, from one 1839 schoolbook—a book that alternately translates *servus* as "servant" and "slave." And this book, published in abolitionist Boston, she actually commends for "its covert attempts to speak against the institution of slavery."[37]

Indeed, Rankine entirely misrepresents the thrust of McCoskey's paper, which is that nineteenth-century Latin schoolbooks were, in fact, better at discussing the brutal reality of slavery than our books are today: "generally more intense and more anxiety-ridden," the older works "tend to treat slaves as autonomous subjects" and "present a more frank portrayal of the institutional forces that perpetuate slavery, such as violence and law."[38] In other words, the alleged failure of contemporary language instructors to teach about the cruelties of slavery is *not* historically entrenched, and "the uneasy connection between race and the Classics" is *not* "iron-clad"—or at least, Rankine hasn't proved that it is.

(3) Philology's Weaponization

Finally, the third charge. Philology, which "pretends to be a neutral and disinvested test of intelligence,"[39] is associated with ideas like merit, rigor, and excellence. It has thus, the naysayers allege, been weaponized to dismiss other approaches to classics, including non-philological work of the kind highlighted above.[40] To quote Eccleston and Padilla Peralta, "the presumptive rigor of philology functioned as much more than a mode or metonym of exclusionary elitism throughout the field. Institutional gatekeepers levied it as a slur that effectively sidelined Black- or Brown-centered methodologies—especially in reception studies—as *me*search."[41]

There is good philology; there is bad philology. There is good reception work; there is bad reception work. We are apologists for philology, but we do not believe that philology is the only worthy way to do classics. We would, however, like to see excellence across the board. This means maintaining high standards in both language instruction and scholarship. Students should be competent in Greek and Latin, and scholars at all levels should be expected to marshal (correctly cited) evidence creatively to make valid—and ideally also sound—arguments in a comprehensible way. This comprehensibility matters precisely because it is what makes the scholarship accessible to non-specialists—that is to say, what makes it *not* exclusionary or elitist. Still, no, not everyone can do it. Excellence necessarily excludes.

But it does not necessarily exclude any one group of people. It is not white supremacy in disguise. According to Rankine, we "have to give the lie to seemingly neutral notions of excellence, just as we unveil the true motives of 'servant' for *servus*."[42] But now we have given the lie to his unveiling, and we see it for what it is: the weaponization of bad philology.

SECTION 3
Cancel Culture

The past decade has witnessed a new and unfortunate phenomenon: the regular cancellation of faculty, including tenured faculty, who are literally removed from their jobs or put in positions where they need to leave for a host of reasons that infringe upon academic freedom, freedom of speech, and due process. The list is so long that one could literally devote a complete volume to just these instances. It is impossible within the context of this anthology to properly represent all the violations of traditional academic norms, not to mention fairness itself, involved in these cases. Many would seem impossible to believe if they hadn't actually happened. All of them are tragic.

Included here are three individuals whose reflections illustrate both the depth of the problem and the diversity of circumstances that lead to dismissal or resignation.

Frances Widdowson had been a tenured associate professor in the Department of Economics, Justice, and Policy Studies at Mt. Royal University in Calgary, Alberta. The initial offenses that raised the specter of dismissal included the temerity to question both the process of "indigenization" at her university and the empirical evidence behind the claims of genocide associated with the supposed site of over two hundred unmarked graves. Ultimately, her questioning of whether "Indigenous Knowledge" was competitive with modern astronomical knowledge led to claims that she was a racist who had created a hostile environment for indigenous faculty and students and, thus, a threat to their safety.

Carole Hooven was a beloved lecturer at Harvard who made the mistake of appearing on Fox News to state a biological truth: sex is binary. For that, she was ultimately hounded out of her department, a loss for all future students at the university.

Finally, Christian Ott, a distinguished astrophysicist at Caltech, was forced to resign his tenured position, not for what he said, but for what others claimed about what he did. What makes his case even more poignant is not just the lack of due process and fairness associated with what happened to him at Caltech—typical of so many recent claims of inappropriate behavior by male faculty at universities—but also the subsequent efforts of numerous activists to distort his case to try and ensure that he could never carry out his research at any subsequent institution. His case illustrates a sad but important truth: universities and research institutions at the current time do not act in the interests of their faculty or students or even in the interests of truth when adjudicating such cases but are motivated primarily by what they perceive to be their own self-interest. In retrospect, this is perhaps not surprising, but it is nevertheless a sad reflection on the current state of higher education administrations.

Indigenization Destroys Academic Values: The University of Lethbridge Case

FRANCES WIDDOWSON

On February 1, 2023, I stepped out of the elevator next to The Atrium at the University of Lethbridge and was amazed to find myself in the midst of a crowd of seven hundred people filling the space to the rafters. Many in the crowd were protesters, wearing orange shirts adorned with "Every Child Matters." Placards with slogans like "RACISM IS NOT FREE SPEECH" were ubiquitous.

For people unaware of developments in Canadian politics, the orange shirts were about the contentious issue of indigenous residential schools—a system of boarding schools in existence for about one hundred years. The contentiousness became particularly pronounced on May 27, 2021, when there was a press release by an indigenous group announcing that "the remains of 215 children" had been "confirmed" at the Kamloops Indian Residential School [KIRS]. In spite of the fact that no evidence had been provided for this claim, the country was flung into a state of mourning for five months with flags flown at half-mast. The hysteria that ensued also led to the imposition of the belief that the residential schools were "genocidal" and that any questioning of this amounted to "residential school denialism" (which was equated with denying the Holocaust).

The slogan "RACISM IS NOT FREE SPEECH" was referring to me. This was how a number of professors, students, indigenous activists, and their "allies" at the University of Lethbridge were characterizing the talk that Paul Viminitz, a professor in the philosophy department, had invited me to give. The talk—"How 'Woke-ism' Threatens Academic Freedom"—had its university space rescinded after administrators received angry emails from powerful university "stakeholders." I showed up anyway to give my talk in a public space to defend academic values.

It should be of no surprise that my talk was shut down, as activists often are not interested in the academic value of actually understanding and analyzing the arguments they oppose. What was more surprising were the revelations that emerged later that showed a university completely estranged from its intellectual purpose. These circumstances came to light because of the court challenge that I was mounting against the University of Lethbridge with Jonah Pickle, a student who wanted to listen to my talk. As a result of this court action, which maintains that the University of Lethbridge violated our freedom of expression rights, 260 pages of emails were released by the university.[1] These emails exposed that a great deal of this academic disintegration was due to the institution's official adoption of what has been called "indigenization."

What Are "Indigenization" and "Decolonization?"

While "indigenization" and "decolonization" often appear in statements together, they refer to different processes. Indigenization refers to *adding* indigenous culture to all aspects of campus life. The emphasis is on making campuses more "inclusive" of the indigenous population.

Indigenization began in earnest with the acceleration of diversity, inclusion, and equity initiatives in universities about fifteen years ago. At first, this seemed innocuous in that it was perceived as recognizing the inequitable results of the discrimination that indigenous peoples had experienced historically. Indigenization, however, became contrary to academic values when demands for equity and inclusivity resulted in standards being lowered for indigenous students and faculty so that their numbers could be artificially increased.

This interpretation of equity and inclusion became even more destructive to academic values when it stressed "belonging," where the "adding" function began to concern itself with *removing* elements that made indigenous people feel alienated and less likely to thrive. This is when indigenization was turned into "decolonization." Instead of leaving the open inquiry and critical thinking aspects of universities intact, it was now being asserted that these elements made indigenous people feel that they were "unsafe" and didn't belong.

In the case of the University of Lethbridge, Indigenization began much earlier than in other universities with the development of its Native American Studies program in the 1970s.[2] Headed by an indigenous activist with a law degree, Leroy Little Bear, the department was formed on the basis of consultations with indigenous communities. In 2011, Little Bear was asked by President Mahon to be involved in developing a strategy to increase indigenous participation in the institution. Recommendations stressed the importance of incorporating traditional cultural features into indigenous educational experiences.

Other indigenization initiatives followed, leading to the compilation of the *Blackfoot and Other Indigenous Peoples Handbook*. This handbook developed the following official "Territorial Statement" for the university:

> Our University's Blackfoot name is Iniskim, meaning Sacred Buffalo Stone. The University is located in traditional Blackfoot Confederacy territory. We honour the Blackfoot people and their traditional ways of knowing in caring for this land, as well as all Aboriginal peoples who have helped shape and continue to strengthen our University community.

The invocations of the supernatural hinted at the threat to Enlightenment values that was to come, but the fact that these events concerned indigenous people usually discouraged those with private concerns from publicly objecting. Indigenization's focus on including indigenous cultural features, however, began to morph into decolonization with the release of the Truth and Reconciliation Commission's (TRC's) report in 2015. Particularly relevant was the TRC's statement that "[t]he education system itself *must* be transformed into one that rejects the racism embedded in colonial systems of education and treats Aboriginal and Euro-Canadian knowledge systems with equal respect [emphasis added]."[3]

As a result of accepting the TRC's advocacy, university administrators began to see including indigenous "knowledge systems" and "ways of knowing" as one of their primary goals. This goal led Mahon to invite Leroy Little Bear to give a talk on February 8, 2016, where he argued that universities should move beyond the "Age of Reason" and embrace subjectivity and the Animistic spiritual beliefs of indigenous peoples. The release of the TRC's report also resulted in a renewed emphasis on "building relationships" with indigenous communities. The most significant development in this regard occurred in December 2019, when the Mastercard Foundation invested $15 million in the university to provide educational support and economic opportunities for indigenous students. This "building relationships" aspect institutionalized activism within the university, which ended up corroding its academic mission.

Indigenization's Challenge to the Enlightenment

When my public lecture was announced on social media, a number of professors sent emails to the administration denouncing me, with many demanding that the talk be canceled. Most significantly, the Department of Indigenous Studies put out a "department statement on the February 1st lecture" that I was scheduled to give.[4] The "department statement" castigated my "anti-Indigenous rhetoric." It then referred to the University of Lethbridge's territorial acknowledgment to delegitimize my visit. In response to the acknowledgment's claim that "We honour the Blackfoot people and their traditional ways of knowing in caring for this land," the indigenous studies department argued that

> This honoring must include a commitment from all faculty to ensure that Indigenous histories, cultures, memories, and lives, past and present, are represented faithfully, truthfully, and safely, on this campus.... It must be a commitment from all faculty to vigorously reject ideologies which continue to propagate violence against Indigenous Peoples through the rhetoric of historical erasure, dismissal,

diminishment, and dehumanization, such as that espoused by Dr. Widdowson.

This department statement was similar to the emails that were sent to administrators by fifteen professors. These emails essentially put forward two arguments as to why my talk should be canceled. The first was that my work was not scholarly, and therefore, the invitation could not be justified on the basis of references to academic freedom. The second was that there were higher values that trumped any academic discussion that might take place. These were values such as preventing psychological harm, prohibiting "hate speech," and the need to "build and be in good relation with Blackfoot and other Indigenous colleagues and staff on our campus."

Although not much space was devoted to refuting my work—to be discussed further below—there were numerous references to the non-academic values that my talk would undermine. The most comprehensive arguments in this regard appear in the correspondence and academic work of the University of Lethbridge's women and gender studies professor, Caroline Hodes. Unlike other professors, who wrote one email or posted a disparaging remark on Twitter, Hodes sent several emails and even wrote up her arguments in the journal appropriately named *Feminist Asylum* (as Hodes identifies as "neurodiverse").[5]

Although Hodes argued that I should not be granted official space because of the unspecified claim that I was "known to distort the facts," a large fraction of her discussion concerned the claim that my views created a "psychologically unsafe environment" and constituted "hate speech":

> Residential school denial is genocide denial and this is a violation of Indigenous rights. This is not a form of speech protected by either the Alberta Human Rights Act or academic freedom as understood by our collective agreement or our Harassment and Discrimination Policy. Both Drs. Widdowson and Viminitz are engaging in forms of hate speech. The university itself may not have invited this speaker but the university is responsible under the Occupational Health and Safety Act to provide an environment

free of psychological hazards for faculty, staff, and students. Dr. Viminitz has created a psychologically unsafe environment on campus not only through his own conduct but by inviting this speaker. I need not remind you that many students, staff and faculty at the University of Lethbridge are residential school survivors whether through direct experience or intergenerational trauma. This is a betrayal of the University of Lethbridge's commitment to Indigenization and reconciliation. I hope that this event is cancelled and that Dr. Viminitz and the department of philosophy are held accountable. Cancelling this event must also be accompanied by appropriate disciplinary action.

Taking the lead from professors like Caroline Hodes, the University of Lethbridge Faculty Association (ULFA) referred to my appearance as a "harmful controversy."[6] It went on to encourage members "to call out and counter unwelcome and hurtful speech and to protect students, colleagues, and community members." It also praised efforts to "call out work we consider to be of poor quality, misleading, or otherwise substandard" and praised the "admirable efforts" of those who were trying "to protect those who are harmed by careless and dehumanizing speech."

While ULFA was obviously not acting neutrally in this dispute between Viminitz and his activist colleagues, it, at least, didn't call for the cancellation of my talk. This was not the case with respect to the president of the University of Lethbridge Students' Union (ULSU). On January 29, President Kairvee Bhatt wrote a letter to President Mahon stating that "the ULSU governance body, consisting of twenty-one, voting, elected, student council members of the General Assembly condemn the hate that is being perpetuated by the speaker, and in turn, the platform that is being provided for them to continue it, by the organizers." Although recognizing the importance of "freedom of speech, academic freedom, and freedom of expression" at a university, she maintained that "there is a responsibility from academics to ensure that a safe and productive environment for such is created."

The Corporatization of the University of Lethbridge Administration

This opposition contributed to the decision made by a number of administrators to cancel my talk. After putting in place a strategy to counter the "backlash" that was expected from the Alberta government for violating its free speech directive, President Mahon released a statement officially rescinding space on January 30. This statement asserted that "it is clear that the harm associated with this talk is an impediment to meaningful reconciliation."

Although faculty and student opposition were significant in this reversal, the role played by activists was particularly pronounced. The University of Lethbridge had hired a number of them as administrators, who then served as a conduit for community pressure. These included vice-provost positions such as "Iniskim-Indigenous Relations" (staffed by Leroy Little Bear), "Equity, Diversity and Inclusion" (the position of Martha Mathurin-Moe), and "Students" (Kathleen Massey presiding). Activism also had been institutionalized through initiatives such as the Mastercard EleV program.

The vice-provost administrator-activists were intensely involved in the flurry of emails exchanged when my talk was announced. Martha Mathurin-Moe, for example, talked about the need to "manage the narrative" and respond to a "difficult situation." Replying to numerous emails, she reassured people that "[w]e as an institution are committed to continue the work to respond to the calls to actions for Truth and Reconciliation and the principles of Equity, Diversity, and Inclusion in a meaningful and respectful way." Nowhere was there a robust defense of the academic values of the university.

Of all the administrator-activists, however, the most significant by far was Leroy Little Bear. He was particularly influential through his engagement with the community activists hired as part of the Mastercard EleV program. When the navigators received President Mahon's original statement that my talk would go ahead, for example, one of the administrator-activists, "Education Navigator: Siksika" Leroy Wolf Collar, sent an email to the members of the EleV program, "wondering where the University's heart truly sits with their relationship with Indigenous peoples." After relaying the extensive efforts that the University of Lethbridge had made to include indigenous people, he lamented that this relationship was now threatened because my talk would "[bring] harm to the Indigenous students, employees, elders at the University and their relationship to the Blackfoot Confederacy (Kainai, Piikani and Siksika) [emphasis in original]." As a result of this opposition, Little Bear met with Wolf Collar and the other "Navigators" and relayed their concerns to the administration.

All of this pressure was not resisted by President Mahon, who had built his career on pandering to indigenous activists. His most cringe-worthy moment occurred after he received a number of messages from activists demanding that the talk be canceled. In response, he told fellow administrators that it "seem[ed] heartless" to "balance...freedom of speech and respecting and supporting our indigenous and BIPOC communities...," thanking them for giving him a "safe space" in which to "express [his] sorrow." He then beclowned himself further when he congratulated the censorious mob that shut down my speech "for conducting themselves in such a peaceful and powerful manner."

Indigenization's Destruction of Academic Values

All of these references to "safety," "harm," "hate," and "building relationships" are intended to obscure the destruction of academic values at the University of Lethbridge. Canadian courts have recognized that academic freedom is a fundamental principle of universities, as this allows the "free and fearless search for knowledge and the propagation of ideas."[7]

As we saw from the earlier discussion involving Caroline Hodes, a position exists at the University of Lethbridge that canceling my talk actually helped to *protect* the university's academic mission. This point, in fact, was made by a number of professors who maintained that my work was unscholarly and, therefore, allowing it to proceed would lower the quality of academic discourse at the university. Sociologist Suzanne Lenon, for example, argued that "Dr. Widdowson's work is not evidence based, rather it is based on ideologies that erase, dismiss and diminish Indigenous peoples."

This was also the position of anthropologist Patrick Wilson, who stated that "people who forward the kinds of positions of the invited speaker are not capable to intellectual engagement, because they do not work in an evidence-based world, therefore, these kinds of talks also do not contribute to vigorous debate around ideas, rather they are disinformation campaigns that do nothing except contribute to intolerance and harm."

What was omitted by these professors, however, was an academic discussion of what aspects of my work were not "evidence-based." The indigenous studies "department statement," however, did specify two of my claims to which it objected—that I rejected the "Truth and Reconciliation's claim that the residential schools were genocidal" and I questioned "the veracity of the unmarked graves."

With respect to my opposition to the use of the word "genocide" to refer to residential schools, it should be noted, in the interest of "truth," that the TRC did not claim that the residential schools were "genocidal." Instead, it stated that they were a manifestation of "cultural genocide," which is not a legal term and refers to circumstances whereby indigenous people were prevented from speaking their languages, converted to Christianity, and had various customs suppressed.

The question of whether the residential schools were actually genocidal is a contentious one in the discipline of history. Two pieces of evidence have been used to support this "truth"—that Canadian politicians and the Pope say that the residential schools were genocidal and that many indigenous children died in the schools. But arguments from authority are not evidence, and child mortality is not the same as showing a concerted effort to exterminate the indigenous population. J.R. Miller, the most significant historian on the residential schools, for example, claims that they were not genocidal because attempts to assimilate are actually contrary to a planned extermination.[8]

The reference to me questioning the "veracity of the unmarked graves," on the other hand, is a distortion of my views. I acknowledge the "veracity" of many unmarked graves; these are the graves in cemeteries that are now "unmarked" because the markers have deteriorated. What I am taking issue with are claims about clandestine burials, and, in particular, those declared to have been found in the Kamloops case. As I have documented extensively, based on the research of Nina Green, there is no evidence of clandestine burials. It is, in fact, highly improbable that there are graves in this case, as not one parent is claiming that their child "went missing." All that we have are GPR "targets of interest" or "anomalies" and the memories of what are called "Knowledge Keepers." The former are most likely the "discovery" of the septic tiles laid in the 1920s, and the latter repeatedly have been shown to be unreliable and subject to implanted memories.

What Is to be Done: Honesty, Not Placation, Is Needed

In order to combat Indigenization's pernicious threat to academic values, a number of changes can be made. The easiest is the implementation of a policy of institutional neutrality. This was proposed decades ago with the University of Chicago's Kalven Report. The report argued that "a good university, like Socrates, will be upsetting" and that official support for a political cause might discourage members of the university community from critically analyzing it.[9]

Reinstating institutional neutrality is important, but it doesn't address the widespread problems with governance that exist in a corporatized university. In such a structure, a university's "brand" is all-important, and an army of administrators is deployed to protect it. In some universities, administrators are elected, and this prevents them from being captured by corporate goals. In the current "woke" climate, however, asserting academic control over universities is difficult, especially because unions like ULFA are firmly ensconced in the corporate structure and are hostile to academic values.

Changes in governance that fight back against the corporatization of universities, however, will not address the major obstacle to protecting academic values. This is the cultural problem that now exists in universities. As we saw at the University of Lethbridge, many professors and student leaders believe that there are higher considerations than academic values at their institution. This needs to be changed and can occur only with widespread pressure demanding that universities be restored to places of open inquiry.

No one is denying that indigenous peoples have suffered terribly and have endured a litany of abuses. But recognizing this circumstance does not mean that the residential schools were genocidal or that 215 children have been buried clandestinely in the apple orchard at KIRS. Indigenous people deserve the truth just like everyone else, but the actions undertaken at the University of Lethbridge have prevented an accurate historical assessment from emerging. This not only destroys the academic values of this institution; it also deprives us of the common understanding needed for actual reconciliation.

Why I Left Harvard

CAROLE HOOVEN

(Reprinted with permission from the *Free Press*, January 17, 2024)

Since early December, the end of my twenty-year career teaching at Harvard has been the subject of articles,[1] op-eds,[2] tweets[3] from a billionaire, and even a congressional hearing. I have become a poster child for how the growing campus DEI—diversity, equity, and inclusion—bureaucracies strangle free speech. My ordeal has been used to illustrate the hypocrisy[4] of the assertions by Harvard's leaders that they honor the robust exchange of challenging ideas.

What happened to me, and[5] others,[6] strongly suggests that these assertions aren't true—at least, if those ideas oppose campus orthodoxy.

To be a central example of what has gone wrong in higher education feels surreal. If there is any silver lining to losing the career that I found so fulfilling, perhaps it's that my story will help explain the fear that stalks campuses, a fear that spreads every time someone is punished for their speech.

The congressional hearing[7] on December 5, 2023, on the rise of antisemitism at colleges did not go well for the presidents of Harvard, MIT, and the University of Pennsylvania. They were accused of failing to condemn the public antisemitic statements made on their campuses. Their defense, as asserted by then-Harvard president Claudine Gay, was that their administrations were "deeply committed to free expression."

That's where I came in.

As Rep. Tim Walberg (R-MI) said[8] at the hearing, "Carole Hooven, an evolutionary biologist, was forced to resign, because she stated that a person's sex is biological and binary…and so, President Gay, in what world is a call for violence against Jews protected speech, but a belief that sex is biological and binary isn't?"

The world Walberg asks about is that of our colleges and universities, particularly elite ones. While the stated aims of DEI may have been laudable, in practice, DEI culture allows the recasting of certain ideas as "dangerous" or "harmful," which squashes viewpoint diversity and the open, vigorous debate that should be at the heart of a thriving institution of higher education. So, while I was not "forced" to resign, Harvard's culture of intolerance—particularly toward my scientific views on the nature of sex—led me to feel that my only choice was to leave.

At the December congressional hearing, Claudine Gay's response to Rep. Walberg's question was a soundbite that severely distorted the truth:

> From the moment that our students arrive on campus, whether it is to begin their Harvard journey as an undergraduate or at one of the professional schools, the message to them is clear—that we are an inclusive community but one deeply committed to free expression.

Today, Harvard's commitment to free expression is anything but clear. Harvard came in dead last in the Foundation for Individual Rights and Expression's free speech rankings, which detailed[9] some of Harvard's many violations.

But Harvard—and American academia—wasn't always this way.

In 1998, before I started studies for my PhD at Harvard, I spent nearly a year doing field research[10] in Uganda on the behavior of chimpanzees. In retrospect, this work turned out to be useful in understanding dominance hierarchies at the school.

After earning my PhD in 2004, I was fortunate enough to get an appointment as a lecturer in my department, Human Evolutionary Biology (HEB). This allowed me to indulge two of my passions: teaching undergraduates and behavioral endocrinology—the study of connections between hormones and behavior.

I ended up in the unusual position of not being tenured faculty (so, relatively low in the Harvard hierarchy) but having a contract that ran indefinitely. This was possible because I also had an administrative job as the codirector of HEB undergraduate studies. I won many awards for teaching and advising and was repeatedly voted one of the graduating seniors' "Favorite Professors." My Hormones and Behavior[11] class was named to. *The Harvard Crimson*'s "top ten tried and true."

In 2019, I took an unpaid year off to write my book, *T: The Story of Testosterone, the Hormone that Dominates and Divides Us*.[12] It received excellent reviews from the popular press[13] and academic[14] journals.

At the end of July 2021, I made my first live TV appearance,[15] on the *Fox and Friends* show on Fox News. I was invited to comment on an article[16] in the *Free Press* by Katie Herzog, in which I'd been quoted. She reported that medical school professors were backing away from using clear scientific terms such as *male*, *female*, and *pregnant woman*, largely in response to student complaints. I said I thought this trend was a big mistake.

In the brief segment on Fox, my troubles began when I described how biologists define *male* and *female* and argued that these are invaluable terms that science educators, in particular, should not relinquish in response to pressure from ideologues. I emphasized that "understanding the facts about biology doesn't prevent us from treating people with respect." We can, I said, "respect their gender identities and use their preferred pronouns."

I also mentioned that educators are increasingly self-censoring for fear that using the "wrong" language can result in being shunned or even fired.

Some of this censorship comes as a result of the growing DEI complex[17] that has a strong foothold in so many institutions. At Harvard, there are "central" DEI offices[18] run by professional staff, headed up by the chief diversity and inclusion officer (CDIO), whose mission[19] is to create "a campus climate that is welcoming, inclusive, respectful, and free from bias and harassment." There are also many departmental DEI "committees"[20] or "task forces" that are typically staffed by faculty, staff, and grad students.

These committees have a profound influence on department culture and on matters ranging from who should give a talk, what is taught, and even who is hired. All this reaches into daily academic life and gives enormous power to graduate students.

The director of our task force was a graduate student; I'll just use her first name, Laura. Shortly after my appearance on Fox, I learned of a tweet[21] from her that read, "As the Director of the Diversity and Inclusion Task Force for my dept @HarvardHEB, I am appalled and frustrated by the transphobic and harmful remarks made by a member of my dept in this interview with Fox and Friends." Attached to the tweet was a clip of my *Fox and Friends* appearance.

I was shocked and distressed to see this public attack on my character, especially from someone representing herself as speaking on behalf of the university and my department. In response, I quote-tweeted[22] her tweet: "I appreciate your concerns. Could you let me (and the Twitterverse) know exactly what I said that you consider transphobic or harmful to undergrads? I think you know that I care deeply about all of my students, and I also care about science. How about a discussion? @HarvardHEB."

My quote-tweet brought a lot of attention to her original tweet, mostly in support[23] of me, including from people with large platforms like journalist Glenn Greenwald.[24] Laura also received supportive[25] tweets, some that attacked[26] me.

Looking back, I can see how naive and unprepared I was for what followed.

Our Twitter interaction received media coverage not only from within Harvard, but in international news outlets such as the *Daily Mail*,[27] the *New York Post*,[28] and the *Australian*,[29] often with clickbait headlines: "Harvard professor who refuses to use the term 'pregnant people' and insists on 'woman' is accused of transphobia by her woke Ivy League colleague." (I had said nothing about these terms.) Current[30] and previous members of my department publicly accused[31] me on social media of intentionally harming Laura, "punching down," and threatening the safety[32] of the "community," particularly undergraduates.

Then, an article[33] appeared in *The Harvard Crimson* about the imbroglio. In it, Laura told the paper, "I also want to reiterate that I respect Carole as both a scientist and valuable member of our department." When asked for comment, I said, "We can be caring and sensitive to the needs and identities of everyone, while also sticking to biological reality."

But the article also linked to[34] a "Statement of Harvard Graduate Student Union Members in Solidarity With Laura [L.]" It claimed that Laura had suffered a "multi-day deluge of personal harassment, racist abuse, and threats of physical harm after her response was amplified by [Hooven]." But my department administrators told me they could not publicly support me because the "optics" of the situation prevented it.

Over the next few weeks, things quieted down, and my fears that my reputation was permanently damaged began to abate. Then, I got an email forwarded to me by a faculty friend in another department. It was sent by the department chair (who was on that department's Diversity, Inclusion, and Belonging committee). When I read the first few lines, I knew my troubles were far from over: the department's "Diversity, Inclusion, and Belonging Committee wants to express our unwavering support for transgender individuals, in our community and everywhere.... We would like to share with you the following commentary from a member of our community on the topic of this week's public lecture."

The "commentary" was a complaint from a graduate student (who happened to be a transgender woman on their DIB committee), objecting to the fact that the department had advertised a talk[35] I was to give on campus discussing my book.

The student wrote: "Carole Hooven, one of the speakers, has had some history of speaking against the interests of transgender and gender diverse people—including a notable recent interview in Fox News.... It is hard to imagine how these sorts of comments are supposed to foster a respectful, academic dialogue around transgender identities and experiences."

The complaint then went on to list my other "offenses," based on quotes taken out of context from podcasts and newspaper articles. It ended with a notice that there would be a "brave space for discussion" after the talk.

While campus community members are entitled to publicly disagree with each other, university administrators (or those representing themselves as administrators) should not publicly disparage the character of those with whom they disagree; doing so damages reputations and inhibits free expression on campus. I knew that this was not going to blow over unless those in charge in my own department, or perhaps the dean of the Faculty of Arts and Sciences—Claudine Gay—spoke out on my behalf. Not to weigh in on the accuracy of my statements, but to strongly defend my right to make them and to reiterate the collegial norms of campus discourse.

So, I was happy to see an email to me from the dean of science offering to meet. He wrote that he was sad about my "turmoil" and that he wanted to explore how we could "foster a climate of collegial intellectual discussion." I should have read the email more carefully. He spoke only about establishing new task forces and initiatives. I eventually interrupted to ask about what he was going to do to address *my situation*. In response, I got a stern lecture on not interrupting. That felt like the end of the line.

But my hopes were again raised when an administrator shared a draft of an email that included this robust defense (and a reference to my place in the Harvard hierarchy): "[The complaint letter] made serious accusations against a much beloved and respected staff member and lecturer in my department, Dr. Carole Hooven…the person hurt by this was not some renowned professor, protected by academic tenure."

I wrote back: "To have you standing up for me and seriously, standing up for science, means the world to me and helped me feel more secure and supported in the work I do for the department and for Harvard.… This insane narrative of my work is being created that has no basis in reality and it is being perpetuated by university administration. And this is appalling. Your strong support means the world to me."

In the end, however, that email absolving me of any wrongdoing was not sent. Another was sent instead. I'd also seen a draft of that one (written by a committee of at least five administrators, including a DEI dean, and cosigned by two of them) and pleaded against it ever being sent.

It contained sentences such as: "It was not the intent of our DIB Committee to cause damage to Dr. Hooven's reputation, but to raise important issues about best practices in science communication and respect for the serious impact our scientific opinions may have on others. We apologize." As an apology, it failed. The clear implication, now broadcast to hundreds of people in my own and the other department, was that I had failed to respect the "serious impact" of my "scientific opinions." It was the opposite of what I'd hoped for.

I just wanted to disappear—and that's pretty much what I did. When I asked the faculty member with whom I worked most closely (also a good friend) for help, explaining that I was concerned about my mental health, he told me to "put your head down, grit your teeth, and be professional." I got the message that I was to just drop it and move on. I did not, and I lost my old friend.

Eventually, I was diagnosed with severe major depression, which included intrusive, persistent, and unwanted suicidal thoughts. I began seeing a therapist and was prescribed medication.

I had someone take over my administrative job and reduced my employment to half-time. This would enable me to continue teaching my beloved Hormones and Behavior course. But for the first time ever, in the spring of 2022, no graduate students would agree to serve as teaching fellows for my course (which typically had sixty to ninety students). And without a high-status position, and a research lab and grad students of my own, there wasn't anything I could do about it.

So, instead, the last class I taught at Harvard (or anywhere) was a small undergraduate seminar. It was a great success,[36] and it received the highest possible course evaluation score from the students.

As of January 2023, although I assumed I'd remain in my job until some ripe old age (I'm fifty-seven), I retired from Harvard. When I made it clear to the powers that be in my department that I felt I had no choice but to leave due to the lack of public support, nobody stepped up to provide it. People were behaving in ways that were invited[37] by the culture of DEI. (This was particularly true of Laura, the graduate student immersed in that culture, and I do not blame her.) Even if they knew the right thing to do, especially for those at the top, there was just too much to lose.

A few brave, compassionate faculty members reached out with support, and I'm indebted to them. I am especially thankful to psychology professor Steven Pinker, who has made it possible for me to have an (unpaid) associate position in his department. And my case was an impetus for the formation of the Council on Academic Freedom at Harvard.[38] Our focus is to promote "free inquiry, intellectual diversity, and civil discourse at America's oldest university." I'm an active member.

As a sign of the political polarization that characterizes the US today, my supporters have tended to come from the Right—although I am a lifelong Democrat. I was happy to accept a position as a senior fellow[39] at the center-right American Enterprise Institute, where lively debate reigns.

The Harvard motto is *Veritas*—truth. But the truth is that the message that members of the Harvard community receive every day—in emails, trainings, posters, pamphlets, and meetings—concerns DEI. The message is that what matters most, certainly above the search for truth, is how people's words affect groups deemed "marginalized."

I care about ensuring that everyone feels welcome at Harvard and believe that diversity on campus strengthens the institution. But this must not come at a cost to free speech. Harvard has a long way to go[40] to restore an environment where those on its campus feel free to teach, write, and speak without fear. Such a culture supports knowledge production, dissemination, and preservation—the true mission of the university.

Delicious Moral Treats

CHRISTIAN D. OTT

"The surest way to work up a crusade in favor of some good cause is to promise people they will have a chance of maltreating someone. To be able to destroy with good conscience, to be able to behave badly and call your bad behavior 'righteous indignation'—this is the height of psychological luxury, the most delicious of moral treats."

Aldous Huxley, *Crome Yellow*

I was a tenured professor of astrophysics at the California Institute of Technology (Caltech), a private nonprofit corporation and accredited institution of higher learning in Pasadena, California. My research focused on computational simulations of stellar explosions and collisions of double black holes and neutron stars. I was also part of the LIGO team that made the first direct observation of gravitational waves in 2015.

In September 2015, I was found responsible for having violated Caltech's policies and Title IX. I was put on administrative leave without pay for an academic year, which was extended by another year with pay. I was the subject of BuzzFeed News articles, and I was pilloried on social media. I ultimately resigned my tenured position voluntarily on December 31, 2017.

This contribution is an account of my cancellation and of what I've learned from it. As is the case for all cancellations of academics, my case is full of complicated details and nuance but is generally reduced by outsiders to what is sensational and misleading.

This is very personal, and it's been hard to write. It has now been more than a decade since the events occurred that ultimately led to my cancellation. It's time that I write about this phase of my life and the events that upended my career. My experience has been nothing short of traumatic, and all of it has had a huge impact on many people in my life and, in particular, on my wife, who stuck with me through these hard times. Writing about it is essential processing.

Backstory

I've always been ambitious, hardworking, and unafraid of obstacles. I was raised by postwar-generation German parents who owned a small business. My upbringing was about finishing what you started, being reliable, trustworthy, and truthful. Nothing could be half-assed, and judgment was harsh but loving. I was a freckled ginger super nerd in school, and I suffered from bullying. This hardened me emotionally and made me determined to achieve what the bullies weren't even capable of dreaming of.

I received my undergraduate degree from Heidelberg University. As a graduate student at the Max Planck Institute for Gravitational Physics in Potsdam, I relentlessly pursued my research goals and fought for the resources I needed. My doctorate advisor had left for another institution, so I was working independently with him and a set of local and remote collaborators. When a new research group leader was brought in, I was not afraid or shy to challenge him when he tried to curtail my resource access (in my case, access to the local supercomputer) or meddle with my research goals. I was thoroughly unafraid of his authority. In fact, I've never been afraid of authority figures and have always known that you always have recourse, certainly in academia. If a professor treats you poorly or unfairly, you have a multitude of options to complain and be heard. You're always protected by the institution and the broader field. That's certainly true anywhere in North America and Europe.

In 2009, two and a half years after I had received my doctorate, I was appointed assistant professor at Caltech and began building my research group while also inheriting a large research effort when a senior leader retired. I got exceptionally lucky with my first students and postdoctoral researchers. We were productive and successful as a team.

I ran into problems when my own research team grew so much that I needed interpersonal and managerial skills I didn't possess. At the time, my approach to managing team members was to simply project my own motivation and goals onto them, holding them to the same punishingly high standards I held myself to and demanding dedication and hard work. I lacked the empathy to understand that not everybody functioned like me or had the same goals in life. When a new team member joined, I made a large time and resource investment in them, but then often grew resentful when they didn't return my investment. Because I had invested in them, their success became entangled with my own. This is a relatively common problem for early-career academics who are under high pressure, but this doesn't excuse my managerial failures.

By the fall of 2014, my team had reached critical mass for bad things to happen.

What Did I Do?

My case involved two graduate students as complainants and a third graduate student who was agitating and driving complaints behind the scenes.

Student A joined my research group in the summer of 2012, and I was her formal graduate advisor. Not long after she started working with me, she confided in me that she was dealing with serious mental health issues that impeded her research and coursework. I'm not a mental health professional and certainly wasn't qualified to help. But I tried to help, listen, and accommodate. Student A and I became close and frequently discussed personal issues. Student A relied on me as her psychological support person.

Student B joined Caltech in the same year as student A. She had already published research broadly relevant to my research. I had met her previously at a summer school program at her alma mater and had worked hard to recruit her to Caltech. I wasn't her formal graduate advisor, but she decided to join my research group and carry out research with me for her first year at Caltech. Student B and I got along well. I felt that there was positive chemistry between us. Our interactions sometimes touched on personal topics.

Fast-forward a year. It's now late summer 2013. Student A and I continued to have a positive relationship, and she was making progress. Student B, on the other hand, struggled and appeared to be unenthusiastic about the research she was doing. I sought advice from faculty colleagues about what to do. In the annual faculty meeting in which the research progress of all first-year students is discussed, I got a clear message from my colleagues: Student B should switch research topics and research advisors. This is a rather common thing. Many students in Student B's graduate program switch topics and advisors after their first year. Overall, not a big deal.

However, it was a big deal for me. I had given her a research project that didn't work out. I had worked hard to recruit her, invested time and energy into her research, and was deeply invested in her success. I couldn't make things work out. I felt like a failure. After meeting with her to discuss the decision that we should stop working together, I felt horrible. My emotions were complex. It was a mix of failure, self-pity, and a sense of loss combined with affection for Student B.

I was lost and didn't know how to deal with my feelings. The right thing, of course, would have been to talk to a counselor and friends outside of Caltech. I felt I needed to talk to someone who would understand the situation. In a desperate moment, while I was on foreign travel and feeling horrible about the situation, I confided in Student A about my feelings. Student A became my confidant and helped me deal with my emotions. She consented to all of this.

A Storm Is Brewing

Student C may be the most important actor in my story. She joined my research group in the summer of 2014. What I didn't know at the time was that she was less interested in physics than in political and social activism. She was quite charismatic, a leader, and she knew how to rile up people for a cause. She also had a keen understanding of how university administrations work and how she could use them to get back at people. When some undergrads she was working with as a teaching assistant criticized her teaching, she complained, and the students were investigated. After a breakup with her boyfriend, Student C was so upset with him that she cajoled multiple women into filing complaints against him. The ex-boyfriend had to take Caltech to court. He prevailed.

Student A became fast friends with Student C and a group of other new students and postdocs who had joined my broader research group. This was a positive development for Student A, and she became less dependent on my personal support.

Early 2015 was an intense time for me at Caltech. I had agreed to take on a class for a colleague who just had newborn twins. I was teaching at twice my normal load. I was also managing a growing research group of more than twenty researchers and graduate students. I was literally at my limit. I was stressed out, irritable, and impatient with students and researchers. Student A, in private conversations, called me out for my poor handling of various situations with students and postdocs. A negative feedback cycle developed where various team members of the group around Students A and C were constantly offended by something I said or by a decision I made. This went so far that we stopped having team meetings since they had become so contentious. Eventually, both Student A and Student C decided to quit my research group. They started working with other advisors.

But they weren't done. Student A had told Student C about our relationship and, importantly, that I had confided in her about my feelings regarding Student B in late 2013. As I learned years later, Student C, the activist and expert complainant, talked Student A into reaching out to Student B to jointly file complaints against me with Caltech's administration.

The Investigation

On June 8, 2015, I was turning on my phone after landing at LAX from a trip to Europe when I received an unexpected email from my division chair. I was to appear in his office for a meeting the next day. I replied, inquiring how I should prepare. The response: "No preparation necessary." The understatement of the century.

I went to the chair's office the next morning to be presented with a letter stating that Student A and Student B had filed harassment complaints against me. No details were provided, not even what I was accused of exactly. The two investigators that Caltech had appointed to look into the case were my chair (and direct superior) and an engineering professor and former vice provost. I was told that there was to be a sequence of fact-finding meetings with the investigators. I wasn't allowed to bring representation. I was also handed a copy of Caltech's policies, which laid out the rules for the kind of investigation I was about to be experiencing. They would be using the preponderance of evidence standard, there wouldn't be any hearings or cross-examinations, and, importantly, no audio recordings of fact-finding meetings with investigators were allowed. I wouldn't be able to access the evidence against me or review the investigation report.

The one right thing I did right away was to get a lawyer experienced with cases like mine. My lawyer was most experienced with student-on-student complaints since they were much more common. I didn't see this as a problem at the time. The first thing I learned from my lawyer was that Caltech couldn't prevent me from having him present at the fact-finding meetings, but that he couldn't actively engage during the meetings.

Naive as I was, I believed that the investigators would seek the facts and base their findings on the facts. Based on the facts that I was aware of, I hadn't violated Caltech's policies. With my lawyer's endorsement, I decided to be completely transparent. I provided the investigators with all emails, instant message chats, and other communications I had had with the students since 2012. I also compiled and provided to the investigators a total of six extensive reports detailing my interactions and relationships with the students.

I met with the investigators four times. In the first meeting, they verbally recounted the complaints to me. I never received the complaints in writing. My best recollection from the meeting (in my notes from early July 2015) is as follows:

> Student B alleged that I stopped working with her because of my attraction to her and that I had created a harmful

work environment for her by talking poorly about her to colleagues.

Student A alleged that I had treated her (Student A) differently from other students because she was a woman. She further alleged that I treated her as a daughter, told her how to dress, told her that she would not succeed without me, and wanted her to be dependent on me. She alleged I used her as a therapist for my issues regarding Student B against her will and that she felt pressured and obligated to listen and give advice because I had power over her.

My response to these allegations was: I stopped working with Student B because my colleagues recommended I do so. My conversations with Student A about my feelings regarding Student B were welcomed by Student A with enthusiastic consent. I provided a trove of exculpatory evidence regarding Student A's and Student B's allegations.

Given the evidence I had provided, I thought all was good. The investigators would seek the truth and go by the evidence. If I were to be found responsible for violating Caltech's policies, my lawyer thought that in the worst case, I'd get a one-quarter suspension and would then be able to return to Caltech.

Verdict and Appeal

According to Caltech's policy at the time, the investigators submitted a recommendation to the provost, who then determined the outcome and possible sanctions. By Caltech's policy, investigations should conclude within sixty days of the initial complaint. In my case, it took almost ninety days, partly because one of the investigators disappeared for an extended Hawaiian vacation. Finally, in early September 2015, I was summoned to the provost's office to learn about the outcome of the investigation.

I received a two-page decision letter telling me that the provost had determined that I had violated Title IX and Caltech's policies by committing gender-based harassment and discrimination against students A and

B. I was placed on administrative leave without pay for an entire academic year. I was banned from Caltech's campus during that time. I had to work with an executive coach to improve my interactions with students. I was not allowed to interact with students and postdocs unless these meetings were moderated by a designated individual and conducted in a predetermined setting. Caltech also effectively placed me under a gag order, which later made it impossible for me to defend myself when my case became public. A new division chair had just been appointed and was put in charge of supervising my leave and interactions with my students.

My lawyer and I were both shocked by the findings and the punishment. It was hugely disproportionate to what was alleged. My lawyer told me that the last time he had encountered such harsh punishment was in a case in which a University of California professor was having drug-fueled sex parties with his students.

Many of my faculty colleagues, my scientific collaborators, and my research group members were similarly shocked at Caltech's ruling. Due to the secrecy around the investigation, almost everybody who wasn't close to me was completely in the dark about what I had been accused of. Naturally, the severity of the sanctions contributed to the rumors about what I had done.

My thoughts and worries turned to my students and researchers and the many ongoing research projects we were pushing forward. Caltech's sanctions would have a hugely negative impact on their work and their ability to carry out research with me. I pledged to myself to make every effort to minimize negative effects on my team. I accomplished this by continuing to carry out full-time work while not getting paid during my leave. An additional factor was Caltech's nearly complete failure to follow through with the supervision of my interactions with my team. More about this later.

About a week into my administrative leave, my new chair met with me via video conference to debrief me on the investigation report and the provost's decision. Since I wasn't allowed to see the report and also hadn't seen the evidence submitted by the complainants, I was eager to learn about the findings and the underlying evidence. The chair verbally summarized what she called "key aspects of the report." Astonishingly to me, the investigators had come to their conclusions by literally cherry-picking evidence and by blindly adopting hypotheses of postmodernist intersectional social theory.

Here is an example: The truth was that student A was a twenty-three-year-old adult at the time. She consented freely and in writing to having personal conversations with me. In fact, she actively sought them out. In contrast, the investigators claimed that she could not have possibly consented to having private conversations because I was her research advisor and, thus, formally, her superior. Thus, they claimed my power over her rendered her an unwilling victim.

At this point, I was still quite naive about academic institutions and administrators. Clearly, there must have been a mistake. The investigators were academics, after all, and used to being in the pursuit of the truth. How could this have gone so wrong?

I was convinced I could make right the wrong by compiling an appeal that addressed the provost's decision letter and what my chair had told me about the investigators' report. The uphill battle here is to be appreciated: I was appealing a decision based on investigative findings that I wasn't allowed to access or review in writing.

I compiled new evidence. I got a supporting statement from a colleague who had been involved in the faculty decision that Student B should switch to a different advisor and research topic. I reached out to Student A's roommate and partner at the time that Student A and I were having our private conversations. He gave an affidavit stating that Student A enjoyed her private conversations with me and benefited from our student-advisor relationship. I also compiled quantitative evidence by running analyses on the seventeen thousand instant messages that Student A and I exchanged over the years. I showed that only a small fraction of them were about my personal problems and that these conversations were equally often initiated by Student A and by me. I put all this together in a forty-five-page appeal.

The provost had decided on the outcome of the investigation, and now that same provost was tasked with deciding about my appeal too. The conflict of interest was clear. About a month later, the provost responded via email. My appeal was denied. The new evidence I had provided wasn't deemed new evidence, and Student A's boyfriend's affidavit was excluded because Student A had asked Caltech to exclude it. My procedural complaints about the investigation were ignored.

At this point, I had exhausted all my administrative remedies. The next step would have been to sue Caltech. When I was considering my next steps, I listened to advice from my lawyer and from colleagues. The overall opinion was that I should just lay low, accept the unpaid administrative leave, and return to Caltech the next year. After all, they hadn't fired me, and I was still a tenured professor. Suing Caltech would have involved first winning a writ of mandamus, which is an administrative case in which I would have had to prove that Caltech violated its own policies. Only in a second step could I have sued for damages. Given the high cost of two separate court cases and the path to a return to Caltech, I decided not to sue.

Public Shaming

While I wasn't paid by Caltech and had to work from home, the first few months of my administrative leave were bearable. There hadn't yet been any publicity about my case, and while there were rumors, my overall reputation was intact. This was about to change.

Student C and the complainants weren't happy with the outcome and were enraged by my audacity to use my right to appeal the outcome of Caltech's investigation. Instead of appealing themselves, they decided to go the far easier and far more effective route. They went public.

By the fall of 2015, BuzzFeed News had made a name for itself for its coverage of various cases of alleged misconduct and harassment at colleges and universities in the US. In all cases, BuzzFeed's reporting was sensationalized, superficial, and biased towards the perceived victims. BuzzFeed was generating revenue from these sensational stories, which put me squarely in their crosshairs.

The complainants kept Caltech in the loop about their plans. My chair told me that there would be a BuzzFeed article coming out in January 2016. In an effort to get ahead of BuzzFeed, Caltech released a vague statement about the case, saying that a professor had been sanctioned for "unambiguous gender-based harassment of two students." That didn't help at all.

Within hours of the statement, Student A had taken to Twitter to identify me as the professor in Caltech's statement. The BuzzFeed article came out a few days later. To her credit, the BuzzFeed reporter had reached out to me with a bulleted list of things she was going to cover in her article and a request for comment. Caltech's gag order prevented me from defending myself. As expected, the BuzzFeed coverage was one-sided and sensationalized. The academic social media, mostly Twitter and Facebook, was in an outraged feeding frenzy. Rumors and speculations about left-out details were abundant. On Twitter, I was quickly called a predator and even a rapist.

Now that my case was public, the larger scale fallout that is now so familiar to so many canceled colleagues was happening to me. I was uninvited from lectures and events, removed from collaborations and editorial boards, and I couldn't visit colleagues at other US institutions since they followed Caltech's campus ban. Colleagues publicly or privately distanced themselves from me with or without hearing my side of the story. Funding proposals were denied because peer-review panels found me unqualified to lead on the basis of what was published in BuzzFeed and on social media. I developed a dedicated following by a group of activists who went out of their way to track what I was doing. They policed conference programs and social media. For example, when I was attending a conference in Japan, they threatened the organizers, who then removed me from the list of invited speakers.

Initial Administrative Leave

When Caltech came down with its decision and I was first put on leave, I felt that Caltech had improperly sided with the complainants. Caltech's overall message, relayed by their public statement and by a private conversation with my chair, was that their primary concern was the well-being and success of their students. While I thought they made the wrong decision in my case, I generally bought their message.

This changed as my leave became more routine, and I again started interacting with students and postdocs. My chair deputized two of my faculty colleagues to monitor and moderate these interactions. I had to copy the monitors on all my email communications with my team, but that was the only monitoring that happened.

I was free to meet in person or virtually with my students, and none of these interactions were moderated or monitored. My monitors were supposed to follow up with my students on our interactions regularly. Yet, often, many months went by during which the monitors did absolutely nothing. Caltech systematically failed to monitor my student and postdoc interactions.

It now finally dawned on me that all that had occurred wasn't about the students at all and actually not even about me and my poor judgment that exposed me to the complaints. I now had striking evidence that Caltech clearly didn't worry about my interactions with students.

I came to the conclusion, which I hold to this day, that the outcome of my case was determined by what Caltech deemed to be the most favorable outcome for Caltech. In the general media and social landscape of 2015/2016, with the beginnings of #MeToo and an increasing rate of cancellations in academia, it was clearly advantageous for an academic institution's public profile to rule against the faculty member in a case involving student complaints.

Looking back, everything in the investigation was set up so that Caltech could arrive at whatever conclusion it wanted: (1) no access to evidence, (2) no hearing or cross-examination, (3) no recordings of meetings with the investigators, and (4) no access to the investigation report.

The investigation was run by high-ranking Caltech officials, and the provost decided on the outcome. Everyone involved had a tremendous stake in Caltech's reputation and, by definition, had a conflict of interest.

Another Year of Leave

In the spring of 2016, about eight months into my leave, my chair and some Caltech colleagues started bringing up that I should take another year off, in the form of a sabbatical, after the end of my unpaid leave. I wasn't on board. I felt that having to work remotely and not being able to have daily interactions with my team had slowed me down, and I wasn't excited about spending another year away. Caltech didn't like it. I suspect that Caltech was, at the time, already under threat from student activists, most likely led by Student C. In addition, neither Student A nor Student B were graduating in 2016, so my return to campus would create a headache for Caltech since it had vowed to keep us separate.

Student A went on a Twitter rant in April 2016, accusing me of deleting her research data and taking credit for her work. None of it was true. I didn't respond to her directly, but, in a weak moment, I posted on my own Twitter feed a summarized copy of Student A's tweets with commentary. This was a mistake, but it was great for Caltech. They accused me of having violated the terms of my administrative leave and happily extended my leave by another twelve months, though this time with pay. Again, I was discouraged from suing Caltech by colleagues and friends. At least, this time, I received my full salary.

More Protests and No Return to Caltech

The one thing about my extended leave that I thoroughly enjoyed was the work with my executive coach. He was a brilliant communicator who normally gets hired to smooth the edges of aspiring C-level executives in Fortune 500 companies. While there's little that I can do about my natural tendency to be frank with people, my coach gave me tools to improve how I come across and manage people.

My coach also helped me prepare for my scheduled return to Caltech in the summer of 2017. Caltech asked that I prepare a self-assessment and meet with a committee that would provide input to the provost about my readiness to return. My coach prepared me extensively for the meeting, and according to my chair, the meeting went very well. Her "only problem" now was the students (her words). In late April 2017, Caltech was so confident about my return that my chair wrote a letter to inform the National Science Foundation.

The students, that is, Student C and her activist friends, didn't want to see me return to campus. They organized protests and a sit-in, all again sensationally covered by BuzzFeed. All was tolerated by Caltech. July 1, 2017, was the date my administrative leave was set to end, and the students had threatened a large protest should I be allowed to return.

Caltech was now stuck between a rock and a hard place. I had fulfilled all my obligations during my leave, had the green light from my coach, and even a psychological assessment that declared me fit to return to Caltech. I had just spent two academic years away from my research group, and it was very clear to Caltech that I wanted to return. The students, on the other hand, made it very clear to Caltech that they would raise hell if I did.

So, Caltech did what academic institutions nowadays always do in such situations. They made the problem go away. I agreed to resign. Before the students could begin to protest on July 1, Caltech had released a statement that I would resign at the end of 2017 and would work off campus until then.

Students A and B both graduated from Caltech with a PhD a few years later. Student A is now an assistant professor, and Student B works as a manager in the tech industry. Student C left Caltech and is pursuing her PhD at another institution.

The Saga Continues

After the decision to resign from my position at Caltech, I still, very naively, thought that there would be a future for me in science. My chair thought I should be able to get a researcher position at one of the national laboratories in the US. I thought I could try for assistant professor positions and re-prove myself as an academic and teacher. As I should have expected, there was absolutely no interest anywhere in the US.

I was still determined to find a way to continue my research. Science was my life. I started applying for postdoctoral researcher positions in and outside the US. Finally, in late December 2017, I was offered a postdoctoral position at the Nordic Institute for Theoretical Physics (Nordita) in Stockholm, Sweden, under the supervision of a female assistant professor, who was fully aware of the backstory as was Nordita leadership. They were all in agreement that I should be given a second chance. I was elated. I would be working with my new boss and her team on a different aspect of astrophysics than I was used to, but I was ready and eager to learn new things and build new collaborations. I was happy to accept the career setback from a tenured full professor to a two-year term postdoc. I was confident I could rebuild my career.

That didn't last. Activists are everywhere, and they never miss an opportunity to punish someone. Within weeks of accepting the offer from Nordita, its director called me to rescind the offer. He said activists had threatened Nordita with protests should they not cancel my contract. It didn't make any rational sense—my new position didn't involve any teaching or student mentoring. But, as I've had to learn, these things rarely make rational sense. The argument against my presence at Nordita was the common "they are trying to pass the harasser" rhetoric. It didn't matter that I applied and was hired by a female assistant professor. There was no old boys club helping me out. There were no backroom deals.

I was smarter this time around. I immediately lawyered up and resisted. By January 2018, I was in Stockholm, visiting Nordita. Nordita knew that they couldn't legally just dump me, so they, with the help of colleagues at the University of Turku, Finland, came up with a new idea. Instead of at Nordita in Stockholm, I would be hired on a two-year contract as a researcher at the University of Turku, paid, in part, by joint funds between Nordita and Turku. Everybody, including the university rector at Turku, was on board. The thought was that Turku could withstand extreme activists better than Nordita.

I tentatively agreed to the deal, but I was wary about what would happen at Turku. I didn't sign any paperwork in Stockholm. I took the overnight ferry from Stockholm to Turku and signed the employment contract the morning I arrived. As at Nordita, my work wouldn't entail any teaching or supervising of students.

By the time I had signed the contract, articles had already been published in Finnish media about a professor accused of sexual harassment in the US being hired by the University of Turku. The fact that Caltech didn't find me responsible for sexual harassment didn't matter.

In Finland, I met new tormentors yet: Syksy Räsänen and Till Sawala. Both were physics researchers at Helsinki and activists extraordinaire. They organized an online petition against my employment at Turku. They also sent a letter to the University of Turku falsely stating I had committed sexual harassment and was a danger to students and that I had not demonstrated sufficient remorse for my actions to be granted a second chance.

Not only was this false, but it was also totally ridiculous. They made themselves judges and determined that because I had not made any public assumptions of guilt or apologies, I had not learned from my mistakes (which were, of course, also misrepresented by them). It didn't fit into their activist brains that I was not allowed to make public statements at the time and, more importantly, I didn't have any obligation whatsoever to do so about a private workplace matter between me, the complainants, and Caltech.

There was, of course, international social media outcry about the University of Turku trying to hire me. BuzzFeed reported. This went on for a few days, then, as I had expected, the University of Turku caved and canceled my contract. I lawyered up again.

I ended up suing both Nordita/University of Stockholm and the University of Turku. In both cases, I won the contract amounts but didn't win any damages. In Finland, defamation can be tried under criminal law. I filed police reports against Räsänen and Sawala, and the State of Finland did indeed indict and try them. This took many years. Ultimately, in the fall of 2022, four years after their alleged crimes, they were acquitted by a court in Turku.

This was still a victory for me. The leading judge stated in the judgment that they had indeed defamed me in their letter to the University of Turku. In addition, they will never get back the many hours they had spent preparing for their trial and being tried in a Turku criminal courtroom in my presence. As their final victory strike, Räsänen and Sawala reached out to *Science Magazine*, which, of course, happily ran a story about my case and their acquittal in Finnish court. *Science* claimed in their article that I had been found responsible for sexual harassment. One easy legal threat later, they removed that falsehood from their article, naturally, without clarifying the changes made.

Done, but Not Done

After the ordeal in Sweden and then in Finland, I was done trying to resurrect my career as a scientist and educator. I pivoted. Given my extensive computing experience and my broad training as a problem solver, I already practically had all the skills to become a software engineer and data scientist.

To become more formally qualified, I participated in a data analytics certificate program at the University of Southern California (USC). I was accepted as a student there. In order to probe the state of my activist following, I added a USC affiliation to my Twitter profile. Sure enough, without delay, activists took time out of their day to call offices at USC to warn them about me. They weren't shy and promoted their actions on Twitter. USC didn't do anything since I was just a student. It was an interesting experiment that taught me an important lesson. Since then, I never list my current employer anywhere that would be discoverable by my activist followers.

I've since found employment in various software engineering and system operations roles with a few companies over the years. They all knew how to google but hired me because of my abilities and because I've been primarily an individual contributor in those roles and not a manager.

I'm happier now. I have fewer responsibilities and much less stress in my day job while still getting to work on interesting problems. I've never been able to reach the same level of compensation I had at Caltech. I'm making up for this by working additional hours as a consultant and building my own business that provides public disclosures software to government agencies. Growth is slow, but I'm enjoying it, and it's meaningful work in public service.

Learnings and Advice

While thoroughly unpleasant, my experience has made me a better person, and it has taught me much. I underwent two years of counseling and executive coaching during my leave, and I've continued working with counselors to process my experience to this day. It was a devastating loss. I got back up. Here is what I learned.

The vast majority of US academic institutions today are in the business of looking good to donors, funding agencies, students, parents, and the public (probably in this order or close to this order). They will ruthlessly distort the truth and railroad faculty, students, and staff if they believe it is in their favor. Universities don't have a tangible product. They only have their reputation. In the case of internal investigations, they will find the outcome that they believe is most favorable to the institution and not to the complainants or respondents. Their goal isn't the pursuit of truth, but the protection of their reputation and, in turn, of their donor and federal grant revenue streams.

The outcome of a university investigation is accepted blindly by everyone at the level of a court ruling. The vast majority of academics, and the general public, don't know that these university investigations are mostly devoid of due process, conducted by amateurs, and contaminated by a severe conflict of interest.

Because of the corrupt nature of these investigations, I would advise to not collaborate with them or, rather, be exceedingly careful about what information to provide to them and what to tell the investigators. I would demand a copy of all evidence submitted by the complainants and then organize my defense around that evidence, providing only the additional information necessary to prove the case. Never give them more than is absolutely necessary.

Once an academic institution has made a final ruling against you, the only recourse is to sue the institution immediately. There is no other way to ever come back from an institution's ruling. Virtually everybody will accept an institution's ruling as if it had the validity and integrity of a court decision. Unless a court overrules the institution's decision, you will be branded forever a "confirmed harasser." The only way to repair one's reputation is through the legal system. I learned this far too late.

The most important advice is to never resign. Ever. Over the past few years, there have been a substantial number of high-profile figures, mostly men, who refused to resign their positions after some scandal. Most of them succeeded in retaining their positions. The storm can be weathered. Academic institutions have yet to understand this. The activists move on when they get bored and find easier targets for their mean-spirited, hateful power.

In dealing with activist attacks on social media, I've found, perhaps surprisingly, that cease-and-desist letters are effective. These days, I maintain a healthy distance from social media, but when I sign on and find things about me that are clearly wrong, I write a nicely but sternly worded cease-and-desist letter, and thus far, this has always worked for me. The other person usually deleted their post.

You cannot expect any open support from any of your colleagues. They will be paralyzed with their fear of the mob. I was so focused on my research that I essentially had no social life outside of science. I lost all of it except for a few true friends who stuck with me. When going through something like this, you learn who your true friends are.

You cannot expect any nuanced understanding of your case from outsiders. Anything that cannot be condensed into a headline will be ignored. The vast majority of the activists in my case weren't even familiar with the details of what I had done or was accused of. The exceptions are those people who went through a similar experience. I've found community with fellow scientists who went through similar mobbings. Their support, and the support of my wife and family, have been invaluable.

SECTION 4

The Impact of Diversity, Equity, and Inclusion Bureaucracies on Scholarship in Academia and Beyond

The unchecked growth of diversity, equity, and inclusion bureaucracies within academia, industry, and government has become perhaps the biggest threat to academic freedom, free inquiry, merit, and fairness in the West today. To those unfamiliar with the current situation, this may seem paradoxical. After all, these three labels reflect what appear to be noble goals. How could anyone argue that working to achieve them is pernicious?

As the four authors in this section demonstrate, however, these goals have been perverted in an attempt not to ensure equality of access but equality of outcomes while infantilizing scholarship, attacking academic excellence, and generating a climate of suspicion and fear in institutions throughout the Western world. Not to mention the diverting of huge economic resources away from the primary mission of universities—to educate students and support the generation of new knowledge—toward the building of a new policing arm of universities that has grown exponentially like a cancer, brushing aside anything that threatens its growth and continued existence.

Janice Fiamengo begins this section with a personal take: as a former equity hire, she provides a compelling argument about the fundamental challenge to both equality and merit that equity hiring represents and how, in many fields, even when it is not stated explicitly, in practice, males simply need not apply. Beyond this, research initiatives that support DEI goals are rewarded at the expense of supporting research of intrinsic scholarly value.

Next, Geoff Horsman describes his experiences at his own university in Canada with its burgeoning DEI infrastructure and his own efforts to counter them. These ranged from questioning the ethics of creating a position in his own department exclusively for an indigenous scholar (which, as it happened, resulted in the department not being able to recruit a very talented black scholar) in the absence of any evidence for previous racial biases to fighting the requirement of producing DEI oath statements in order to receive federal research grants and the demoralizing effects this has on research efforts.

Alessandro Strumia is a scientist, formerly of CERN in Geneva, whose own research arguing against the empirical basis of discriminatory DEI hiring quotas in physics led to his being first censured, then censored, and then expelled from that organization. His chapter in this volume describes his experience and then his perspective on the broader DEI problem in academia.

Husband and wife academic team at Penn, Roger Cohen and Amy Wax describe what ultimately seems a more harrowing new intrusion of DEI—this time in medicine. Here inserting ideology into education and research can literally have deadly consequences. They argue cogently that the changes proposed in health care in the name of DEI should be subject to the same gold standard of evaluation and scrutiny as the rest of science. A modest and rational proposal, but alas, one that they argue appears to fly in the face of the dominant direction in medicine.

Finally, philosopher Peter Boghossian proposes that the academic crime of plagiarism, which in the past would discredit any scholar, has been overlooked today for high-level academics and administrators at Harvard because of the rise of postmodernist critical social justice activism via the incursion of diversity, equity, and inclusion bureaucracies.

Equity Hiring, by an Equity Hire

JANICE FIAMENGO

As a woman hired on an equity ticket in 1999 and again in 2003, I am well qualified to speak about the effects of equity hiring on recipients and on the academic community generally. I was given two tenure-stream positions for which well-qualified men were not invited to compete at a time when, if there had ever been an underrepresentation of women in English departments, it hadn't existed for decades in a field avowedly feminist, queer, anti-racist, and postcolonial. As a woman, I was given many opportunities that were denied to my male colleagues, who were expected to accept (and who, at least outwardly, did accept, some enthusiastically) that their disadvantage was a good and necessary corrective. Even at the time, the exclusion of male applicants struck me as bizarre and shameful. Decades later, equity hiring is even more firmly entrenched in the academy, with an even more elaborate rationale for its long-term necessity. For how long will white (and now Asian) men tolerate such unjust discrimination? How long can the general population ignore what is going on? These questions remain open.

Equity hiring has become more overt and more unapologetic in the two and a half decades since it brought me and thousands of other preferentially hired (mostly female) candidates into academia. In the year 2020, the Irish government announced that it would use taxpayer money to create twenty women-only professor positions to close the alleged academic gender gap.[1] Designed to raise the percentage of female professors at Irish postsecondary institutions, the move was hailed as a "game-changing moment" by the minister of state for higher education, Mary Mitchell O'Connor, who enthused that she was "incredibly proud that this intervention will ensure a swifter gender re-balance."

It's not clear what there is to be proud of in (at least implicitly) admitting that, a full century after being granted legal and political equality with men, women still aren't capable of competing with men on merit and that the government must ride to their rescue with imposed quotas and exclusions. O'Connor stressed with patent dishonesty that "appointments to these posts will be subject to the highest standards and rigorous assessment processes as currently adopted by the institutions for prestigious posts at these levels."

It doesn't take a PhD to recognize that choosing applicants on the basis of sex (or race, as we shall see), and thus excluding many applicants who might well have superior qualifications, will inevitably lower the intellectual standard allegedly being upheld.

This is just one example in one country of a practice that has been widely pursued for many decades, in which white men's taxpayer money pays for significant professional advantages for female, racial minority, or sexual minority academics who then, more often than not, make a career out of claiming to be oppressed and advocating for more professional advantages. My focus here will be on Canada, in particular, and North America, in general.

Anyone who doubts how widespread and well-entrenched equity hiring has become—now with an elaborate specialty language to affirm and justify its discriminatory proceedings—should take a look at the archives of the Society for Academic Freedom and Scholarship, a volunteer organization that lists and protests preferential hiring advertisements in Canada. We shall have cause to examine some of these advertisements in detail. In the academic year 2022–2023 alone, equity searches were conducted for dozens of academic positions.[2]

A small sample includes: the University of New Brunswick ("Only applicants who self-identify as racialized individuals will be considered for this opportunity"); University of Saskatchewan ("We encourage applications from members of equity groups"); Lakehead University ("A lived experience or worked experience of any of these [feminist, Indigenous or racialized] issues is preferred"); University of Calgary ("This position is only open to qualified Black [...] scholars"); Acadia University ("This is a focused hire for three African Nova Scotian and/or African Canadian Scholars"); University of Ottawa ("Only qualified applications from racialized or Indigenous peoples will be considered and evaluated for the position"); Carleton University ("We especially encourage applications from candidates from groups underrepresented in Computer Science and/or with experience in mentoring students from such groups"); Simon Fraser University ("Preference will be given to equity-deserving candidates"); Nova Scotia College of Art and Design ("These opportunities are restricted to applicants who identify as Black, Indigenous, or as a person of color"); Dalhousie University ("This initiative will support inclusive excellence by appointing five Black scholars to the university across multiple disciplines"); UBC Okanagan ("We encourage applications from members of groups that have been marginalized on any grounds enumerated under the B.C. Human Rights Code, including sex, sexual orientation, gender identity or expression, racialization, disability, political belief, religion, marital or family status, age, and/or status as a First Nation, Metis, Inuit, or Indigenous person"); and others too onerous to list here. The SAFS archives regarding restricted hiring date back to 1999.

Equity hiring was already common when I was first on the job market in 1999. Some job applications required candidates to fill out forms declaring their identity categories. The advertisements to which I responded did not explicitly exclude male candidates, but they made it clear, as many advertisements did, that female candidates were preferred as part of the university's diversity agenda. I was part of two shortlists of four candidates each, each shortlist entirely female.

During the time I taught at the University of Saskatchewan, my English department conducted four further job searches using an equity framework. The stated rationale was that where two or more candidates were equally qualified, the candidate whose hiring fulfilled the university's equity goals would be selected. In reality, of course, candidates are never "equally qualified," and qualifications soon cease to matter once the goal has been established to hire a woman or a racial minority. Equity, by its nature, involves gross discrimination at all stages of the hiring process.

White male candidates who applied for these positions were not told that they were wasting their time, but they were. In most cases, their applications were put into a pile that did not receive serious consideration. Members of the hiring committees were not explicitly instructed, and certainly never admitted, that even outstanding white men should be overlooked, but they were. We *were* explicitly told that we should not discuss the person hired as an equity candidate; on the contrary, we were to insist that the candidate had been chosen for her outstanding academic qualifications. In this way, the overt rejection of merit was hidden from general view.

Such gerrymandering has been commonplace since the 1980s, as Martin Loney demonstrates in *The Pursuit of Division: Race, Gender, and Preferential Hiring in Canada* (1998).[3] Rooted in equal rights legislation from the 1960s, government initiatives to provide positions for certain groups were aggressively expanded in the early 1980s, mainly at the behest of feminist activists in both the United States and Canada. In 1983, the Canadian federal government introduced an affirmative action policy to promote equitable representation of women, aboriginal people, and the disabled, which was expanded in 1986 to include visible minorities. The idea was given impetus and legitimacy by Judge Rosalie Abella's Report of the Commission on Equality in Employment, published in 1984. In this report, Abella developed the concept of *employment equity*, which was to significantly influence the academic discourse on equity hiring.

Some years later, Abella rationalized the overt discrimination required by equity hiring in a manner that was to find favor with many social justice activists. She attempted to explain as follows, distinguishing equality from equity:

> The reason in human rights that we do not treat all individuals the same way is that not all individuals have suffered historic generic exclusion because of group membership. Where assumptive barriers have impeded the fairness of competition for some individuals, they should be removed even if this means treating some people differently.

> Otherwise, we can never correct disadvantage, chained as we would be to the civil libertarian pedestal of equal treatment of every individual. There is nothing to apologize for in giving the arbitrarily disadvantaged a prior claim in remedial responses (Abella, qtd by Loney, p. 9).

Abella's statement is riddled with logical fallacies and non sequiturs that, in an era that valued quality, would not be acceptable even in a first-year undergraduate essay. They are even more appalling given the considerable influence held by this feminist judge, who was eventually the chief of the Supreme Court of Canada. In her explanation, Abella refers to "individuals" who "suffered historic generic exclusion because of group membership." By definition, individuals today have not experienced *historic* exclusion. Her next sentence is even more logically tangled. "Where [...] barriers have impeded the fairness of competition for some individuals, they should be removed...." This part of the assertion sounds reasonable: indeed, barriers that impede fairness should be removed. But the sentence ends "...even if this means treating some people differently." Why is it not sufficient to *remove barriers*? Why set up *new* barriers against other groups? Such is not explained in the following sentence, which claims melodramatically, "Otherwise, we can never correct disadvantage...." Here is sad nonsense indeed. Removing barriers *does* correct disadvantage to the extent that the law can correct disadvantage.

Abella implies that groups treated unfairly in the past are still experiencing disadvantages in the present, but she makes no attempt to prove this with evidence (or even to argue for it logically), and the implementation of laws to prevent discrimination would seem clear evidence to the contrary. Abella doesn't even try to justify why individuals today must be compensated for discrimination experienced by previous generations, and she certainly never explains why able-bodied white men, some from relatively disadvantaged backgrounds, should pay a penalty for historical injustice, where it existed, that they themselves never practiced and never benefited from. Her simplistic racism and sexism are, in her own words, *unapologetic*.

As Abella's discussion suggests, merit is not a primary consideration in equity hiring. Sometimes, as is evidenced by numerous recent job advertisements, it seems to be hardly a consideration at all. Many of these advertisements are lengthy and bombastic celebrations of diversity ideology, with very little stress on individual excellence or achievement. Where excellence is mentioned, it is qualified as *inclusive excellence*, whatever that might be, as in an advertisement for an assistant professor in data science and analytics being sought by the Faculty of Science at the University of Calgary.[4] This position is open only "to qualified Black (e.g., Black Pioneer, African, Caribbean) scholars" and is part of an "Inclusive Excellence Cluster Hiring Initiative." The description of the position makes clear the revolutionary ideological impetus behind the initiative. It is described as a "key step to ameliorate the under-representation of members of equity-deserving groups at the university, to build a critical mass of scholars who reflect the diversity of our student body and broader community, to advance epistemic pluralism and the creation of new knowledges and ways of knowing, and to broaden the culture of inclusivity and engagement." However many black scholars are hired within this initiative, it will be hard to say when the goals have been achieved.

Implied in the assertions but never proved (and in fact, unprovable) is the allegation that "equity-deserving groups" are deserving precisely because they have experienced discrimination, have been unjustly excluded from positions they were otherwise qualified to attain. Notably, the university's statement does not mention the importance of rigorous academic standards, the advancement of science, or the pursuit of (unqualified) excellence. Such considerations are notably secondary to the god-term *representation*.

The statement also refrains from mentioning any such qualities characteristic of the classical liberal university as a commitment to free speech, tolerance of diverse viewpoints, respect for academic traditions, love of learning and discovery for their own sakes, the preservation of the Western inheritance, or the valuing of objectivity in research and judgment. It is not at all clear that many candidates, tutored in contemporary grievance politics and disdain for Western culture, would embrace these. The advertisement makes it clear that the ideology of perpetual revolution is an end in itself, productive of a benefit that is asserted in grandiose language but not clearly delineated.

The advertisement for three Mi'kmaw/Indigenous scholars at Acadia University is even more non-academic in its stated commitments and values.[5] After a lengthy land acknowledgment that affirms Mi'kmaw priority and claim to the land on which Acadia University was built (with no mention, of course, of the Western achievements that made the university possible or the white men who built it from the ground up), the advertisement emphasizes its intention to "decolonize" the Acadia campus by "listening to and committing the campus to following through on the needs and interests articulated by a range of Indigenous Elders, community leaders and campus members to advance the aspirations of Indigenous students, scholars and staff, and to support Indigenous-driven collaborative opportunities."

A host of questions is raised but not answered by the statement: how are the needs of Indigenous students at university different from those of non-Indigenous students, and does the prioritizing of Indigenous needs require the neglect of other needs? That such needs and aspirations do not include being taught by individuals with expertise in traditional disciplines is made clear in the university's stated indifference to "standard academic criteria" as qualifications for the positions. The advertisement clarifies that folk knowledge, "community relationships," and "lived experience" may all be sufficient. "The hiring committee encourages Indigenous individuals with varied academic qualifications and professional experience to apply. For example, we welcome applications from: Elders and Knowledge Keepers with experience in research, teaching/mentoring, and community engagement, with or without formal post-secondary credentials [...]." The implied contempt for traditional university knowledge and training methods is startling in its totality. Why require PhDs—or indeed *any post-secondary education at all*—from anyone if the knowledge gained thereby can be so easily replaced, or even eclipsed, by oral traditions and lived experience? Or is only Indigenous wisdom of value? Left unanswered is the question of how a Knowledge Keeper's research or scholarship can compete with the work of those trained in academic methods and held to the highest academic standards.

As previously noted, hundreds more advertisements for preferential hiring on non-academic grounds are archived on the website of the Society for Academic Freedom and Scholarship. The abandonment of the merit principle is well evident.

As the language of these advertisements makes clear, equity hiring is not only a matter of increasing the number of women, blacks, the Indigenous, or other groups that have been affirmed to matter to the academic community. DEI is an all-encompassing program designed to reshape what counts as research and, indeed, what counts as knowledge itself. As the reference to "epistemic pluralism" in the University of Calgary job advertisement makes clear, much of this knowledge will be subjective, ends-oriented, and narrowly defined. Diversity applicants must not only fit particular ethnic and gender criteria but are expected to practice the ideology of DEI in every aspect of their teaching and research, from the words they use in the classroom to the types of research projects they pursue and the students they choose to mentor and hire for their labs. It is not enough merely to be a female art historian or a social scientist of Indigenous ancestry; each equity representative must be seen to advance the specific political, cultural, and ideological interests of their group, as defined by contemporary feminist and left-progressivist ideology.

Research that does not advance these narrowly defined interests is discouraged or at least seen as less urgent and less socially significant. Such research will necessarily be looked upon less favorably by publishers, editors of scholarly journals, those who vet scholarly articles, grant-dispensing agencies, and university administrators, all of whom have been thoroughly schooled in DEI and many of whom answer to government bureaucrats or others who have pledged fealty to it. An evident hierarchy of academic research subjects has been created, which has all but guaranteed the death of curiosity-driven research.

Many subjects of general discussion are likewise off-limits in the classroom and out of it, including alternative views of the experiences of men and women, whites and racial minorities—and even including, of course, the justice and usefulness of equity hiring itself.

It's hard to imagine how the atmosphere of febrile complaint, grievance-mongering, victim one-upmanship, and melodramatic assertion is helpful for anyone in academia, even including the beneficiaries of DEI largesse. These beneficiaries inhabit a queasy no-man's-land of unearned privilege as standard-bearers for the oppressed, their resentment as victims coexisting awkwardly with their awareness of material advantage. They may tell themselves that they and their people have suffered discrimination, but such self-justification wears thin in the face of abundant sympathetic attention and reward. In compensation, many beneficiaries become ever more sensitive to perceived slights, microaggressions, or bigotry and more inclined to see condescension, doubt, or dismissiveness in their colleagues' words and actions.

A few years ago, accusations of racism at the annual meeting of the Society for Classical Studies, held on January 5, 2019, highlighted the volatile mix of grievance, bullying, and hypocrisy that has become so common in academia and is not likely to be decreased by further equity hiring. In the Q&A following a panel called "The Future of Classics," an independent scholar, Mary Frances Williams, was reported to have uttered a racist remark, accusing a black classics scholar, Professor Dan-el Padilla Peralta of Princeton, of having "got his job because he is black."[6] How exactly this was a racist statement when, as has been seen, many job advertisements clearly announce their determination to hire black scholars because they are black was never made clear, but disciplinary action against Williams was announced by the president of the society, Mary Boatwright, in the society newsletter. Williams was informed that she had committed racial harassment of Peralta and was no longer welcome at conference sessions.

If it had been true that Mary Frances Williams had told Dan-el Padilla Peralta that he was hired because he was black, it would have been an awkward and uncivil accusation (though not different from what feminist activists of various stripes have been saying with impunity about white men for many decades). Yet the response of the Society for Classical Studies would have been extreme and unnecessary. Peralta held a plum position at an Ivy League college. If privilege and power matter in these circumstances, as we are always assured they do, he certainly had them over Williams. She had no academic position: no prestige, no tenure, no large salary, no job security, no power over others' appointments or opportunities; and she could not in any universe, even a white supremacist one, be seen as a threat to Peralta either personally or professionally. Moreover, the manner in which her comment was reported guaranteed her professional disgrace.

But the kicker is that what was reported in the society newsletter was not what she had said. She had actually said, "You may have got your job because you're black, but I would prefer to think you got your job because of merit." The president of the Society for Classical Studies significantly misreported Williams' words, causing her ousting from the conference on unjust grounds and implying that she was a racist bigot who had denigrated Peralta.

What turns Williams' expulsion into something more than an overreaction or even an injustice based on rumor is the astounding fact that Professor Peralta had just finished making a speech at the panel in which he explicitly *called* for the advancement of black scholars for no other reason than because they are black. I recommend readers watch the entire panel discussion about the future of classics—which is a harrowing spectacle of asinine virtue-signaling—but if you don't have time for the entire thing, you should definitely watch the eleven minutes or so of Professor Peralta lecturing the mostly white audience about the need for racial reparations, specifically in academic publishing.[7] As he put it, white men must surrender the privilege of seeing their words published and disseminated; people of color like himself will "take the place of white men whose words could have or had already appeared" in the classical studies journals.

It was remarkable to note the sea of white audience members sitting respectfully and indeed applauding (though perhaps less than enthusiastically) as Peralta informed them that they should look forward to a future in which, solely because of skin color, they themselves or their children would have to accept racial discrimination and be happy about it. It was in this context that Williams made her comment about merit, not realizing that the merit train had departed the station long ago. That comment was aggressive, perhaps, but it was also conciliatory, an attempt to find common ground with Peralta around the importance of merit. She was only wrong in thinking there could be any such common ground.

In the topsy-turvy world of DEI, the Society for Classical Studies can condone a race-based system of preferences, as advocated by Peralta, and

Williams, a relatively powerless female scholar, is condemned for allegedly suggesting that Peralta benefited from a race-based system of preferences.

If there is any satisfaction to be taken in sorry affairs such as this one, and in the current state of equity hiring, it is perhaps in an ironic prediction: more equity hiring will not bring us to the promised land.

The cycle of blame and promotion will continue, leading to oft-repeated accusations and apologies, expulsions, and acts of contrition. All will depend on where one stands in the victimhood hierarchy, but those at the top of the victim hierarchy today will not always occupy a niche of untouchability. As the episode between Peralta and Williams demonstrates, the white women who drove the culture of victimhood and reparation are no longer always holding the reins, and they will increasingly find themselves forced to apologize and make amends far more thoroughly than they have yet done for their white cis-gendered and able-bodied privilege. Dr. Peralta, as a man, is not permanently invulnerable either. There may be black lesbian scholars who may call out Peralta one day on an inappropriate comment or an exclusionary decision or an oppressive flick of his eyelash.

Show trials will proliferate. The whole lot of politically correct scholars may find themselves forced out for various inadequacies and failures, for neglecting to thoroughly enough purge their discipline of whiteness, or perhaps even for sneaking in merit through the backdoor. Only one thing is for sure: academia as we knew it, that bastion of nerdy specialization and beautiful esoteric knowledge and curiosity-driven research, is dead, and the trendy new socially relevant so-called scholarship that killed it is a sickly thing, likely to be easily picked off by new waves of righteous warriors. Whether anything can be saved is a question best left to more penetrating minds than mine.

Equity, Diversity, and Inclusion: The Dismal Pseudoscience Threatening Science

GEOFF HORSMAN

"You should stop talking about EDI," my colleague suggested to me over lunch. You've got a family, he said. Think about your kids, he implored.

I'll admit that my heart rate increased, because it sounded a lot like, "Nice family you got there, would be a shame if you were no longer able to provide for them."

Only later did it occur to me that the perfect response would have been to quip that I was unaware that I worked for Tony Soprano.

Unbeknownst to me at the time, I was the focus of some consternation in my faculty. People had been talking, according to this colleague, and they were worried about what I was "thinking."

As it turned out, what I had been "thinking" was that we needed to re-normalize a culture of liberal science on campus, especially with respect to radical new ideas like equity, diversity, and inclusion (EDI).

Jonathan Rauch coined the term *liberal science*[1] to describe an open society's epistemic system, which constitutes the third leg of liberalism's three-legged stool along with free-market economics and the liberal democratic political system. Liberal science has just two rules: (1) no one has the final say (knowledge is provisional), and (2) no one has personal authority (knowledge is universal).

Universities, the institutions that should be at the very heart of liberal science, are increasingly ignoring these rules. The first rule is often broken through confident declarations that a controversial topic is "beyond debate" or "settled." The second rule is often subordinated to certain sacred identities—usually regarding race, sex, or gender—and their "ways of knowing" or "lived experiences." As I tell my students, if you do an experiment well and describe it properly, it should be replicable by someone on another continent, of an entirely different culture, centuries into the future. Your personal characteristics are irrelevant.

Here I will outline how EDI itself has become sacred, which undermines liberal science and demoralizes researchers by silencing dissent and corrupting scientific research. Sacralized EDI beliefs are first protected from scrutiny through mechanisms of soft censorship. After stifling dissent, the next stage of demoralization involves coercive displays of support for EDI. Required EDI statements corrupt principled scientists by incentivizing lying, which demoralizes scientists and compromises their capacity to resist groupthink.

Silencing Dissent: Thou Shalt Not Question the Sacred Beliefs of EDI

By the time of this lunch with my colleague, I had done a few things in the spirit of liberal science that seemed to rankle the EDI faithful. First, I questioned why my department was participating in racial discrimination by advertising exclusively for an indigenous professor. This was part of a university-wide "Inclusive Excellence" program to "address systemic racism" by hiring six black and six indigenous professors.[2] I thought barring ethnic groups from faculty employment was immoral. Even if you agree with excluding European or Asian ethnicities from faculty employment, the program required racial discrimination against black and indigenous applicants. For example, Inclusive Excellence positions granted to departments to hire indigenous scholars were not open to black candidates, and vice versa.[3]

I sought to protect our department by requesting information from the administration. The university's announcement of the program was unhelpful; according to the president, "Inclusive Excellence embodies our vision to fully embrace discovery, scholarly exploration, and application of new ideas, while engaging and challenging the world in all its complexity."[2] No evidence was provided to demonstrate hiring bias against qualified indigenous or black applicants. So how could implementing *concrete* racial discrimination address the vague concept of *systemic* racism? Even if data showing discrimination existed, I thought it far preferable to first remove all sources of discrimination before adding more.

Similar concerns were articulated in a letter from the Society for Academic Freedom and Scholarship (SAFS),[4] Canada's preeminent organization committed to academic freedom and the merit principle. In response, I suggested the following motion for our next department meeting: "Our department requests that the administration respond to SAFS in defense of the Inclusive Excellence program. Any data, policies, and metrics should be made available to assure candidates that Laurier is addressing any existing discriminatory hiring practices." After all, I thought, if Laurier had self-diagnosed as systemically racist,[5] would not prospective faculty want to know the problem was being fixed?

I was unable to get this motion on the agenda for a department meeting. Instead, a faculty-only meeting was arranged to decide if my motion should even proceed to a department meeting. After I made my case—which included our department hypothetically rejecting a qualified applicant for being black—my motion was defeated for the following reasons:

- We should not ask the administration to defend this program. They don't owe us any answers. Instead, we need to educate ourselves on EDI.
- SAFS is an alt-right organization, and therefore, we can dismiss their letter. The administration is correct not to dignify them with a response.
- We don't need data to know that indigenous people are underrepresented.
- These positions are new positions carved out after negotiations between our faculty association and the administration.

These are smart and good people. I am quite fond of many of them. But how does one explain such strange justifications? Surely the definition of "underrepresented" involves comparing an observed number to an expected number. One would most certainly need data to achieve this comparison.

In my opinion, this reaction was driven by fear of being called racist for questioning EDI. Because evidence-based critiques might put their profession at risk, EDI apparatchiks exploit politeness and fear to contain thoughtful discussion to small groups. This prevents people from openly debating and confronting EDI-mandated racial discrimination.

I hoped that the administration might respond to the SAFS letter, but my inquiries revealed that they would not even acknowledge receipt of—much less respond to—the letter. I found this deeply troubling. Easily defensible policies don't normally elicit such dramatic demonstrations of intellectual insecurity. I was getting serious cult vibes.

Somewhat prophetically, my concern for a hypothetical black applicant came to pass. A colleague from outside my department casually contacted me about an excellent black candidate for our Inclusive Excellence position. I had to respond that this position was only for indigenous people. Somehow, we had found ourselves doing something that would have been unimaginable a decade prior: excluding a candidate from faculty employment for being black.

As it turned out, we received no applications. Our original Inclusive Excellence position was given to another department, which it used to hire a relatively senior academic from another institution. Tenure-track positions are rare and extremely competitive. Instead of considering tens or even hundreds of applications from budding young scientists, we had advantaged an established academic while depriving another institution of a prized indigenous scholar. I am waiting for someone to explain to me how this helps anyone other than virtue-signaling administrators and the poached scholar, who presumably was financially enticed to move to Laurier.

Another of my liberal science transgressions was facilitating a classroom discussion about EDI in science, which included panelists articulating both benefits and harms. In my view—which, until recently, I would have assumed my university would share—a panel obviously includes those on both sides of a controversial issue. While the panel discussion seemed to go well, strange things began happening. This includes an attempt to remove me from the course and my being asked to apologize for traumatizing panelists who felt "ambushed" by EDI critics. The purge of dissenting opinions on campus is so complete that, to many, it would simply not occur to them that there exist valid critiques of EDI. In an astonishing admission of fundamentalism, an administrator informed me that EDI "is not debatable" while also claiming "I'm not an expert" when personally invited to explain this viewpoint to my class.

These administrators should know that EDI does more harm than good because I have sent them a recent report by my Laurier colleague David Haskell.[6] This report summarizes extensive research literature, including several systematic reviews, showing that EDI does not reduce bias but instead increases prejudice and even activates bigotry.

Administrators should also know—again because I informed them—about Eric Kaufmann's report that the Canadian public rejects EDI and instead favors *cultural liberalism* (free speech, due process, etc.) by a two-to-one margin.[7] Regarding the Inclusive Excellence program, for example, research has found that "[b]y 70 to 30, people prefer a colour blind rather than colour-conscious approach to issues in society." This tells us that EDI is a controversial topic over which the public is divided. In other words, EDI represents a political view, and an unpopular one at that.

But despite—or perhaps because of—the ineffectiveness of EDI and its rejection by the public, administrators offer little more than silence in response to these concerns.

The EDI Statement—Sowing Seeds of Corruption

These examples show how an illiberal culture of silence has come to dominate the university administrative class. It is little wonder, then, that dissenting professors might become demoralized and stop questioning. It's easier to rationalize their own compliance than swim against the current. After all, they might think, how bad can it be to make a written statement committing oneself to EDI if it just means helping people feel included?

I am afraid that my patience for such willful ignorance is fast dwindling. Critically thinking professors can no longer credibly claim to be unaware of problems with EDI; critiques have become too loud, too logical, and too ubiquitous to ignore. In 2024, Harvard and MIT dropped EDI statements,[8] several US states removed EDI from public universities,[9] and even the Washington Post turned on EDI.[10] And this all occurred prior to Donald Trump being elected on an anti-EDI platform, which was quickly implemented in a flurry of executive orders. In Canada, I was one of about forty academics outlining problems with EDI in a policy brief to Parliament.[11] Our recommended removal of EDI from federal research funding was covered in the national media and must have been widely noticed.[12] A subsequent parliamentary study invited several EDI critics to appear as witnesses, prompting the national press to note that the Canadian government is wasting money by funding social justice instead of science.[13]

While willful ignorance is one problem, another is the compliance of silent dissenters. For example, a scientist at another Canadian university told me, "I have made my peace with EDI. I will lie about my most deeply held beliefs or convictions on paper in order to get funding."

This comment was in reference to the EDI statements increasingly required for federal government research grants. For example, the New Frontiers in Research Fund (NFRF) made EDI a priority from the outset, stating that it "has formally embedded EDI requirements in its program design."[14]

Justification for EDI in Canadian research grants is hard to find. The three main funding agencies, known as the Tri-Council, have published a Tri-Agency Statement on EDI asserting that "Achieving a more equitable, diverse and inclusive Canadian research enterprise is essential to creating the excellent, innovative and impactful research necessary to advance knowledge."[15] The NSERC Guide for Applicants states: "The Evidence is clear. Equity, diversity, and inclusion strengthen the scientific and engineering communities and the quality, social relevance and impact of research."[16] The citation for this statement links back to the Tri-Agency Statement, which itself provides no clear evidence that EDI improves research. Perhaps the upper-case "Evidence" in the NSERC statement is a Freudian slip of sorts; much like a belief in God, the "Evidence" for EDI is clear only to the faithful.

When applying to NFRF, I took the EDI section very seriously because it is evaluated pass/fail; a great research proposal means nothing without

EDI compliance. A document titled "Best Practices in Equity, Diversity and Inclusion in Research" stated that the applicant must demonstrate a strong commitment to EDI, understand microaggressions, and identify and address systemic barriers.[17] As I learned, "systemic barriers" are neither concrete policies nor remediable actions. Instead, they are vague and controversial ideas like implicit biases, "whiteness," and disparities like wage gaps. In preparing for the NFRF application, I read enough Thomas Sowell[18] to know that discrimination rarely contributes to racial disparities in modern Western societies like Canada.[19] Furthermore, it is unreasonable to expect a biochemist to tackle such complex social science problems in a grant application on an unrelated topic.

These concerns led me to an online course called "Inclusive Research," which my university offered to educate scientists about EDI.[20] Shockingly, this course abandoned the liberal tradition of not judging books by their covers; it even reinforced sexist stereotypes. For example, in a video workshop about writing EDI statements,[21] participants were asked to critique a job advertisement for a postdoctoral fellow. Herculean efforts to problematize the advertisement included speculation that the term "highly motivated" is male and might put off female applicants. Apparently, "hard words" like "skilled" should be avoided because they don't appeal to women. This was not accompanied by any evidence to support these surprising claims, nor were the benefits to science of infantilizing women clearly articulated.

This is all very confusing to classically liberal or conservative scientists. Not long ago, we prioritized equal treatment without regard to race, sex, or country of origin. Nowadays, such liberal sentiments might sink your grant application. For example, Patanjali Kambhampati believes his grants were rejected because he stated his meritocratic desire to "hire the most qualified people based upon their skills and mutual interests," even though officially he was only informed that "the Equity, Diversity and Inclusion considerations in the application were deemed insufficient."[22]

This further highlights the problem that EDI bureaucrats seem loathe to provide clear written feedback. As another example, one researcher was informed that they "did not sufficiently describe concrete practices that would be put in place to ensure that EDI is intentionally and proactively supported." Because so often EDI practices involve discrimination, it's understandable that bureaucrats might favor vague language. A paper trail of grant rejections for scientists favoring non-discrimination might become a problem.

Without documentation, we are driven to anecdotes. For example, colleagues who had recently served on a grant review panel recounted how an EDI statement proclaiming the importance of holding all students to a high standard was, to them, obviously insufficient. Perhaps after being subjected to the unchallenged barrage of assertions in EDI training courses like Laurier's "Inclusive Research," one might reflexively take offense at such forthright adherence to merit.[23]

While my own NFRF application was successful (for which I now feel some shame), I worried that EDI statements might unnecessarily exclude innovative research based on the politics of the applicant. My inquiries revealed that, in the 2021 competition, 2.5 percent of applicants failed on EDI alone. Oddly, considering the EDI portion was pass/fail, 29.5 percent of applicants got a "mixed pass" because they passed the EDI evaluation but didn't get full marks. Perhaps 32 percent of applications were not sufficiently ideologically committed but rejecting them all would have brought too much heat.[24] While 2.5 percent isn't huge, it's still concerning that *any* grant would be rejected on ideological grounds.

But this small fraction of grants being culled for political noncompliance is just the tip of the iceberg. It signifies that beneath the surface lies a much larger problem of ideological corruption.

Demoralization and its Effect on Science

It is challenging to articulate exactly why EDI statements are so corrosive, because many people do not recognize a problem. I've been told, for example, not to worry about EDI statements because there are rubrics that clearly describe what to write. Still others have extolled the virtues of ChatGPT for writing EDI statements. That none of these scientists were disturbed by surrendering free thought to government agencies or chatbots is concerning. For liberal science to function, we cannot so casually outsource this basic liberty.

While the scientists admitting to surrendering free thought don't think they're doing anything wrong, how does this differ from those who admit to lying about EDI? Maybe the only difference is that the former agree with EDI, while the latter do not. How would the former view using ChatGPT to write, say, a statement committing oneself to advancing libertarian principles like free markets and limited government? Would they see such statements as the ideological tests they are? Would they pretend to agree just to get a grant? If so, would they recognize that they are lying, as most left-leaning professors would be?

Recognizing self-betrayal may not be straightforward. The EDI worldview requires one to view people not as individuals but as members of groups. As our Inclusive Excellence program—an exemplar of EDI initiatives—showed, identifying into an "equity-seeking" group was a basic job requirement. How, then, might a traditional Christian consider the sacredness of the individual being subordinated to EDI's ethnic or sexual group identities? Would not Christianity and EDI represent incompatible belief systems? In other words, would someone who values the sanctity of the individual be lying if they agree to EDI in any way? In my view, the answer is probably yes. And that most Canadians also disagree with EDI further highlights the cognitive dissonance that must, to varying degrees, confront many scientists.

This brings me to my final thought: what are the costs to scientists, and to science, of going along with the crowd? How does even a "little" lie, like saying you are committed to EDI when you aren't, affect science?

The answer may lie in demoralization from self-betrayal.

Deceitful conformity leads to demoralization through the quiet shame of having transgressed. We become more cowardly when we compromise our integrity. Our self-respect is diminished, which makes us slower to speak up and challenge things we disagree with, including groupthink in our own disciplines. We may more readily relinquish agency and accept the corruption of superiors. When this demoralization touches many individuals, the cumulative effect is an upward transfer of power. Those at the top decide upon "institutional values" and dictate those values to underlings who will feel pressure to adopt them. A vicious cycle of further self-betrayal and demoralization ensues.

The Czech dissident Václav Havel wrote about demoralization under communism in his famous essay *The Power of the Powerless*. He described how a greengrocer would be sent signs from party headquarters to display in his shop. Just as regurgitating a banal phrase like "I am committed to equity, diversity, and inclusion" might seem like the harmless cost of entry to a research grant, hanging a "Workers of the world, unite!" sign might also seem like a small price to pay to stay in business. But these compelled expressions of political loyalty signal our subjugation; the humiliation of surrendering free thought leaves us diminished, defanged, and dull. Havel's story reminds us that totalitarians understand how easily we will agree to these small self-betrayals, because we convince ourselves it's only a little thing. But what totalitarians also know, and most people do not, is that by diminishing ourselves in this way, we weaken our defenses and become capable of agreeing to almost anything.[25]

How this will fully impact science remains to be seen. Anna Krylov has made startling comparisons to the politicization of science in the Soviet era.[26] For now, we are still doing some good science, but that was also true of the Soviet Union. But our defenses are clearly down, and the demoralization felt by many from humiliating themselves with a required EDI statement has doubtless contributed. They say that no besieged city falls without those inside opening the gates to the enemy. Scientists, I'm afraid, are naively welcoming EDI inside the walls of liberal science, and most have no idea what may be unleashed upon them. My faculty of science recently revised its tenure and promotion guidelines to include "indigenous forms of knowledge," in clear violation of Rauch's second rule of liberal science. It will become dangerous to challenge anything proposed by a member of this sacred identity group. Most scientists will use fear as an excuse to accept the unacceptable and, in so doing, will begin to reap the indignities they will have so rightly earned.

A public that remains—at least for now—culturally liberal may not easily forgive us. Confidence in science is falling,[27] and the public may no longer agree to have their tax dollars misspent; scientists may allow *themselves* to be humiliated, but many other Canadians are not so compliant.[28] A hatchet to science budgets may end up being the ultimate—and even merciful—cost of the "innocent little lies" of EDI statements. "After all," as I mentioned in a recent op-ed, "how much confidence can we have in a research ecosystem that incentivizes betraying oneself?"[29]

Despite these dour final sentiments, I do feel some optimism. But things will probably have to get much worse before they get better. Those of us who avoid the political knife may not escape the budgetary axe. We are doubtless in for tough times, and none of us is in the clear. But, as Havel might say, if we remain oriented towards truth, we will at least retain something of true value that can only be willingly surrendered but never taken—our self-respect.

Dedication

This chapter is dedicated to my mother, Carol Horsman, who passed away as it was being written. Mom was a gentle, humble, and hard-working prairie farm girl who quietly modeled what it means to stick to your principles. With a twinkle in her eye, she would softly delight in letting her dissenting opinions slip. She never feared the costs of doing what she thought was right.

Love you, Mom. This is for you.

The Leaning Ivory Tower

ALESSANDRO STRUMIA

In the past decade, we lost some excellent apolitical journalism, media, entertainment, cartoons, search engines, encyclopediæ, scientific magazines, and academic and scientific institutions. I accidentally got involved in these troubles while at CERN thanks to a European Research Council grant for physics. In 2018, CERN decided to host a workshop about a new topic, gender. Having worked on bibliometrics, I had the data needed to test anecdotal claims about why women remain underrepresented in physics.

According to the politically correct mainstream theory, STEM (science, technology, engineering, and mathematics) disciplines conspire to keep women out. This happens despite that Western academia is one of the most progressive environments globally. Why should these champions of diversity seek to exclude women, especially in countries with higher levels of gender equality (Stoet et al. 2018)? Apparently, because we all suffer from sexist "stereotypes" and "unconscious systemic bias" (such as believing that biological gender differences exist) that create "invisible obstacles" via "micro-aggressions." Interestingly, this pervasive bias disproportionally affects STEM scientists and truck drivers, as women long ago reached parity in other fields and professions, including those near to power, such as the judicial system.

To me, this seemed to be a bizarre conspiracy theory. So, I conducted basic bibliometric checks that CERN could have performed before hosting claims that physics discriminates against women. For example, I calculated the number of papers published and the citations received by each author at the time of hiring. Do women need, on average, higher bibliometric indices than men to be hired? The data indicated no evidence of discrimination in favor of men.

Many previous academic studies on gender and STEM similarly found no discrimination (Ceci et al. 2018), yet these findings were ignored at the CERN workshop. Despite the evidence, some physicists cannot believe that they collectively address gender issues in a fair manner. They argue that since there is a gap, there must be discrimination. So, I also presented an alternative interpretation of the gender gap, hoping that an audience of scientists could accept a disagreement from simple-minded egalitarianism. The data could be explained assuming that current female underrepresentation in STEM is dominantly due to two main factors: gender differences in interests and higher male variability (HMV). Gender differences in interests of systemizing activities along the "people vs things" dimension is a large, well-known effect in psychology (Su et al. 2009). HMV is a small effect: bibliometric distributions revealed a 10 percent gender gap in variance, a known trend in biology (Halpern et al. 2007, Murray 2020) originally noticed by Darwin.

I was aware that this science was dismissed by gender theorists and thereby unsayable in academia and that activists were visiting CERN for the gender workshop. So, after my talk, I privately warned some colleagues that trouble was possible. They saw no reason to worry. When trouble arrived, they suddenly realized that my talk was bad. CERN stated, "*CERN stands for diversity,*" while canceling my talk (the only diverse talk) and deeming it "*highly offensive.*" The CERN cancellation backfired, drawing huge visibility to my slides, which freely circulated on the internet while the word "*cernsorship*" trended on social media.

In my opinion, a scientific organization should have stated: "We stand for science; we take no institutional position on socio-political issues (including diversity and gender); we respect individual free speech." Indeed science, in its pursuit of truth, occasionally finds offensive results. This happened centuries ago, when science cast doubts on sacred beliefs, and it's happening now, when activists (instead of bishops) sometimes use claims of offense as a weapon to silence dissent. Discussing natural human differences is indeed considered "offensive" in some sociological circles. A mainstream view, labeled as the standard social science model by some authors, ideologically rests on the blank slate paradigm, which denies human differences. By refusing to accept input from biology, a part of sociology becomes flawed and at odds with important observations. In contrast, the standard model of physics agrees with observations because physicists prioritized being scientifically correct rather than politically correct. An alternative theory, "nuclear democracy," postulated that no particle is more fundamental. But physicists followed data rather than egalitarian ideologies. As Richard Feynman stated, "one of the signs of intelligence is to be able to accept the facts without being offended," and "it doesn't matter how beautiful your theory is...if it doesn't agree with experiment, it's wrong."

Many (including most physicists) are unaware that:

- gender differences in interests and HMV are solid quantitative science that replicates;
- neuro-imaging data correctly tell the binary sex of a person with 99.7 percent accuracy;
- some activists use media to paint real science as discredited and their ideology as science;
- many politically correct studies, promoted on media as scientific, suffer from replicability issues;
- gender is a field where politically correct hoaxes got published (Sokal 1996, Lindsay et al. 2018), while genuine scientific papers are deemed "controversial" and canceled. The table below lists recently canceled papers on gender and STEM.

Authors	Journal	Published	Canceled	Survival
Hill, Tabachnikov	The Math. Intelligencer	accepted		0 days
Hill	NY J. of Math.	2017/11/6	2017/11/9	3 days
Hudlicky	Angewandte Chemie	2020/6/4	2020/6/5	1 day
AlShebli et al.	Nature Comm.	2020/11/17	2020/12/21	34 days
Kormendy	arXiv, PNAS	2021/10/7	2021/11/1	5 days

For example, one apparently compelling claim of discrimination was presented at the 2018 CERN gender workshop: blind auditions increased the probability of women being hired in orchestras by 50 percent (Goldin et al. 2000). However, a check found that this claim was never supported by data (Sommers 2019). Blind auditions are now seen as an obstacle to diversity (Tommassini 2020).

Currently, I am the only participant in the 2018 CERN gender workshop who presented results that got published in a scientific journal (Strumia 2021). Its editors demonstrated that the normal scientific standards can be upheld. With a bit of courage, there is no need to surrender to cancel-culture activists. Despite being published, my paper cannot appear on the arXiv preprint bulletin, which is widely read by physicists. ArXiv claims that my publication is off-topic while routinely accepting politically correct preprints on the same topic, including a heated criticism of my paper (Ball et al. 2021). ArXiv is managed at Cornell University, which, in 2023, proclaimed an initiative to restore free speech. I contacted the Cornell free speech committee. Nobody responded.

The worst attacks against my results and me come from the US, where a physicist known for she/her discovery that "Black women must, according to Einstein's principle of covariance, have an equal claim to objectivity" (Prescod-Weinstein 2020) and for activism (Kay 2024) formed a "Particles for Justice" (P4J) academic mob that petitioned my CERN *superiors* to get me punished (P4J 2018). My "superiors," after canceling my data, imposed gender quotas onto CERN (CERN gender program). They should have known that gender is part of a partisan political ideology that was damaging US academia. Then, my "superiors" could have avoided unnecessarily importing a disease of the US system on a tolerant French-style excellence. Since 2018, CERN has lost around 40 percent of its bibliometric output, probably mostly because of scientific problems. I got support from (Anonymous 2018) and (Anonymous 2019), while our courageous CERN colleagues moved a workshop to support a physicist at CERN who was arrested and later sentenced for plotting Islamic terrorism (Butler 2016).

In 2022, I was invited to talk about dark matter at a CERN school. Suddenly, my lessons got canceled, as some organizers still disliked my data about gender. Other organizers refused to collaborate with the cancelers, and the whole school was never held. In 2024, I could not attend a physics workshop in Germany featuring a US "diversity coordinator." My talk was instead welcomed in China, where physics workshops remain about physics.

In 2024, CERN hosted a new Gender Meeting (CERN 2024). My published data on objective bibliometric indices of seventy thousand physicists worldwide in the past fifty years were not allowed. In contrast, the CERN organizers allowed other speakers to report about "interviews with 27 self-identified progressive white-male physics" interpreted via the "epistemology of ignorance," finding that "White men's ignorance leads to their complicity in racism and sexism even when they are well-meaning" (Dancy et al. 2023). This scientific discovery was showcased on a slide titled "White Cis Men Have Limited Awareness" at (CERN 2024). The well-meaning physicists at CERN did not find this presentation to be "offensive science." I half agree: is this science?

Discriminatory Quotas

The 2024 CERN Gender Meeting openly discussed the real issue: quotas. Back in 2018, the goal of gender workshops seemed to be writing job ads in inclusive language to avoid subtle gender bias. Many did not consider that a plausible goal of false victimization narratives was obtaining "positive" preferential discrimination. Merely suggesting this possibility was punished as unethical in 2018. By 2024, discrimination became explicitly and officially declared in various job advertisements worldwide: only the specific minorities elected as "equity deserving" by identity politics are eligible to apply. Common official terms used to restrict academic job qualifications are: *identify* as a woman, as black or racialized, as "2SLGBT+." For instance, more than half of Canada Research Chairs opened in 2023 enforce such exclusionary criteria (Burgess 2024). This means that Einstein today could not apply to such physics positions because he *was* male, white/jewish, and heterosexual. Einstein had similar problems with the early-Nazi movement, *Deutsche Physik*, and did the right thing: he left Germany preemptively, without waiting for discrimination to become official.

The current examples of discrimination imply political discrimination, as members of "equity deserving" groups who view themselves as human beings and support universal liberalism don't want to be hired in discriminatory positions. These discriminations disproportionally impact young physicists, as hiring decisions are managed by senior physicists who don't step down from their own positions to make way for minorities more equal than others. Rather, senior physicists discriminate against young colleagues to seek absolution for sins they haven't committed by granting privileges to individuals who have not suffered discrimination.

Now many people notice that something went wrong. I next propose my perspective on why we find ourselves in a historically bad moment for science and academic institutions. Being a physicist, I over-simplify, focusing on what I consider the main factor, and present a coarse-grained brief summary.

The Political Origins of the Problem

A century ago, Marxists believed that workers were poor and thereby discriminated against and thereby morally superior. In some countries, this seemingly benevolent belief led to revolutions, dictatorships, and massacres. In Western countries, Marxism led to a strategy known as the "long march through the institutions." This meant gradually infusing progressive anti-authoritarian ideas in academia and other institutions, according to the perspective of those who favored such ideas. According to the opposite perspective, this strategy led to gradually occupying institutions to subvert them from within. The term "cultural Marxism" remains controversial.

By the '60s, it had become apparent that capitalism was enriching workers, while communism resorted to repression. Different ideologies had been tried, and the socioeconomic model inspired by Marxism had failed. Western Marxist intellectuals responded to this reality check similarly to how the Church responded when science started being "offensive." They retreated into dogmatic absolutism, developing an ideology known as *postmodernism*, which ended up justifying disregarding reality. In its current version, postmodernism rejects reason, science, individual merit, and equality before the law, seen as tools of oppression, and claims that only power exists.

Given that a class Marxist revolution had become impossible in the West, political activists tried exploiting extra divisions beyond class. To trigger resentment among people who were improving their conditions, activists switched from absolute poverty to relative inequity.

"Gender" was invented in this context, paraphrasing Marxian dialectic conflict theory, with "patriarchy" replacing "capitalism," and so men replacing the bourgeoisie as the oppressor. Political power was used to replace women's rights with gender. To (mis)interpret gender differences as discrimination, gender denied natural differences. Up to the point that some scientific institutions write "*identify* as women": Einstein could become eligible for a physics position by changing his "gender assigned at birth." Contrary to what was predicted by gender ideology, however, gender differences grew in more liberal countries like Sweden (Herlitz et al. 2024).

"Critical race theory" (CRT) was invented as a Marxian variant adapted to target the US in its Achilles heel, replacing "capitalists" with "whites." Following the assassination of Kennedy, in 1965 US voters accepted reverse discrimination as a temporary tool to address racial gaps due to past discrimination. Contrary to what was predicted by the ideology, however, certain racial gaps persisted. So, the emergency laws become permanent, effectively establishing unconstitutional discrimination as an alternative constitution and curtailing free speech under the guise of political correctness. According to CRT, the problems of blacks must be attributed to "pervasive systemic racism," despite contrary evidence: some groups outperform whites. Reasonable anti-crime measures are deemed racist and illegal due to their "disparate impact," up to the point that shoplifting and fare evasion got de facto legalized in some US states. Crime spread, and some US towns became no-go areas.

"Decolonialism" is the variant adapted to the third world. Contrary to what was predicted by the ideology, however, many decolonized countries regressed economically and socially, sometimes brutally. Even in South Africa, hailed as a beacon of peaceful, positive change, corruption escalated, basic services such as electricity are in jeopardy, and murder rates are about one hundred times higher than those in Italy. In Western countries, many politicians welcomed immigration, expecting that any problems would dissipate by the second generation. Instead, the opposite is happening.

When the Soviet Union collapsed, class Marxism mostly ended. But history did not end. These Western variants of Marxism inherited its resources and its media.

I emphasized examples where reality contradicts social constructivist ideology. In my opinion, the root of the problem is that a big part of current mainstream political thought is based on this flawed premise, which originated in academia. This is particularly evident in the uncritical interpretation of gaps as discrimination. A different interpretation is that gaps due to social nature have been progressively addressed up to the point that various remaining gaps have a different nature. To understand current issues, one needs to move beyond the social constructivist postulate and consider that different groups follow (approximatively normal) distributions with varying means and variances for relevant characteristics. Some differences might have biological origins.

However, just arguing for freedom of research on these topics can lead to professional repercussions. Scientific findings that challenge social constructivism are routinely attacked as sexist and racist. For example, an obituary attacked a deceased biologist for using the "so-called normal distribution" (Coyne 2021). Particles for Justice accused me of "belittling the ability and legitimacy of scientists of color," when my 2018 talk was solely about gender. In 2020, P4J switched from gender to the race component of US identity politics, accusing physicists and police officers of "systemic racism." The results of critical race theory policies are now evident: delegitimizing and undermining the police led in the US to a sudden 30 percent increase in black homicides and deaths (Alexander 2022), a death toll higher than that of the 9/11 terrorist attacks.

Allowing free speech and science on these sensible topics would help prevent bigger ongoing mistakes. Social taboos on these topics are selectively enforced by partisan politics, as illustrated by the next two examples.

First, individual differences are not considered an offensive taboo. We can say that somebody is taller or somebody is smarter without being attacked by academic mobs. We can say that biology contributes to such differences. However, we cannot say that, consequently, genetic drift can result in related differences at the group level.

Second, let's consider criminality rates: data indicate three big gaps depending on the age, gender, and race of the offender. Nobody pretends to be offended by the age gap. The gender gap is politically correct. The race gap is an unsayable taboo, enforced by the same political area that highlights the gender gap, coining sexist terms such as "femicide." Consider the sentence: "According to recent US statistics, black women commit homicides at a rate comparable to white men." Half is politically correct; half is offensive.

These and similar examples reveal the hypocrisy of political correctness: a selective taboo enforced by partisan politics, painted as moral superiority.

The Political Takeover of Academia

The goal of academia was truth. So, academia was a place where sensitive topics could be discussed in a civil, intelligent way, where tenure protected freedom of research, and where experts reasoned on what statistical results really imply instead of getting offended and fighting with anecdotes and cancellations.

Now, US academia (and consequently Anglo academia) has become a main source of the problem. Here is my understanding of what went wrong and risks repeating in other countries.

Half a century ago, private donors funded seemingly well-intentioned "studies" rooted in the postmodernist ideology. Activists entered academia and, rather than seeking truth, pursued their political agendas, gave academic credibility to victimization narratives, presented their ideology as science, and gradually indoctrinated students, who embraced the label of "woke."

Next, the movement could *individually* target and silence dissenters within academia, imposing their ideology on the humanities. Professors and students found it personally convenient to self-censor and adapt to cancel culture. Some graduates of "grievance studies" started being hired as administrators. The number of administrators grew so much that, to pay their salaries, student fees in the US increased, exceeding the economic value of higher education: paying loans became impossible for many students.

Many US universities started imposing an explicit political discrimination: DEI statements. DEI means diversity, equity, and inclusion, but in an Orwellian sense. These words, designed to obfuscate old illiberal ideas and make them sound like a kinder version of liberalism, hide a precise political meaning. Equity means replacing equal opportunities with equal outcomes. Diversity means replacing individual merit with group-based discrimination. Inclusion means excluding those who disagree. DEI statements are used to exclude applicants with "bad" political opinions without even considering their science. The physicist Steven Weinberg, while well-established and a Nobel laureate, could state: "I will seek out the best candidates, without regard to race or sex." I agree with Weinberg, but his statement is now considered a microaggression in some universities.

The phase transition completed: STEM came under attack and was submerged by the DEI wave. Most scientists who sought institutional leadership roles found it convenient to follow the prevailing politics. Minorities placed as heads of institutions were motivated to advance the DEI ideology that advanced their careers. So, STEM started adopting discriminatory practices, such as DEI/gender/race hiring. Standardized tests aimed at selecting excellence were removed because they helped poor people irrespective of their race and gender. Some courses, such as "Black Holes: Race and the Cosmos," blend DEI with science. Just naming scientific instruments, such as the James Webb Space Telescope, now often results in sad disputes (Kay 2024). Some scientific prizes started to be assigned to activists and to minorities without strong qualifications. Some academic journals started publishing ideological content without allowing replies (Reichhardt et al. 2023); others enforced ideological restrictions. Some scientific magazines became political. Various conferences started having diversity quotas, gender pronouns, land acknowledgments, and codes of conduct that forbid expressing specific opinions. Some physics conferences forced colleagues to "contribute to the culture of inclusivity, equity, diversity" and excluded those with different political opinions, such as preferring fairness, equality, and merit.

At this stage, the "long march through the institutions" has reached its destination: the political takeover of US academia has been achieved. Perhaps this outcome was not the original intention and labeling it as "cultural Marxism" is inaccurate. In practice, the result is evident: about 90 percent of US professors lean towards leftism (NAS 2018). I had no issues in working in academic environments when the Left was tolerant and scientific. With dissenters canceled, the intolerant and anti-scientific institutionalized woke ideology developed unchallenged.

People started noticing that part of what is presented as science is nonsense. Polls find that the majority of people outside the left-leaning spectrum started perceiving *US higher education as negative*. Ben Shapiro writes: "Our elite universities began as places to teach eternal truths and seek actual knowledge. Now they're left-wing activism training centers. Defund them." The Atlantic writes, "The ideas of a radical vanguard are now the instincts of entire universities—administrators, faculty, students." "Group identity assigns your place in a hierarchy of oppression. Between oppressor and oppressed, no room exists for complexity or ambiguity. Universal values such as free speech and individual equality only privilege the powerful. Words are violence. There's nothing to debate" (Packer 2024). The opposite political area took notice and started to push back. When the Left loses elections, DEI discrimination moves from being considered a universal human value to being outlawed.

Harvard used to be the pinnacle of academic excellence in all rankings. But its president, Larry Summers, resigned in 2006 after having mentioned higher male variability to gender activists. In the same incident, the physicist Lubos Motl also left Harvard. Recently, the biologist Carol Hooven left Harvard after daring to write about biological sex differences (Hooven 2024). The Harvard economist Roland Fryer found no race bias in US police shootings and dared to publish, despite warnings that this could ruin his career (Fryer 2024). The Harvard temple of knowledge (if we still wish to call it that) now hosts many Particles for Justice advocates and occupies the bottom position in the FIRE free speech rankings, with an "abysmal" below-zero rating (FIRE 2023).

In the Soviet Union, a bell signaled when people could safely stop clapping after Stalin's speeches. In the US, the bell rang when the de-colonialist component of DEI attacked Jews and the Harvard president (among others) argued that, depending on the context, *their* speech is free speech. The double standard became too evident: "Harvard is now the place where using the wrong pronoun is a hanging offense but calling for another Holocaust depends on context" (Pinker 2023). People started to question the academic credentials of the Harvard president despite her power and intersectionality score as a Black woman.

Various authors stopped clapping to DEI and started checking its claims, finding that they don't replicate. Nevertheless, DEI claims keep being heavily promoted as scientific by mainstream media. As a result, some people have lost trust in science and have started generically viewing academia as politicized.

What Can Be Done?

If academia recovers, the current period will be remembered as historically dire, akin to the legacies of Lysenko and McCarthy. What makes the current failure more shameful is that it's self-inflicted within democracies. It could have been avoided with a bit more courage. Now, many recognize the negative aspects of the DEI ideology. But now, DEI has become institutionalized. It will not go away spontaneously.

Internal reforms of captured institutions seem difficult. Moderates within academia could have prevented the problem but didn't want to fight. Perhaps they now think that they have more to lose by remaining silent.

External reforms provide a more realistic hope. This path is not without difficulties: I concluded my 2018 CERN talk warning that politicized scientific institutions risk being involved in messy political battles. These are now unfolding. In some states, political authorities started forbidding by law some worst aspects of DEI and/or removing "grievance studies." These actions risk conflicting with academic freedom. Nevertheless, academic freedom comes with responsibilities: it should have been safeguarded by never allowing "studies" to masquerade as academic fields and politicize academia. "Grievance studies" are systemically flawed. Canceling politically incorrect but scientifically correct results seems as bad as scientific fraud. Simply upholding the standard norms of academia would allow the removal of the "studies" that attacked all other fields as biased and political while producing biased politics.

Removing "grievance studies" might sound extreme, but their absence remains the norm in many countries outside the Anglosphere, where "studies" are only now attempting to enter academia. In these countries, moderate actions would suffice to avoid repeating the US mistakes. Academic institutions could still convincingly adopt institutional neutrality and uphold the freedom of speech of their members.

A final possibility is that the Anglo academic system will not return to the meritocratic, open, apolitical excellence that I had admired. In the international panorama, it was an exception: the special product of a rare trust-based society. This exceptionalism might be lost.

The previous text was written before the 2024 US elections. Politicized academia lost, and the elected US president is taking the expected actions to "reclaim America's educational institutions from the radical left … removing all Marxist DEI bureaucrats." His words, and his revocation of the 1965 temporary order that started DEI discriminations, indicate that his goal is reverting the political takeover that started from academia and expanded largely unnoticed through decades of top-down bureaucratic policies. If successful, the US will once again demonstrate its strength to first try new ideas and next to reject the failed ones. Achieving stability will require using free speech to mainstream the reason of the failure: differences don't imply discriminations.

DEI in Science and Medicine: Missing Metrics and Measures

ROGER B. COHEN AND AMY L. WAX

The process for developing and approving new drugs and treatments for cancer is rigorous. As overseen by the federal Food and Drug Administration, it involves multiple rounds of testing and systematic and precise analysis of patient outcomes, including comparisons to how patients fare with standard therapies. A key part of clinical testing is to define the metrics and measures that determine success or failure. Either a new therapy effectively treats the cancer, improves symptoms and signs, or prolongs life, or it doesn't. Those effects must be demonstrated by substantial amounts of reliable data. And, of course, assessments of new drugs also include careful monitoring for side effects, which can be life-shortening in themselves or otherwise intolerable and can defeat any positive effects. Most experimental drugs for cancer, often hailed with great fanfare and with millions of dollars invested, fail the tests. For those that do pass and are approved for licensing and marketing, data collection and monitoring continue to ensure that the initial claims of benefits are accurate and that the drug is performing as expected in the real world and outside of the clinical trials process.

The clinical trial process in medicine is meticulous and unforgiving. Nothing is taken for granted, and all assumptions are designed to be falsifiable. Impressionistic observations and wishful thoughts and feelings don't count. The supporting data must be detailed, statistically valid, and demonstrably reliable.

This article is written by a specialist in cancer clinical trials (Roger Cohen) and a law professor (Amy Wax, his wife). In this essay, the authors argue that the gold standard for clinical trials should apply to all interventions and changes in health-care fields that are planned, proposed, or implemented. In particular, the changes proposed for medicine in the name of so-called "diversity, equity, and inclusion" should be subject to rigorous evaluation and scrutiny. Proposed DEI initiatives range widely and affect all aspects of the medical system, from how new doctors and other personnel are selected and trained to the design and selection of treatment methods and protocols and to research design and funding. The injection of DEI principles into medicine warrants no less scrutiny than any other intervention in the health-care field. Before implementing any proposed program or modification of existing health-care practices, key questions must be considered and defined. What are the goals? What are the metrics, measures, and "deliverables"? What precise improvements or benefits do we hope to generate, and how will they be demonstrated and monitored? Once the intervention has occurred, what plans are in place to assess whether the goals were accomplished and whether health outcomes are improved or degraded? What kind of data will be gathered, and how? Are the methods for analyzing the data rigorous and unbiased? Are there plans to replicate any initially promising results before widespread acceptance and implementation occur?

Currently, these questions are rarely being asked, pursued, or analyzed in a sustained and systematic way. To the extent that DEI is being evaluated at all, the efforts are scattershot and fall well below any scientifically defensible standard. In other words, DEI initiatives are being implemented in medicine on every front with little sustained effort to subject the DEI-inspired changes to the gold standard of scrutiny and assessment that prevails in every other area of modern medicine. Examples abound. Last June, for instance, a document was released to the community by the leadership of the main teaching hospital of the Perelman School of Medicine at the University of Pennsylvania, one of the best medical schools in the country. The document described how Penn is signing up with enthusiasm for a new Joint Commission health equity certification. The Joint Commission inspects hospitals once every three years to determine whether they meet defined standards of sound medical practice. The additional health equity accreditation by the

Joint Commission, a new part of the certification process, is still technically voluntary. But it is understood that declining to seek such certification would not only be unwise, but downright unthinkable. A hospital's ability to operate legally and bill for its services rests on private accreditation by the Joint Commission. Proving "equitable" bona fides—which means adopting "diversity, equity, and inclusion" principles, priorities, and practices in all hospital activities—is therefore effectively mandatory. No hospital, even the most prestigious and venerable, can afford to refuse.

The Joint Commission has gone full woke: "*This new, voluntary advanced certification for [a] health care equity program provides the structure to guide your organization's journey to achieving health care equity. It guides forward movement in imbedding health care equity in all aspects of care, treatment, and service delivery.*" The Joint Commission's ukase on DEI does not even try to explain or justify the new equity certification in the traditional terms of better health outcomes for patients, which is the Joint Commission's traditional ambit. Will "health care equity" make the hospital safer and more efficient, or the patients better off? Will fewer patients die or develop complications if this new certification occurs? Will it improve the services that hospitals exist to deliver? Radio silence on these issues.

We are now about fifteen years into the grand medical DEI experiment. Its grip on the field only grows stronger. The infiltration of DEI into every aspect of medicine, starting with undergraduate (medical school) and graduate (residency and specialty fellowships) admissions, education, and training, and now extending into hospital management, hiring and personnel management, human resource policies, medical and treatment resource allocation, and medical research priorities and funding, rests on a series of unproven and often dubious assertions. Each step into the DEI abyss either rests on no credible evidence at all or on a small number of oft-cited but questionable and poorly designed studies that rely on crude and unexamined assumptions. These studies have never been subjected to the painstaking and searching scrutiny that has long been the standard at the top of the American medical establishment. This wholly inadequate evidence, which would never be tolerated in any other scientific field, is used to justify sweeping changes in every aspect of our profession. Validation, replication, control variables and groups, critical analysis and re-analysis, the application of well-established statistical standards, and diligent vigilance against the elementary error of conflating correlation with causation are conspicuously missing from the literature that is driving these changes. Vague jargon and buzzwords (the ubiquitous parlance of "diversity, equity, and inclusion," "systemic racism," "structural racism," and "social determinants of health") are routinely deployed without ever being precisely defined. These egregious methodological flaws, and specifically the absence of well-defined outcome metrics and their systematic, careful measurement, are the focus of this essay.

The shortcomings described here are grievous and ought to concern everyone and not just people in the field. Modern medicine, which has matured over centuries to rest on a sound scientific foundation, is based on constant, unsparing examination and scrutiny. Like any biological or physical science, the search for truth and the process that leads to progress requires constant testing and debate, often leading to the outright rejection of treatments and approaches that don't work or fail to meet well-defined metrics and expectations. Nothing like this is happening today in the newly hatched field of "health equity." Two centuries of methodological and scientific progress in medicine have been jettisoned, or at best ignored, in the name of enshrining the DEI woke transformation of a proudly precise and disciplined field.

The return to rigor and standards should start with medical education. One of us has written previously about the dramatic shifts that have recently occurred in how we select and train future doctors and has urged that we approach such reforms with great caution. Deciding to place a major and sometimes primary emphasis in medical admissions and training on social justice and diversity, equity, and inclusion, rather than on candidates *demonstrated and measured* (not, as DEI acolytes would have it, so-called latent) ability, is nothing short of reckless.

Before making such changes, we should, at the very least, subject departures from longstanding practices to objective and unbiased analysis of their effectiveness. Without such an assessment, we imperil the overarching goal of achieving the very best outcomes for our patients based on the highest quality medical care. And we also jeopardize future advances in basic and medical science that depend critically on the quality of the next generation of scientists. Future therapeutic advances will necessarily be sacrificed if diversity initiatives take priority over finding, admitting, and training the best and the brightest people who have demonstrated the intellectual attributes and abilities for scientific achievement at the highest levels. Once students are selected, what they are taught is also crucial. Medical school and physician training is brief. More focus on social justice and health equity means less time and attention to the technical, demanding, and vastly time-consuming efforts to acquire medical knowledge, including learning scientific methods and techniques and performing scientific research. Advocates for curricular changes and a reorientation of medical training to prioritize social justice and equity rather than the scientific and medical fundamentals rarely (never, really) claim that this radical reform project will promote scientific excellence and sustain the scientific innovation that has long characterized American medicine. Rather, they speak vaguely of a "new excellence," which is never actually defined.

Currently, empirical support for the goal of *"embedding health care equity in all aspects of care, treatment, and service delivery"* is, when properly examined, practically nonexistent. Apart from a few studies, repeatedly cited but easily faulted, the main justifications for DEI in health care are rhetorical and based on politically popular, feel-good assertions that are far from proven but that people in the medical field are very reluctant to challenge. It is worth pointing out that the diversity in DEI has nothing to do with diversity of thought or political point of view, including any ideas that challenge DEI paradigms and the basic assumptions behind a social justice focus. The diversities valued by the AAMC (American Association of Medical Colleges), as customarily presented in public diversity talks, are socioeconomic status, race, ethnicity, language, nationality, sex, gender identity, sexual orientation, religion, geography, disability, and age. Conspicuously missing are the categories of diversity of ideas, thought, political position, and point of view that would encourage, if not demand, that DEI initiatives stand on more solid ground before they are allowed to transform long-standing admissions, training, and treatment practices. In fact, the lack of proper skepticism towards the unscientific, irrational, empirically ungrounded nature of the changes presently proposed and implemented under the banner of DEI is demonstrated by the attitude towards nonbelievers. In most health-care institutions and medical schools, even the most prestigious, questioning the tenets of DEI and the "bias narrative" at its heart or positing any alternative explanation other than racism or prejudice for minority group ills is a form of punishable heresy. The mere existence of such heresy indicates the need to double down. It is taken as proof positive that even more DEI is needed.

What kinds of assertions and evidence are currently adduced to justify the DEI-based transformation of health-care practices?

> Numerous articles and presentations in the medical literature assert: "Diversity, equity, and inclusion (DEI) are essential principles that physicians must embrace to enhance rapport with patients, increase patient compliance with treatments, improve patient outcomes, increase patient satisfaction, and build trust."

This set of claims is one variation of an oft-stated assertion: DEI efforts produce better health outcomes, and especially for minority patients. How exactly do DEI efforts produce this result? The focus is on increasing the number of underrepresented and minority physicians and health-care workers. Patient outcomes will improve if their doctor is the same race, it is claimed, because patients are more likely to adhere to and trust treatment recommendations, and minority providers will be more effective and astute caretakers of their minority patients. The so-called "racial concordance" thesis, which posits positive effects from doctors and patients sharing a common group identity, has also been extended selectively to other categories, such as gay people—but not to whites or South Asians. The bald and sweeping assertion in the statement quoted above is typical of the genre. But it is backed up by little or no solid evidence. What evidence should we be looking for? The physicist Wolfgang Pauli famously stated that a hypothesis cannot be established as true unless and until an experiment can be devised to prove it wrong. It must be "falsifiable." If it is impossible to conduct a test to disprove a claim, then that claim is not one that belongs to science but rather to the realm of ideology, fantasy, dogma, and wishful thinking. The racial concordance fad reveals the current state of the practice of DEI in health care. No studies have been proposed, or carried out, in any attempt to disprove the hypothesis.

Instead, this idea continues to exert a pervasive influence despite weak and limited evidence to support it. The so-called "Oakland study" (Alsan et al. 2019), purports to show that black patients with black doctors have better health outcomes. Despite being repeatedly cited, this study is riddled with basic methodological flaws that effectively render it useless. These include the lack of an adequate control group, the projection of facially preposterous lifetime mortality benefits from a single point-in-time observation, and conclusions based only on patients' expressed willingness to engage in preventative care without any quantitative evidence of increased use of preventative services or of actual health benefits, from whatever source, in the short- or long-term.

Another widely touted and reported study claims that newborn black babies are more likely to survive if they are cared for by black doctors (Greenwood et al. 2020, 117 (35)). The single study that is the basis for this claim is also fatally defective. For one thing, it uses an administrative database of the general racial composition of caretakers in the studied health-care facilities as a substitute for determining the actual race of the doctors caring for individual babies—a methodology that is entirely inadequate and unreliable by any standard of sound medical social science. It also fails to control for the birthweight and medical condition of newborns as an important determinant of the doctors who end up caring for them. (Borjas and VerBruggen 2024).

These shoddy articles are repeatedly and uncritically cited to justify dramatic overhauls in policies and practices in health care delivery and training. What is remarkable is how the two studies just described have achieved such hallowed status, with eight hundred-plus citations in the literature for the Alsan et al. study and approximately three hundred for the Greenwood et al. study. The few academic critiques that do exist are generally ignored. Among the DEI bureaucrats, proponents, and so-called "experts" who now abound in medical schools and other medical establishments, attempts to replicate, disprove, or even systematically critique such "research" almost never occur. Instead, presentations of the "concordance" idea are routinely accompanied by vague references to "many studies" or "studies show" without specific citations. Because DEI initiatives in health care are motivated overwhelmingly by political priorities rather than actual evidence, there is no reason to believe that research that comes to the "wrong result" will ever get reported or published. "Another widely cited DEI pillar is the assertion that 'teams with diverse perspectives achieve higher scientific and economic impact.'"

In 2015, 2018, and 2020, the venerable consulting firm McKinsey released three reports purporting to show that greater racial and gender diversity in large public companies' executive ranks results in higher profits. Reliance on these studies has become a staple of DEI presentations, in which speakers regularly draw a straight causal line from more personnel diversity to higher profits. The problem here, of course, is that correlation is not causation. Maybe the causal arrow, in fact, runs the other way, with more profitable firms choosing to prioritize hiring a more diverse workforce or being better able to afford to do so.

There are other reasons to doubt the bona fides of the McKinsey study claims. A paper published earlier this year by economists Jeremiah Green and John Hand (*Econ Journal Watch* 2024) tried to replicate the McKinsey data. They failed. They found no link between racial and ethnic diversity and the financial performance of S&P 500 firms. To be sure, there are differences between the Green and Hand study and the original McKinsey reports, including the types of firms examined and their location, which could explain the discrepant results. Unfortunately, McKinsey's refusal to disclose the raw data behind their study makes a more detailed comparison impossible. And there are additional metanalytic and quasi-meta-analytic papers in the literature that find insignificant, negative, or mixed effects on productivity and profits based on workforce age, gender, and cultural background.

There are also other sources on business performance that could be analyzed to test the McKinsey reports' conclusions. For instance, at least two prominent mutual funds that invest in "diverse" businesses have been created: iShares Refinitiv Inclusion and Diversity UCITS ETF and SPDR® MSCI USA Gender Diversity ETF. The performance of these funds can be readily examined; they grossly underperform their peers. The experience with these so-called "diverse" investment funds is in tension with the conclusions of the McKinsey studies. These facts should be front and center in any discussion of the relationship of personnel and leadership diversity to firm profits. Yet they are not.

The story for scientific team performance is just as weak. There are summary analyses in the literature, referred to above, that have shown that the link between team diversity and aspects of science team performance is decidedly mixed and sometimes negative. But the literature on this topic—which is admittedly sparse—is short on specific metrics and measures of scientific success, quality, and contributions. Some possible parameters to examine might include publications, patents, citations, H-indices (which reflect how often papers are cited), intradepartmental collaborations, interdepartmental collaborations, and number of invitations and oral presentations at major meetings. The literature does not fully explore the available data.

One metric of outcomes that is currently useless, unfortunately, is the number of NIH or other prestigious government grants awarded to an investigator or team. That is because affirmative action is now rampant in federal science funding. The NIH FIRST Grants Program is a typical and lavishly financed example of this phenomenon. Scathing exposés of questionable practices in this program can be found on several sites[1]. The stated goal of NIH FIRST and programs like it is to achieve more "diversity" in science, with an emphasis on race and skin color. In describing and justifying this objective, the relationship of diversity to desirable outcomes and "success" is both assumed and touted in vague, jargon-laden terms that have nothing to do with well-established, concrete measures of the outcomes that should matter: greater human health and longer life. An emblematic statement can be found in the 2023–2027 NIH-Wide Strategic Plan for DEIA: *"To achieve institutional and research excellence, NIH must foster and sustain an inclusive and equitable culture that embraces DEIA, both in the workplace and in the pursuit of biomedical and behavioral science. The true measure of success for cultural change is belonging—the feeling and knowledge of being included in the NIH mission."* Another passage in the 2024 NIH government grants guide[2] states: *"Research shows [no references given] that diverse teams working together and capitalizing on innovative ideas and distinct perspectives outperform homogeneous teams. Scientists and trainees from diverse backgrounds and life experiences bring different perspectives, creativity, and individual enterprise to address complex scientific problems."* How those "different perspectives" result in specific health improvements is neither addressed nor elaborated. Once again, the standards of basic scientific rigor that inform other areas of medical science are conveniently forgotten.

DEI principles have now become a scientifically unproven priority for government science funding. This distorts which scientists and projects are supported. But DEI considerations also now dictate how the research is performed! One egregious example can be found in research on HPV-related head and neck cancer. In recent years, there has been a pronounced uptick in this disease in the USA, Europe, and Australia. This potentially fatal type of head and neck cancer is not an equal opportunity affliction. Whites are affected almost ten times more often than blacks, and 80 percent of patients are men. Epidemiology studies have established that the incidence patterns are in large part related to sexual behaviors that differ among groups, including the self-reported performance of male-on-female oral sex.

For a recent clinical study in HPV-related head and neck cancer conducted at the University of Pennsylvania, the National Institutes of Health (NIH) program officer, reviewing the study's progress, asked the investigators why there were no blacks in the study. In response, the investigators pointed out that blacks seldom get this disease and that this well-known fact had been noted and discussed in the original grant proposal. That answer did not satisfy the DEI commissar: "That answer is non-responsive and unacceptable, find and enroll them or we will stop funding your grant." How the cessation in funding would help in finding better treatments for this type of head and neck cancer was not addressed.

To illustrate the absurdity and destructiveness of this reflexive bureaucratic reaction, consider an analogy to sickle cell anemia. Yes, some whites are afflicted with sickle cell disease, but they are few. The ratio of blacks to whites is around fifty to one. A demand that all research on sickle cell disease involve some "adequate number" of white patients is not "equitable." It is irrational. And such a demand would mean that important lifesaving research on sickle cell anemia, which mostly affects blacks, would be seriously impeded.

A core DEI claim is that adopting DEI principles and practices is the only way to effectively address health disparities.

Obamacare dramatically improved patient access to health care after 2011. More people acquired personal physicians and visited their doctors, including minorities who were previously underserved. But better access has not improved racial and other group disparities in health-care outcomes. As one national DEI proponent, Dr. Consuelo H. Wilkins from Vanderbilt University Medical Center, recently pointed out, *"[t]he absence of progress [on the impacts of racism and inequities in health care] compared with the survey from nearly 3.5 years ago is striking.... Both the extent of the disparities and the perception of how patients are impacted is virtually unchanged. This is despite all the discussions we have had and the new programs and increased funding for equity initiatives."* See NEJM Catalyst in 20243. One possible explanation for these disappointing results might be that the diagnosis and prescription are wrong. DEI proponents routinely point to "structural" or "systemic" racism as the main cause of intractable group disparities, and especially for poor health outcomes for blacks. The proposed solution is "anti-racism," a concept that is never precisely defined. Alternative explanations of health disparities (including choices and behaviors of the patients themselves) are rarely proposed, let alone investigated. Indeed, suggestions of any personal responsibility for inferior health outcomes are quickly shut down under the angry rubric of victim-blaming. And the suggested solutions are always the same: more resources, money, programs, and "accountability." That is obviously an inadequate and dogma-driven approach to health disparities. Perhaps, for example, weak and unstable families and a relative lack of family support might contribute to the failure to see a doctor, comply with treatment, or make lifestyle adjustments conducive to positive health outcomes. At the very least, the claim that DEI in health care might mitigate or remedy the effects of the lack of familial comfort and support should be studied and verified. Specific interventions to compensate for family failure should be proposed, and their effects scrutinized and measured. The data must be made available for evaluation. None of this is currently happening. Such tunnel vision would never be tolerated in any other area of medicine. The routine failure to consider alternative hypotheses and causal factors, even glaringly obvious ones, would be regarded as scientific and empirical dereliction.

> DEI proponents are ardent supporters of funds for disparities research, especially between racial groups. What is the evidence that health disparities research has resulted in better health outcomes?

The increase since 1990 in government funding for research on so-called "health disparities" is staggering. As of 2020, a third of all National Science Foundation (NSF) grant awards included one of the following terms: "equity," "diversity," "inclusion," "gender," "marginalize," "underrepresented," or "disparity," up from 3 percent in 1990 (data summarized by the Center for Partisanship and Ideology 2021). Young investigators not surprisingly follow the money, so the number of grant requests to support disparities research has exploded. In the same vein, applications for prestigious residency or fellowship training programs in medicine, even from MD/PhD candidates being trained to do basic research, increasingly mention and highlight the applicant's interest in disparities research. Since 2015, the National Institutes for Health (NIH) has required the inclusion of "score driving" (NIH's own minatory words) diversity priorities for all projects seeking funding within its seventy-two National Cancer Centers, including research on preclinical models of basic biological processes in organisms such as worms and zebrafish! The money diverted into disparities research, or used to contort studies where diversity priorities have no place and add nothing, is money that cannot be used for other projects. Resources are scarce, so disparities research and priorities should at least be scrutinized, tested and evaluated, and called upon to demonstrate efficacy. Nothing like that is happening now. Indeed, the standards for disparities research are embarrassingly low and would not pass muster in any other area of medical investigation. Most disparities research is highly descriptive, and the vocabulary is vague, ideologically informed, and stereotypical. The terms structural racism, systemic racism, and anti-racism provide little guidance on how to address exceedingly complex, multifactorial societal problems, let alone scientific ones.

> DEI supporters assert that diversity in medicine benefits everyone, not just racial and ethnic minorities. The claim is that white physicians in racially diverse medical schools are more culturally responsive and report feeling more comfortable treating diverse patient populations.

Even if, as often claimed, physicians and patients trained in DEI "feel more comfortable around diverse people," there is effectively no evidence that these feelings lead to better health outcomes. It should be obvious that feeling better is not the same as making people better. Cultural sensitivity and cultural humility are different from competence; they may or may not go together. The goal of creating more diverse medical schools and promoting a "sense of community" has generated a cottage industry of trainers, handlers, and consultants and an endless plethora of training sessions, workshops, and presentations on microaggressions, implicit bias, unconscious discrimination, and similar topics. But there is no reliable evidence that these activities help us get along better, let alone that they improve health care or medical results. In fact, some troubling evidence has emerged that these training courses may have negative effects, promoting cynicism and resentment rather than insight and cooperation.

One especially favored training tool is the implicit association (IAT) test, also called the implicit bias test, which claims to help people discover potential prejudices that lurk beneath their awareness—and correct them! The flaws of this test are well established. It does not even measure what its aficionados say it does (racism), is not replicable (based on test-retest data), and its well-respected academic inventors have publicly decried its routine use in institutional anti-racism training. Yet, it has become the centerpiece of a profitable industry. Health-care institutions such as Penn Medicine now routinely subject their employees to regular IAT training and testing and boast of exceptionally high compliance rates.

Fifteen years in and with increasing recognition of its pernicious effects on institutions and scientific excellence, the time has come to recognize that ideologically driven DEI initiatives have no place in institutions of higher learning, especially in medical science, unless they are validated by hard evidence and data. Metrics and outcome measures for all diversity initiatives must be defined, and the results of all interventions must be precisely measured and demonstrated. Some universities have as many as one hundred staff in the DEI office. In business parlance, there must be "accountability" and "deliverables." Medical science is adept at measuring—that is its stock-in-trade. Its tried-and-true practices should be extended to investigating the value of DEI.

There are some simple, specific steps that could be taken. The weak studies underpinning many sweeping diversity initiatives need to be sunsetted, starting with the Oakland adults and Florida newborns studies. Neither article is worthy of respect even under the basic standards of social science. In science, mediocre and flawed papers get replaced by better papers. (See, for example, the most recent PNAS study quoted earlier on black infant mortality.) Older treatment paradigms in medicine are regularly abandoned in favor of better treatments. Drugs are retired or have their FDA approval rescinded. There is nothing wrong with this. Quite the contrary—it is essential to medical quality and progress. Without new data and disruptive thinking, we would still be bloodletting. These insights should be applied to DEI, and as soon as possible.

Understanding Why Plagiarists Are Not Fired: From Veritas to DEI

PETER BOGHOSSIAN

You are the provost of a major US research university. After a fair administrative proceeding, it has been conclusively determined that one of your faculty has plagiarized large sections of their dissertation.

You have the following options:

A) Terminate their employment.
B) Give them a written warning.
C) Terminate their employment and revoke their licensure.
D) Do nothing.

If this question were asked fifteen years ago, people would find it bizarre—the answer is obviously either A or C. Yet today, university administrators have answered unequivocally: D.

Why?

There is an ideology undergirding decisions to turn a blind eye to academic plagiarism, but to better understand the degree of the problem, let's first look at a few high-profile examples. The highest profile case of academic plagiarism, that is, theft of another's ideas without giving credit, is Claudine Gay's. Gay is the former president of Harvard and a serial plagiarist. In addition to plagiarizing her doctoral dissertation, she also plagiarized four peer-reviewed papers. In equal parts astonishing and impressive, she even plagiarized her dissertation's acknowledgments. Harvard responded with a variation of Option D: She stepped down as Harvard's president but retained her $900,000 a year salary, professorship, tenure, and title as "professor of government and of African and African American studies."

Harvard's chief diversity officer, Sherri Ann Charleston, is also a serial plagiarist. According to independent (i.e., non-Harvard) reviewers, she "plagiarized more than 40 passages of her 2008 dissertation, 'Converging Frameworks: Examining the Impact of Diversity-Related College Experiences on Racial/Ethnic Identity Development.'" Harvard Extension School administrator Shirley R. Greene is a serial plagiarist as well (forty-two instances of plagiarism in her dissertation alone!), and so too is the prominent professor of sociology Christina J. Cross.

Plagiarism scandals are not confined to Harvard. Plagiarists have been exposed at Columbia's medical center, the University Wisconsin-Madison, MIT, and elsewhere. At MIT, two of six high-level DEI officials are serial plagiarists. One DEI official, Tracie Jones-Barrett, "copied an entire section on 'ethical considerations' from a classmate in her Ph.D. program."

In any other period of academic history, plagiarists like Claudine Gay, Tracie Jones-Barrett, and others would have been fired and disgraced. The difference now, however, is that an invasive ideology has overtaken universities. Helen Pluckrose, coauthor of *Cynical Theories*, terms this new orthodoxy "applied post-modernism." Applied postmodernism is the view that "society has been constructed in the service of dominant groups by their having created and perpetuated discourses—ways of speaking about things—that exist to marginalize people of color and particularly women of color, both actively and passively." Once one comprehends the ethos of applied postmodernism, it will become clear why academic administrators universally chose to not fire plagiarists.

Understanding the applied postmodernist rationale requires a fundamental shift in one's thinking about the purpose of the university and the peer-reviewed system. It is to shift from Harvard's official motto, "Veritas," to the unofficial moral maxim of diversity, equity, and inclusion. In the applied postmodern worldview, the purpose of the university is not to find truth, but rather to oppress marginalized people. Universities and the peer-reviewed system (along with the police, capitalism, the judicial system, and every other system) *cause* disparities in outcomes. Specifically, universities and their literature canons are oppressive structures whose goal is to keep minorities (African Americans in particular) disenfranchised and disempowered while actively working to preserve the power of dominant groups. That is, quite literally, their raison d'être.

Applied postmodern academic Alison Bailey refers to this as "privilege-preserving epistemic pushback." That is, the phenomenon that privilege (and those who wield privilege, like white people, men, heterosexuals, and the able-bodied) will seek to preserve or maintain itself at the expense of all else. The goal of DEI, which is a manifestation of applied postmodernism, is to disrupt and dismantle instruments of privilege maintenance. Privilege maintenance machines, like the university system, which is "deeply embedded in the fabric of society," are instruments the dominant groups use to retain power and, thus, must be disrupted and dismantled because they are the main drivers of inequalities.

Once these systems and the discourses that flow from them are disrupted, marginalized peoples—black people, women, homosexuals, disabled people, fat people, trans people, midgets—can be liberated from oppression. But as privilege seeks to preserve itself, the tools of oppression will not be voluntarily relinquished by the dominant groups. They must, therefore, be subverted, disrupted, and dismantled. In fact, this is the only proper use of power—to remediate oppression by placing power in the service of advancing the needs of the downtrodden.

There are two primary ways to subvert the academic engines of oppression: (1) for marginalized people to hold positions of power and authority, and (2) for oppressed people and their allies to create and institutionalize lines of peer-reviewed literature that change discourses and actively promote certain values. In the first case, marginalized people need to occupy faculty and administrative positions. As such, they will be better situated to promulgate certain values, advocate for other marginalized peoples, and disrupt and dismantle the system. In the second case, marginalized peoples need to contribute to the literature that actively undermines oppressive power structures and then codify that literature as university policy. Specifically, diversity (uniformity of viewpoint in ideological matters among people who are marginalized), equity (equality of outcome), and inclusion (making oppressed people feel welcomed by prohibiting hurtful speech).

Once the principles of applied postmodernism are made explicit, it is easy to understand how academic administrators have *only* one option, D—do nothing when confronted with plagiarism. Any other response, like terminating the plagiarist's employment (A) or giving a written warning (B), would actively work against *the* objective of disrupting the system—and disrupting the system is the moral mandate of applied postmodernists.

This is why African American plagiarists who advance the cause of DEI are not fired. And it is why President Gay was found innocent of plagiarism before her investigation had begun. And once there came to light indisputable evidence that she had repeatedly plagiarized, her actions were termed "duplicative language" instead of the stigma-carrying term "plagiarism." Gay was not terminated from employment because the objective was to keep Gay, a black woman whose academic work actively promoted DEI and who successfully instituted said policies, in the highest position of authority in one of the world's most elite universities. Retaining Gay's position as a full professor at Harvard satisfies applied postmodernism's primary goals: one, marginalized people holding power and authority, and two, contributing to peer-reviewed lines of literature that disrupt dominant discourses.

Implications of tolerating plagiarism in academia are profound, as they signal a shift away from accountability and toward a prioritization of ideological alignment over time-honored ethical standards. By allowing plagiarists to retain their positions, universities not only undermine their credibility but also compromise the integrity of the educational system. This raises significant concerns about the value of academic credentials and the trustworthiness of peer-reviewed literature. As academic institutions grapple with conflicting demands of social justice and academic integrity, they must consider the long-term consequences of their decisions. Ultimately, the academic legitimacy hinges on prioritizing Veritas over DEI.

In conclusion, the reason plagiarists are not fired stems from the dominating influence of applied postmodernism and its emphasis on DEI. This ideology prioritizes the preservation of marginalized voices within institutions over the adherence to traditional academic integrity. As demonstrated by high-profile cases, the ramifications of this shift lead to a disturbing tolerance of grossly unethical behavior when it aligns with the broader goals of dismantling perceived oppression. The decision to not fire plagiarists highlights the urgent need to reevaluate principles that govern academic institutions and their ethical commitments. As we move forward, the academic community must decide whether it values truth or DEI.

SECTION 5
Gender, Race, Ideology, Science, and Scholarship

This section follows up in detail on a hot-button issue that has been raised in this volume. It merits further detailed exploration, however, because perhaps nowhere is there a more divisive scientific issue with significant popular impact at the current time than the relationship between sex and gender.

Two areas, in particular, have resulted in lawsuits and legislation that are affecting millions of people around the globe every day: (*a*) the availability and advisability of "gender affirming care" (GAC), especially for minors, and (*b*) the question of how society should accommodate self-identifying transsexual individuals, and whether, for example, public spaces previously reserved for females, such as dressing rooms and toilette facilities, should be opened up for individuals who identify as women, and whether concerns about such practices is equivalent to transphobia.

The section begins with a chapter by Drs. Lauren Schwartz and Arthur Rousseau, who compellingly argue that too often claimed research studies associated with gender-affirming models are based in ideology and not science. As they put it, "Given the positive connotation of 'affirmation,' GAC *should* have been the compassionate and comprehensive care that considers all the complexities that transgender, nonbinary, and gender-diverse individuals face in their lives. Instead, it has become a caricature of health care, calling for trainees, physicians, and mental health providers to uncritically accept a patient's stated gender identity and to rubber-stamp treatment based on their self-diagnosis. Consequently, even the youngest and most vulnerable patients are subjected to unproven and experimental medical interventions with lifelong and debilitating consequences."

Next, Alex Byrne and Moti Gorin discuss the divide between the US and many other Western countries associated with GAC and the failure of both philosophers and bioethicists to seriously address many of the substantive issues associated with GAC, as well as the shoddy state of much of the research that is currently being used to justify GAC interventions. They conclude with some recommendations for improving academic integrity in this area.

Judith Suissa and Alice Sullivan turn to the chilling effect in academia of restricting free and open discussions of issues associated with gender, where merely stating that sex exists as a meaningful category distinct from an individual's self-declared "gender identity" is now often branded as transphobic. They describe the potentially disastrous consequences for learning and knowledge production of this new form of censorship and, more generally, its broader implications for democracy.

Finally, Karleen Gribble explores the implications of the desexing of language in women's health care, such as replacing "women" with terms such as: "people who menstruate" or "bodies with vaginas"; "mothers" with "parents"; and "breastfeeding" with "chestfeeding." She argues that dehumanizing and often scientifically incorrect semantic replacements, sometimes now required by scientific journals, are not innocuous and can have deleterious effects both on the design and implementation of human research studies and on the ability to effectively provide health care to those segments of the population who are not attuned to such new awkward terminologies. In particular, she argues that this imposition of new language in the West is not appropriate for export to much of the rest of the world.

Gender-Affirming Care: The Abandonment of Medical and Academic Standards

LAUREN H. SCHWARTZ AND ARTHUR W. ROUSSEAU

The Infiltration of Transgender Ideology

The foremost fundamental responsibility of the physician is to protect the safety and well-being of every patient. First, we do no harm. Ethical medical practice demands that we keep in the forefront well-established gold standards and avoid experiments with unproven standards of care promoted by self-proclaimed experts. Patients seeking care related to identity and gender deserve to receive the same evidence-based, safe, comprehensive health care that we strive to provide for all patients.

The gender-affirming model has upended every sacrosanct tenet of health care, infiltrating academia, research, medical schools, residency programs, general practice, and public policy. Gender-affirming care (GAC) fails to meet even the most basic expectations demanded of medicine through the insidious conflation of experience with clinical expertise and civil rights with patient rights. This becomes appallingly clear when contrasted with the widely accepted, proven standards of care and evidence-based treatment protocols in child development, pediatrics, and psychiatry, scientifically established over a century or more. In place of scientific principles, gender-affirming providers offer their "lived experiences," an extensive collection of anecdotes that import every imaginable bias into allegedly scientific literature (Goetz and Keuroghlian 2023).

All aspects of medicine have been influenced by the ideological overshadowing of care for individuals seeking gender-related support. Yet GAC does not have the medical, pediatric, or psychiatric consensus that US professional associations insist it does. The outcome of this misrepresentation is an on-demand, vending machine approach to the treatment of gender dysphoria, espousing patient rights to bad medicine.

Given the positive connotation of "affirmation," GAC *should* have been the compassionate and comprehensive care that considers all the complexities that transgender, nonbinary, and gender-diverse individuals face in their lives. Instead, it has become a caricature of health care, calling for trainees, physicians, and mental health providers to uncritically accept a patient's stated gender identity and to rubber-stamp treatment based on their self-diagnosis. Consequently, even the youngest and most vulnerable patients are subjected to unproven and experimental medical interventions with lifelong and debilitating consequences.

It is unthinkable that these radical medical interventions are being carried out overwhelmingly on children and young adults, at an age when they do not have the mental capacity to either understand the enormity of these interventions or their consequences. However, here we are. GAC instructs health care providers to maintain focus on gender distress, even as patients present a host of complex comorbidities. Consequently, for most patients, the symptoms of their co-occurring challenges get worse. Instead of eliciting any second thoughts among gender-affirming providers, a patient's deterioration often leads them to double down on affirming interventions (Grossman 2023).

This chapter does not deny that there are individuals who struggle with devastating distress related to an incongruence between their physical bodies and their psychological sense of gender. Additionally, this discussion does not attempt to pathologize identity variance, diversity of self-expression, or experience. It simply asks which medical approaches are best for each young individual.

Several nations have published their own independent systematic reviews, including Finland, Sweden, and the United Kingdom (Cass 2024). In response to the outcomes of these and other investigations, many countries have adopted more cautious approaches, especially for children. Others are under significant pressure from public health agencies and professional associations to reevaluate current medicalized models.[1]

The medical and mental health communities in the United States have access to these same international reviews, so why have academic institutions, national medical associations, and health care communities championed GAC without question? Why are hundreds of gender clinics and clinicians across the United States still providing puberty blockers, cross-sex hormones, and sex-trait-modifying surgeries for thousands of young patients? By doing so, they are complicit in the exploitation of significantly distressed children and young adults while hijacking the compassion and empathy of their parents. Bullied into silence, parents watch helplessly from the sidelines while their children suffer trauma after medical trauma. Children and young adults deserve less harm-inducing, less sterilizing, more flexible, and more wellness-preserving support.

In Name Only: Pseudo-Scientific Guidelines

Transgender-identified individuals, a cohort that has already faced difficulty accessing health care, are now victims of an extreme version of misguided medicine. Understanding how gender-affirming guidelines infiltrated the US medical community requires an understanding of the self-proclaimed transgender health care authority, the World Professional Association for Transgender Health (WPATH).

Previously known as the Harry Benjamin International Gender Dysphoria Association (HBIGDA), the organization was founded in 1979. At that time, standards for medical research were only just emerging from a history of patient testimonials, expert opinion, and highly selective, biased case reports.

In 2007, HBIGDA changed its name to WPATH, a rebranding that spoke to its ambition to be the world leader in guiding care for transgender people. The name suggested a level of medical sophistication that never existed in the organization. Yet the world accepted the self-designation at face value and submitted to its "Standards of Care" (SOC). To date, its influence is ubiquitous, and its pseudo-scientific guidelines are entrenched worldwide (Grossman 2023).

A quick search provides an interminable number of associations that either directly refer to WPATH or heavily rely on their guidelines. Many of them rely solely on WPATH's best practice recommendations for gender-distressed patients. Although far from comprehensive, the following list includes some of the largest and most influential medical and mental health associations in the United States (list continued in footnote):

> American Medical Association, American Psychiatric Association, American Academy of Child and Adolescent Psychiatry, American Psychological Association, American Academy of Pediatrics, and American Academy of Family Physicians[2]

Despite the unethical absence of evidence-based medical and psychiatric practices throughout WPATH's guidelines, the authors' own professional organization, the American Psychiatric Association (APA), champions them. The APA frequently cites WPATH standards in its guidelines and resources for mental health professionals working with transgender individuals. The APA further shocked its rank-and-file members in November 2023 by publishing a "textbook," *Gender-Affirming Psychiatric Care*. Peer-reviewed and marketed by the APA's own publishing house, it references WPATH no less than thirteen times (Goetz and Keuroghlian 2023).

The Evolving GAC Landscape

In April 2024, Dr. Hilary Cass, a British pediatrician and former president of the Royal College of Paediatrics, published the *Final Cass Report* and, in doing so, revealed WPATH's central role in the unfolding GAC scandal. In 2020, Dr. Cass was appointed by the National Health Service of England to chair an *Independent Review of Gender Identity Services for Children and Young People*. The investigation evolved into a four-year international review to comprehensively examine the current state of (trans) gender-related care for young people worldwide.

Dr. Cass and her team commissioned the University of York to conduct multiple systematic reviews, which included an extensive investigation of international guidelines and recommendations (Cass 2024).[3] In her 388-page landmark report, she provided the results of the largest independent review of transgender health care for children and young people to date. Her review included the following statement:

> The World Professional Association of Transgender Healthcare (WPATH) has been highly influential in directing international practice, although its guidelines were found by the University of York appraisal process to lack developmental rigour (Cass 2024, p. 128).[4]

Dr. Cass's report was not limited to an investigation of WPATH's guideline development. All twenty-three existing international transgender health guidelines were reviewed by Dr. Cass and the University of York's research team. Her report revealed that, apart from Finland and Sweden, all appraised guidelines[5] were heavily influenced by two earlier guidelines, WPATH's SOC 7th Version and the Endocrine Society's 2009 gender-affirming guidelines. As was the case with SOC 8th Version (SOC-8), these documents lacked developmental rigor and were based on a paucity of evidence (Cass 2024, p. 130).

WPATH's Overt Demand for Transgender Sub-Standards of Care

In 2017, WPATH selected a research review team at Johns Hopkins University (JHU) to perform systematic reviews addressing dozens of critical questions regarding gender-related care. In doing so, WPATH gave the appearance of sponsoring independent, nonbiased investigations of current literature to support their newest transgender guideline recommendations and revisions. But commissioning these reviews and stating that they did so publicly appear to be the end of WPATH's commitment to international standards for guideline development.

The details that follow shed light on one of the largest attempts to extinguish academic freedom to date. In May 2024, a federal court unsealed documents from *Boe v. Marshall*, a case in which the State of Alabama's law limiting unproven medical and surgical transition interventions for transgender youth was challenged.

Court documents reveal that WPATH did enter into a sponsored research agreement with JHU. However, in the months leading up to the final contract, WPATH pushed for authority over what the JHU research team could publish (Society of Evidence-Based Gender Medicine [SEGM] 2024).

These court documents included an email from Karen Robinson, PhD, dated March 6, 2018 (*Boe v. Marshall* 2024). Dr. Robinson, the SOC-8 Evidence Review Team's principal investigator and director of the JHU Evidence-Based Practice Center, explained:

1. First, Hopkins as an academic institution, and I as a faculty member therein, will not sign something that limits academic freedom in this manner. In other words, a sponsor cannot change or suppress publication of research.
2. Second, I will not sign off on language that goes against current standards in systematic reviews and in guideline development. It was my understanding that WPATH wanted to move toward the current standards for guideline development. To do so, the review team needs to be independent (*Boe v. Marshall* 2024, Exhibit 167, p. 26).

Despite its contentious beginnings, the research agreement was finalized in April 2018. By 2020, JHU had reportedly completed dozens of evidence reviews for at least six chapters[6] in the new SOC-8 (SEGM 2024).

When results began to emerge, WPATH determined that JHU's findings conflicted with their ability to continue recommending current hormonal and surgical interventions. WPATH leadership pressured JHU to withdraw systematic reviews already submitted for publication. Then, in August 2020, after blocking the publication of the submitted reviews, WPATH created a new policy. The policy provided for an approval mechanism over any future publications related to the sponsored body of work. It allowed WPATH control over the final verbiage of proposed conclusions and over manuscript publication approval. It also forced JHU to include a disclaimer that they were not in any way influenced by WPATH (SEGM 2024).

After establishing the new research policy, WPATH only allowed JHU to publish one narrow review examining cross-sex hormone "therapy." Results of the review do not appear directly in SOC-8, but the review is cited in support of recommendations for treatments that the review itself had determined lacked conclusive evidence. The published review states:

> The strength of evidence for these conclusions [across the outcomes investigated] is low due to methodological limitations.... It was impossible to draw conclusions about the effects of hormone therapy on death by suicide (Baker 2021).

What appears in their SOC-8 cannot be reconciled with the above-stated conclusion. WPATH's published narrative, which brazenly references the above review, states:

> There is strong evidence demonstrating the benefits in quality of life and well-being of gender-affirming treatments, including endocrine and surgical procedures, properly indicated and performed as outlined by the Standards of Care (Version 8).... They are safe and effective at reducing gender incongruence and gender dysphoria (Coleman et al. 2022, p. S18).

Reflecting JHU's compliance with WPATH's new research policy, the 2021 review included WPATH's deceptive disclaimer. The publication states that WPATH did not influence the review despite its involvement in the design, draft, and approval of the final manuscript (SEGM 2024).

To date, WPATH's efforts to interfere with the publication of what they consider "unfavorable results" have been successful. According to unsealed emails, the limited evidence JHU has published from this large body of work is a small fraction of the actual research JHU completed. It is unclear whether these studies will ever be published. From Dr. Robinson's emails, what *is* clear is that JHU's systematic reviews "found little to no evidence about children and adolescents" (*Boe v. Marshall*, Defendant's Motion for Summary Judgment and Brief Support 2024, p. 16).

In summary, JHU, one of the most trusted evidence-based research centers in the United States, knowingly or unknowingly gave WPATH unprecedented control over their research. This included allowing WPATH the ability to approve any proposed studies and, once completed, provide final approval of any manuscript publication "to ensure that publication does not negatively affect the provision of transgender healthcare in the broadest sense" (*Boe v. Marshall* 2024, Exhibit 167, p. 91).

Ideology Overshadows Science

What WPATH's guideline development lacked in rigor, it made up for in political and legal influence. This was explained by James Cantor, PhD, a clinical psychologist, world-renowned neuroscientist, and researcher. Dr. Cantor provided an expert report for the State of Alabama in *Boe v. Marshall*. In the Appendix A to his supplemental report, he stated:

> Members of the WPATH Guideline Development Group went so far as to explicitly advocate that SOC-8 be written to maximize impact on litigation and policy even at the expense of scientific accuracy (p. vii).[7]

SOC-8 ignores actual fundamental standards of care in medicine, scientific research, and mental health that must always be considered when treating any patient. Additionally, unsealed court documents in 2024 have shed light on more nefarious intentions.

Age restrictions for medical and surgical sex-trait modifications for minors have been at the center of several lawsuits, many involving the medical community. *Appendix A to supplemental expert report of James Cantor Ph.D.* also revealed that the WPATH Development Committee was significantly pressured by the American Academy of Pediatrics and by Admiral Rachel Levine, assistant secretary for health of the US Department of Health and Human Services, to remove all age restrictions for medical transition interventions from SOC-8.[8,9,10]

> (But) AAP issued an ultimatum to WPATH: Should WPATH not delete the age minimums, AAP would not only withhold endorsement of SOC-8, but would publicly oppose the document.
>
> As a result of this additional pressure, on top of that from Assistant Secretary Levine, WPATH capitulated and removed the text in violation of its own process despite the preference of its own committee members to retain the age limits. (p. x)

It is disturbing to know that politics, not evidence, led to the elimination of age restrictions for gender-affirming interventions from SOC-8 guidelines, essentially reducing decisions influencing children's lives to nothing more than a political football.

The Corruption of Medical Education

Despite the lack of utilizing comprehensive systematic reviews, rigor of guideline development, and professional consensus, GAC is taught in medical schools and residencies and is widely practiced across the United States today. Many physicians, trainees, and mental health providers are led to believe that GAC is accepted clinical practice, completely unaware that there is no actual medical consensus. Instead, there is a growing divide between countries seeking better care for gender-distressed youth and those doubling down on affirmative approaches (Posner 2024).

United States medical schools and residency programs have barreled forward with GAC, regardless of a patient's age or comorbid mental or medical challenges. Physicians-in-training are taught to provide affirming medical care at any patient's request. This includes patients identifying as transgender, nonbinary, or nonconforming, whether or not affirming medical or surgical interventions are mentally or physiologically necessary. Questioning or exploring other options is rarely tolerated.

Medical students and residents have spoken out about being taught to ignore evidence and biological facts when they do not align with a patient's current experience of gender. For example, residents recently reported to the *New York Post* that they "are instructed that gender-questioning children will kill themselves if not prescribed puberty blockers and cross-sex hormones. They are barred from talking to their young patients about other treatment options" (Posner 2024).

Because training programs are now heavily influenced by gender ideology, students are taught a harmful conflation of inclusion and falsely labeled best practices. The current academic state of affairs is a direct result of the infiltration of social justice into medical education. This pernicious influence appears in the APA's textbook, *Gender-Affirming Psychiatric Care*, which instructs trainees and practitioners to ignore evidence-based practices:

> TNG[11] patients may not desire what the medical industrial complex considers evidence-based. In listening to our stakeholders, we needed to stray outside of evidence-based cognitive-behavioral therapy or dialectical behavioral therapy groups and provide nonclinical groups, imagining nontraditional mental health services, to foster empowerment, resilience, and gender euphoria (Goetz and Keuroghlian 2023, p. 361).

Making health care decisions based purely on a patient's request and teaching trainees to do the same is grossly negligent. The foundation of medical training is anchored in evidence-based research and the medical community's actual standard-of-care practices. In contrast, the concluding chapter from *Gender-Affirming Psychiatric Care*, which boasts "evidence-informed" versus evidence-based methods, states:

> Cisgender people in power have traditionally created systems that bar access to quality gender-affirming care. Instead, TNG communities should lead the way: reimagining the bounds of health care and fostering an activist health care environment that provides comprehensive care.... We need

systems that are dedicated to advancing gender-affirming care.... We have a duty to fight injustice in every encounter, to ensure people are affirmed.... The right care does not just mean treatment or prevention, it also includes affirmation (Goetz and Keuroghlian 2023, p. 356).

This is not the practice of clinical medicine. This is the practice of clinical activism. Repetition of medical fallacies does not make them suddenly true. The above statements are dangerous for patients and the providers who continue to trust these self-proclaimed gender authorities.

The Suicide Myth

As countless parents have disclosed, that single dreaded word, "suicide," pushed them into giving their consent to medically transition their children, even as every instinct told them something was very wrong. This is a cruel but nonetheless extremely effective version of bullying that is being perpetrated by the gender-affirming establishment.

One individual who has courageously voiced her personal experience is Chloe Cole. Chloe's testimony to Congress, given on her nineteenth birthday, publicly exposed the emotional coercion experienced by not just her parents but thousands of other parents across the United States (Barnes 2023, Olohan 2024). Chloe testified that when she was twelve years old, prior to any medical affirmation,

> They asked my parents a simple question: Would you rather have a dead daughter or a living transgender son? The choice was enough for my parents to let their guard down, and in retrospect, I can't blame them.
>
> This is the moment that we all became victims of so-called gender-affirming care (Cole 2023).

Chloe is one of many children who felt she had no choice. She made life-altering decisions out of fear after her caregivers warned her of looming suicidality if she did not medicalize with off-label puberty blockers and cross-sex hormones. Off-label use of medical treatments is neither unethical nor uncommon. What is unethical is to present off-label interventions as first-line, lifesaving, or as the only available option, especially when there are less invasive alternatives that protect a child's development and future adult functioning.

The 2024 Cass Review, as previously referenced, warns:

> This is an area of remarkably weak evidence, and yet results of studies are exaggerated or misrepresented by people on all sides of the debate to support their viewpoint. The reality is that we have no good evidence on the long-term outcomes of interventions to manage gender-related distress (Cass 2024, p. 13).
>
> It has been suggested that hormone treatment reduces the elevated risk of death by suicide in this population, but the evidence found did not support this conclusion (Cass 2024, p. 33).
>
> …the Review would recommend an extremely cautious clinical approach and a strong clinical rationale for providing hormones before the age of 18. This would keep options open during this important developmental window, allowing time for management of any co-occurring conditions, building of resilience, and fertility preservation, if required (Cass 2024, p. 34).

Lack of Informed Consent

Legally and ethically, physicians must give patients the ability to make informed decisions about treatment. Therefore, it is in the hands of the medical and mental health communities to develop safe, effective standards, demonstrate clear medical necessity, provide transparency of risks, and improve mental health for these young patients.

As emphasized by Dr. Cass:

> (C)onsent is more than just capacity and competence. It requires clinicians to ensure that the proposed intervention is clinically indicated as they have a duty to offer appropriate treatment. It also requires the patient to be provided with appropriate and sufficient information about the risks, benefits and expected outcomes of the treatment (Cass 2024, p. 34).

SOC-8 recommendations regarding informed consent for a child do not align with WPATH members' discussions behind closed doors. For example, their guidelines advise that physicians discuss fertility and the long-term consequences of blockers and cross-sex hormones. However, even WPATH members admit that the majority of minor patients cannot comprehend these consequences of treatment.

In March 2024, *Environmental Progress* released an investigative report written by Mia Hughes titled *The WPATH Files*. The report included leaked videos of a WPATH panel discussion and members' internal messages. Many shocking revelations regarding the true nature of WPATH were revealed. Informed consent was one damning issue that came up repeatedly. For example, Dianne Berg, a child psychologist, member of WPATH, and coauthor of SOC-8, disclosed that children considering puberty blockers and cross-sex hormones can discuss adverse outcomes with providers; however, for the purpose of informed consent, these discussions are "out of their developmental range to understand the extent to which some of these medical interventions are impacting them" (Hughes 2024, p. 47).

A Canadian endocrinologist, Dr. Daniel Metzger stated:

> …it's always a good theory that you talk about fertility preservation with a 14-year old, but I know I'm talking to a blank wall…. They'd be like, Ew, kids, babies, gross…. (Hughes 2024, p. 192–193)[12]

Gender-affirming professionals are curiously uninterested in participating in an open discussion about obtaining informed consent. Most recently, Dr. Hilary Cass pointed out that doctors are afraid to raise concerns about the evidence and whether or not this approach truly represents best practices when it comes to transgender-identifying youth (Cass 2024). However, there is no more vulnerable subset of patients than those who are unable to make their own medical decisions, not the least because they have years of development in front of them. They are unable to conceptualize lasting, life-altering interventions.

When internal discussions about the informed consent of minors became publicly visible, it was evident that WPATH leadership repeatedly discussed that children cannot consent. Yet this admission led to no consideration of postponing, not recommending, or not performing medical and surgical treatments.

The Way Forward

There is much work to be done. The education of parents, patients, medical schools, residency programs, medical associations, psychologists, therapists, and clinicians regarding the fallacy of GAC is paramount to a better path forward. The international community is waking up. Globally, we are seeing changes in real time. It is time for the United States to heed what the rest of the world has concluded.

For those of us who provide care for transgender-identifying youth, now is the time to demand excellence from our associations, our academic institutions, and our licensing boards. American professional associations have been challenged to address their complicity and, to date, have either irresponsibly remained silent or doubled down on their GAC endorsements (Posner 2024, Doctors Protecting Children 2024). We must continue to demand our academic institutions and associations maintain the high standards required for excellence in patient care.

The argument claiming that politics and legal bans are preventing children from accessing lifesaving care is misdirected. It is the lack of evidence supporting GAC, a disregard for safe clinical practice, and an unfounded push for irreversible interventions in children that have spawned these types of legislative bans. Legislators have no choice but to rein in unethical, non-evidence-based medical treatments that conflate healthy identity exploration and gender nonconformity with medical necessity determined by social justice.

Our young patients and their families deserve better research, better guidelines, and better treatment options. The fundamental chain of trust between physicians and their patients can be restored, but the medical community must acknowledge harm has been done. As the lack of evidence behind the affirmation model comes to light, medicine cannot remain complicit.

Addendum:

On September 3, 2024, twenty-four US medical and mental health organizations submitted an amicus brief ahead of oral arguments made before the Supreme Court in the case of *U.S. v. Skrmetti (2024)*. The case examines whether or not Tennessee's law to limit the availability of gender-affirming medical interventions to minors is constitutional (Sapir et al. 2024). These organizations, purporting to represent over 1.2 million medical professionals, include the American Psychiatric Association, the American Medical Association, the American Academy of Pediatrics, and the Association of American Medical Colleges. It should be noted that the twenty-fourth organization listed was the WPATH itself (American Academy of Pediatrics et al. 2024). In the brief, these associations repeatedly argue that GAC, including medical interventions for minors is "critical, medically necessary, evidence-based care for gender dysphoria." At the time of this writing, the Supreme Court has not yet ruled on the case.

Ignoring a growing number of national and international systematic reviews (see Cass 2024, Miroshnychenko 2025, and Miroshnychenko et al. 2025), our professional organizations continue to cite the scientifically discredited Endocrine Society and WPATH SOC-8 as *the* "established, evidence-based clinical guidelines." Associations and members therein that remain silent are complicit in the harm of thousands of children and young adults. Every child has a right to the healthy developmental process of unhindered puberty. These medical associations have a duty to examine the evolution of evidence-based literature regarding the treatment of gender dysphoria, acknowledge the damage that has been done, and protect their patients from further harm.

On January 25, 2025, the President of the United States issued an executive order highlighting many of the issues raised in this chapter. The executive order calls on federal agencies to rescind policies that rely on WPATH guidelines, specifically mentioning SOC-8. It directs the Secretary of Health and Human Services to review all existing scientific literature regarding the best treatment options for patients struggling with gender dysphoria and halt affirming sex-trait modification interventions. Interventions specifically mentioned include puberty blockers, cross-sex hormones, and surgeries, all of which are recommended by the WPATH guidelines (Trump 2025).

This Executive Order addresses the substandard tenets of GAC as it pertains to patients struggling with gender-related distress. However, without the involvement of medical associations, licensing boards, research and academic institutions, the inexcusable risk of harm remains. The infiltration of gender ideology at the expense of patient safety cannot continue. The abandonment of evidence-based care and guideline standards must be addressed by the very medical communities that have, to date, promoted and heavily relied on WPATH's gender-affirmation model.

A Deafening Silence: Bioethics and Gender-Affirming Health Care

ALEX BYRNE AND MOTI GORIN

The "affirming" health-care model for gender-distressed youth is endorsed by the medical establishment in the United States, but many European nations have retreated from it. Coverage of this issue in major media outlets in the US has been poor for years; a lengthy 2022 article in the *New York Times Magazine* by Emily Bazelon marked something of a turning point. The subhead of Bazelon's article was, "More teenagers than ever are seeking transitions, but the medical community that treats them is deeply divided about why—and what to do to help them" (Bazelon 2022). This controversy would be expected to attract the interest of philosophers and bioethicists, with a diverse range of opinions appearing in academic articles. However, when philosophers and bioethicists have ventured into print, they have almost invariably endorsed the affirmative approach, which involves life-changing medical interventions on children with psychological problems. As we will explain, this is a sign that the process of academic research and writing is not functioning as it should.[1]

The Gender-Affirming Model

Gender dysphoria is an extreme aversion to one's sexed body. Or, as the latest version of the *Diagnostic and Statistical Manual of Mental Disorders* (DSM) explains it, "the distress that may accompany the incongruence between one's experienced or expressed gender and one's assigned gender" (APA 2022: 512). Gender dysphoria may be *early-onset*, affecting young children, or *late-onset*, mostly affecting heterosexual men. Starting around 2015, clinicians noticed a new presentation of *adolescent-onset* gender dysphoria, mostly affecting girls. This third presentation was labeled *rapid-onset gender dysphoria* (ROGD) by the physician and researcher Lisa Littman in a 2018 paper, which immediately ignited an explosion of activist-driven controversy. Here we will be concerned with the treatment of gender dysphoria in children and adolescents.

The criteria for a DSM diagnosis of "gender dysphoria in children" are: (*a*) "A strong desire to be of the other gender or an insistence that one is the other gender (or some alternative gender different from one's assigned gender)"; (*b*) five criteria from a list of seven, including "A strong preference for the toys, games, or activities stereotypically used or engaged in by the other gender," and "A strong dislike of one's sexual anatomy"; and, lastly, (*c*) "associated…clinically significant distress or impairment in social, school, or other important areas of functioning." These symptoms should be present for at least six months (APA 2022: 512). Among children with a diagnosis of GD, post-puberty, there is "a high incidence of sexual attraction to those of the individual's birth-assigned gender" (516); i.e., these children grow up to be homosexual.

Until comparatively recently, treatment for GD in minors did not routinely involve medicalization or even a "social transition," where the child wears clothing typical of the other sex, with a name and pronouns to match. The "Dutch protocol," which introduced gonadotropin-releasing hormone (GnRH) agonists ("puberty blockers") as a treatment option around the early 2000s, avoided social transitioning, and prior to medicalization, patients were carefully screened for other mental health conditions and lack of family support. When US clinicians at Children's Hospital Boston began to use puberty blockers around 2007, they relaxed the Dutch protocol. Gone was the requirement that treatment start no earlier than twelve and that any comorbidities should be addressed beforehand. The codirector of Children's Gender Multispecialty Service (GeMS) and colleagues reported that "one of the most striking characteristics of our population [is] the prevalence of psychiatric diagnoses and history of self-harming behaviors" (Spack et al. 2012: 422). Pediatric gender medicine in the US came to favor a much more child-centered approach than that of the Dutch. The psychologist Diane Ehrensaft explained the US approach as follows:

> The gender affirmative model is defined as a method of therapeutic care that includes allowing children to speak for themselves about their self-experienced gender identity and expressions and providing support for them to evolve into their authentic gender selves, no matter at what age. Interventions include social transition from one gender to another and/or evolving gender nonconforming expressions and presentations, as well as later gender-affirming medical interventions (puberty blockers, cross-sex hormones, surgeries). (Ehrensaft 2017: 62)

Ehrensaft is the director of mental health at UCSF's Child & Adolescent Gender Center and a leading proponent of the gender-affirming model. She conceives it as facilitating the child's quest for their "authentic gender self," which may (or may not) involve irreversible hormonal treatments and surgeries. The duty of medical professionals is to help children along their gender journey, as race officials hand out water to marathon runners, not to nudge the child in one direction or another. The destination is set by the child themselves: "When it comes to knowing a child's gender, it is not for us to tell, but for the children to say" (63). Whatever a child's "gender" may be, it is certainly not simple. In her book *The Gender Creative Child*, Ehrensaft proposes a baroque taxonomy of "gender creative children," including "Gender Priuses—Half Girl/Half Boy," "Gender Ambidextrous Children," and "Gender Tootsie Roll Pops" (Ehrensaft 2016: 34–42).[2]

Although they may balk at Gender Tootsie Roll Pops and other animals from Ehrensaft's gender bestiary, nearly all the major medical associations in the US endorse the gender-affirming model, including the American Medical Association (AMA) and the American Academy of Pediatrics (AAP). According to an AMA press release, "the majority of transgender and diverse-gender patients report improved mental health and lower rates of suicide after receipt of gender-affirming care" (American Medical Association 2021). In August 2023, the AAP re-endorsed a 2018 policy statement, which asserts that "pubertal suppression in children who identify as TGD [transgender and gender diverse] generally leads to improved psychological functioning" (Rafferty et al. 2018: 5). Rachel Levine, assistant secretary of health at the Department of Health and Human Services in the Biden administration, says that gender-affirming care is "suicide prevention care," which "improves quality of life," "saves lives," and is "founded on a vast body of medical literature" (Levine 2022).

If gender-affirming care prevents suicide, one might wonder why philosophers or bioethicists should bother getting involved. However, this apparently solid evidence-based consensus starts to fall apart at the mildest prodding.

Take, for example, the AMA's claim that "patients report...lower rates of suicide after receipt of gender-affirming care." This is poorly worded since self-reporting completed suicide is difficult. But at the very least, the AMA is claiming that there is an association between gender-affirming care and lower odds of suicide or (more likely) attempted suicide. There are no citations in the press release; instead, there is a link to a letter the AMA sent to the National Governors Association in April 2021. In that letter, the AMA states that studies of "gender-affirming care...*demonstrate* dramatic reductions in suicide attempts, as well as decreased rates of depression and anxiety" (Madara 2021, emphasis added). Focus on the first part, about dramatic reductions in suicide attempts. That is not quite the same as dramatic (or indeed any) reductions in suicide; still, it is an impressive claim to make about any treatment.

The AMA gives two citations at the end of the quoted sentence. One is a prospective study of patients who underwent sex reassignment, applying for cross-sex hormone therapy and surgery at a mean age of 30.9. Psychological functioning modestly improved at follow-up, around four years later. There is nothing in this paper about suicide, and, more importantly, this was not a study of gender-affirming care.

The AMA's second citation is a systematic review of hormone therapy specifically. This paper does mention suicide attempts: "Suicide attempt rates decreased after sex reassignment but stayed higher than the normal population rate" (Murad et al. 2010: 216). But, again, the paper is not a study of current gender-affirming practice: many of the studies reviewed are from the good old days of "sex changes," with the earliest paper dated 1971. Moreover, Murad et al. state their conclusion extremely cautiously: "*very low quality evidence*" suggests improvements of various kinds (229, emphasis added). Presumably, the AMA picked the most favorable citations. If so, then its claim that studies of gender-affirming care "demonstrate dramatic reductions in suicide attempts" is unsupported.[3]

Alarm bells were obvious, even without chasing down citations. In December 2020, a Divisional Court from the High Court of Justice in the UK ruled on the use of puberty blockers by the National Health Service's Gender Identity Development Service (GIDS), expressing skepticism that young teenagers were in a position to understand the consequences of these medications. The judgment was later overturned on appeal, but GIDS has now been shuttered. The recent Cass Review, the most comprehensive examination of pediatric gender medicine to date, effectively sounded the death knell for gender-affirming care in the UK, as well as the more conservative Dutch protocol. Other European countries have also rejected the adolescent pathway from puberty blockers to cross-sex hormones to surgery. Like England, Sweden and Finland have conducted systematic reviews and concluded that the evidence is too weak to justify these interventions, given the known and unknown risks, the lack of high-quality studies, and major changes in the composition of pediatric patients.

Clearly this important and controversial topic is a goldmine for philosophers and bioethicists, who are always on the lookout for something new to write about.[4] After the thousandth paper on abortion, the law of diminishing returns starts to set in. The dominant affirmative model is grounded in under-theorized and sometimes incoherent or conflicting claims: adolescents and even very young children have a "*transcendent* sense of gender" (Turban 2024: 38); the healthy body may be an impediment to the expression of an individual's "true" self; bodily alienation is a non-pathological example of human diversity and—simultaneously—an urgent problem that requires invasive medical treatment.[5] The upshot of these framings is that liberation, both from psychiatric distress when present and from purportedly oppressive norms governing the function and appearance of the human body, must come in the form of pharmacological and surgical products, offered to minors by medical experts, paid for by public or private insurance schemes, and endorsed and enforced by the state itself.

The ingredients for lively and wide-ranging philosophical discussion and debate—about the nature of sex and gender, autonomy and paternalism, the meaning and moral salience of human diversity, the relation between the mind and the body, the scope of first-person authority, the aims of science and medicine, the nature and normative significance of identity—are laid out before us, impossible to miss. Moreover, insofar as philosophy has a valuable role to play in high-stakes debates of great interest to researchers and the public at large—spotting unstated assumptions, formulating precise arguments, locating and resolving theoretical inconsistencies, explicating normative commitments, and so on—one would expect to see it play that role here. Yet, despite the low-hanging fruit, there has not been much picking. What we find in the philosophical and bioethics literature, as well as in public-facing contributions from scholars, is surprisingly limited in quantity and, crucially, breadth of perspective.

Disciplinary Failure

Sex and gender controversies extend well beyond pediatric medicine. They include issues of social policy, for example, gender "self-identification" laws and rules governing eligibility for female sporting categories or the female prison estate. There are also more "metaphysical" questions, notably "What is a woman?"—notorious for stumping politicians on both sides of the Atlantic. Unfortunately, the discipline of philosophy proved unable to accommodate all opinions, in particular feminists who insist on the social importance of sex and resist the slogan "Trans women are women." (These feminists are often called "gender critical"; here we will use this phrase broadly to label any who are out of step with mainstream philosophical thought on sex and gender.) Ironically, it was the branch of the discipline known as "feminist philosophy" that exhibited maximum intolerance of dissenters, with the go-to allegation for any kind of departure from orthodoxy being "transphobia."

Apart from the usual tactic of "no-platforming," some feminist scholars displayed a related tendency—one particularly striking in the academic context—namely an extreme reluctance to cite or engage with those on the gender-critical side. There are some honorable exceptions, but for the most part, the defenders of orthodoxy write in a "she who must not be named" style. A recent example is the philosopher Matthew J. Cull's *What Gender Should Be*. Cull obliquely alludes to "those who would call themselves 'gender criticals,'" writing that "much ink has been spilled on the inadequacy of contemporary gender critical thought and I will not recite such arguments here" (2024: 160). Of course, it's perfectly fine for Cull not to spill any more ink refuting gender-critical philosophers—and anyway, an author can only do so much in a book. Turning to the notes, though, one expects to find citations to gender-critical philosophers A, B, and C, together with the refutations by X, Y, and Z. That is simple intellectual honesty. But while Cull provides many citations (some decidedly eccentric), there are none to the self-labeled "gender criticals," leaving their identity and writings a mystery.

Mainstream feminist philosophers have generally avoided wading into debates about transgender health care, but when they have touched on this issue, motivated reasoning has clearly been in play. Take, for instance, Kate Manne's 2024 *Unshrinking: How to Face Fatphobia*. Manne is perhaps the most distinguished feminist philosopher of her generation, one of *Prospect Magazine*'s 2019 "world's top thinkers," and will be a leader of mainstream feminist philosophy for years to come. She had declared her allegiances early in the gender wars that roiled philosophy, uncompromisingly tweeting in 2019 that trans women are "women in every sense of the term." Philosophy, Manne claims in *Unshrinking*, is not just "transphobic," but "increasingly" so (2024: 121).

Since one of Manne's aims in the book is to argue that the health hazards of obesity—at least when decoupled from stigma—have been exaggerated, naturally, she turns a skeptical eye on studies purporting to show the opposite. However, her skepticism is replaced by credulity when she compares gender-affirming care to bariatric surgery for severely obese adolescents, both currently endorsed (as she notes) by the AAP. Bariatric surgery, according to Manne, "sets fat kids up for a lifetime of being unable to meet their basic nutritional needs or satisfy their hunger without suffering.... It also seems likely to increase suicide risks...." In contrast, "gender-affirming care...enables trans kids to flourish and be who they are," and "*demonstrably reduces* their risk of depression and even suicide" (198, emphasis added). Manne's citation for the latter statement (Tordoff et al. 2022) did not report on suicide, but rather suicidality, a very different phenomenon. More importantly, the data in that study showed that gender-affirming care (in this context, puberty blockers or cross-sex hormones) was *not* significantly associated with a reduction in either depression or suicidality.[6] Surely, the AAP's woefully misguided recommendations about bariatric surgery (as Manne sees them) should have prompted her to question whether the organization was right about gender-affirming care?

Manne does not expound on gender-affirming care at any length, but in 2019, an instructive exchange took place in the *American Journal of Bioethics*, a leading bioethics journal. The exchange centered on philosopher and bioethicist Maura Priest's paper "Transgender Children and the Right to Transition: Medical Ethics When Parents Mean Well but Cause Harm" (2019), which appeared in the *AJOB* along with twelve "peer commentaries." According to Priest, adolescents should have a legal right to consent to puberty blockers without parental approval, and moreover, the state should play a direct and active role via the public education system in teaching children about transgender identification, gender dysphoria, and the availability of medical interventions. Priest's main argument is that because medical transition prevents serious psychological harms, parents who do not support medical intervention are guilty of neglect and hence that "the state should pay special attention to, and has a duty to protect, transgender minors from psychological harm inflicted via their caretakers" (46).

As Priest notes, guidelines from both the Endocrine Society and WPATH recommend the provision of puberty blockers for the treatment of gender dysphoria beginning at Tanner stage 2, the onset of puberty (48). Tanner stage 2 can be as early as eight years for females and nine years for males. Priest's position, then, is that caregivers who do not support pharmacologically blocking the biological development of their gender dysphoric children are inflicting serious harm and, consequently, that state power should be exercised to prevent this harm. The view certainly is provocative, and Priest's arguments are clearly articulated and defended. This is exactly what we should expect from a philosophically informed bioethics paper addressing a controversial topic in a top journal. So far, so good.

Some of the twelve commentaries were sharply critical. None of those were the four whose authors included philosophers or bioethicists with advanced philosophical training (i.e., graduate degrees in philosophy). The most critical of the four, by Lauren Notini, the bioethicist Rosalind McDougall, and Ken Pang, agrees with Priest that puberty blockers are indeed necessary to prevent serious and imminent harm but argues for an individualized approach, where the decision whether to involve the state in dealing with disapproving caregivers should be taken on a case-by-case basis, rather than via the establishment of a universal legal right to access blockers (Notini et al. 2019). The remaining three commentaries fully endorse Priest's conclusion, with two—one by philosopher Robin Dembroff and another by philosophers B. R. George and Danielle Wenner—arguing that by grounding the minor's right to puberty blockers in the prevention of psychological or physical harm to health, Priest wrongly pathologizes transgender identity and that, consequently, her proposals, while correct, do not go far enough. According to these authors, Priest's argument from harm prevention errs in making medical transition contingent on the protection of health rather than on the right of adolescents to exercise control over their bodies.

Dembroff finds Priest's approach "shocking" in its neglect of non-binary or other gender non-conforming minors, who may pursue body modification not to alleviate clinically significant distress but instead to better express their gender as they understand it (Dembroff 2019: 61). Priest is guilty of endorsing a perniciously narrow conception of what it is to be transgender and, consequently, gives the wrong kind of argument for the right kind of policy.

As we saw with Manne, Dembroff seems to be reading with reality-distorting ideological spectacles, claiming that "Priest notes the vastly increased risk of suicide" in "trans adolescents who are denied puberty-blocking treatment" (60). First, Priest is more cautious: "Such factors [including normal puberty] put transgender minors at high risk for mental health problems and *potentially* suicide" (Priest 2019: 51, emphasis added). Second, Priest gives seven citations at the end of this sentence, only two of which are remotely relevant, and neither of those two comes close to showing that puberty blockers reduce the risk of suicide.

In their commentary, B. R. George and Danielle Wenner argue that because most adolescents are regularly "allowed to proceed" through puberty without first undergoing any assessment by mental health professionals, the requirement that children seeking puberty blockers receive such an assessment prior to the prescription of puberty blockers amounts to a morally objectionable double standard (George and Wenner 2019: 81). On their view, treating puberty as the default course of adolescent development is prejudicial against transgender minors since clinicians must first assess them as "good candidates" for pubertal suppression; non-transgender minors, in contrast, face no such medical gatekeeping prior to going through endogenous puberty. For George and Wenner, the presumption in favor of normal puberty makes sense only if one believes "that trans adulthood is an inherently worse outcome than cis adulthood" (81). They do not address the objection that minors are usually restricted from directing their own medical treatment or that medical transition of any kind is standardly supposed to be neither necessary nor sufficient for being transgender.

Finally, the bioethicist Lisa Campo-Engelstein adds another voice to the choir. Campo-Engelstein agrees with Priest that children have a right to puberty blockers without parental consent but worries that she has not sufficiently emphasized the importance of social support. Health-care authorities, schools, and other organizations should engage in "targeted social interventions aimed at families" and the general public in order to increase awareness of pediatric medical transition (Campo-Engelstein 2019: 86). Like the other philosophers and bioethicists, Campo-Engelstein takes it as established that puberty blockers offer great clinical benefit.

Priest's views and those of the accompanying commentaries are fairly representative of the (small) philosophy and bioethics literature engaging directly with the field of pediatric gender medicine. What explains the absence of contrarian opinions from people paid to write, talk, and argue, who are members of a profession that prides itself on asking thorny questions, challenging the status quo, unearthing and critiquing comfortable assumptions?

Climate Crisis

To be fair, the self-portrait of the modern-day philosopher as a dauntless rebel is unduly flattering—many of us seem more interested in citation counts and our tenure case (if we are lucky enough to be on that track) than we are in playing the role of fly in the ointment. Still, there is an incentive in philosophy to defend the seemingly indefensible, to upend the commonsense applecart, and to boldly go where no argument has gone before. Usually there is little downside to heterodoxy—those who think that tables do not exist, that having children is immoral, or that electrons are conscious are commonly viewed as intellectual provocateurs with unconvincing arguments that are nonetheless challenging and enlightening.

However, philosophers (and bioethicists) need to put bread on the table like everyone else. When heterodoxy brings threats to one's reputation, character, or even livelihood, the pragmatic calculus starts to look very different. If the threats are external, they may be resisted if support from professional colleagues is strong—it is a different story when the threats are coming from inside the house. And, as we will now recount, that is exactly what has been happening in the profession of philosophy. The obvious conclusion is that the inhospitable climate is largely responsible for the intellectual monoculture described in the previous section.

The best-known case in philosophy is that of Kathleen Stock. When Stock—then at the University of Sussex—raised objections to proposed amendments to the UK's 2004 Gender Recognition Act that would have made self-identification sufficient for legal change of gender, she was subjected to a campaign of vilification (including threats to her physical safety) that ultimately led to her resignation in October 2021. Philosophers played a leading role in this campaign, at one point penning an "Open Letter Concerning Transphobia in Philosophy" that convicted Stock of producing "discourse" that "contributes" to various harms suffered by transgender people, including "physical violence" (Bettcher T. M. et al. 2021). Stock was also blamed for her part in restricting "trans people's access to life-saving medical treatments." This bizarre accusation was leveled in response to Stock's reservations about the care provided to gender dysphoric youth at the Tavistock Gender Identity Development Service (GIDS) in the UK, which turned out in retrospect to be precisely on-target.[7] Nevertheless, the open letter, signed initially by twenty-seven philosophers and then by almost eight hundred faculty and students, remains online, and no apologies can be expected.

About six months after Stock was pressured out of her job, another open letter appeared, addressed to Oxford University Press in anticipation of their publishing *Gender-Critical Feminism* by political philosopher Holly Lawford-Smith. That the book was not yet available to read did not prevent the self-described "members of the international scholarly community" from harshly condemning it. According to the letter, written in an all-too-familiar catastrophizing style, Lawford-Smith's gender-critical position reinforced "policies targeting the right of trans people to live freely or at all." Pediatric gender-affirming medical interventions made an appearance here, too, described again as "lifesaving." OUP had already subjected the book to an unusual, additional level of scrutiny late in the publication process, when they required Lawford-Smith to respond at length to a set of comments from a medical expert whom the editors had invited to review the chapter on transgender issues, which addresses (among other things) pediatric gender medicine. Yet another letter, organized by the union representing OUP's New York staff, pushed the hyperbole to stratospheric levels, claiming that "the publication of this book will embolden and legitimize the views of transphobes and contribute to the harm that is perpetrated against the trans community globally" and urging "management to reconsider their decision to publish this title" (Weinberg 2022).

The circulation of these letters via blogs and websites, as well as the usual back channels, ensured that everyone got the message: advocating for (some) single-sex spaces and criticizing the gender-affirming model was unacceptable. By the time Lawford-Smith had been painted as philosophy's witch-of-the-semester, the campaign against Stock had been widely reported in the media. Expressions of support for these two colleagues drew furious criticism.

It can be hard to measure the success of a campaign when its intended effect is silence. As every authoritarian knows, suppressed dissent can be indistinguishable from widespread consensus when observed from the outside. Still, there are many signs that the relative absence of critical discussion is not due to reasoned agreement.

We have heard firsthand from colleagues (both in bioethics and philosophy) that while they think the issues surrounding sex, gender, and health care are interesting and important, they are unwilling to engage with them publicly out of fear of social and professional opprobrium. The fear is palpable and manifests in various ways: as supportive but furtive asides over drinks at conferences; in confidential, supportive emails containing subtly apologetic admissions that the author feels uncomfortable making their agreement known publicly; as nervous requests from colleagues about what they might expect from students or administrators were they to include readings by gender-critical authors on their syllabi; as colleagues' willingness to read and comment constructively on paper drafts followed by hesitation about and in some cases outright unwillingness to accept public recognition for their efforts (for example, in the "acknowledgments" section of the published paper). These anecdata are not exactly the horrors of the gulag, but nevertheless are symptoms of an unhealthy intellectual climate.

A less subtle clue is that some philosophers publicly endorse and implement the "no debate" strategy. When the trans historian Susan Stryker, together with the philosophers Quill Kukla and Robin Dembroff, learned in 2019 that their short essays on philosophy and transgender issues would appear alongside those of Kathleen Stock, Holly Lawford-Smith, and the feminist campaigner and writer Julie Bindel, they retracted their contributions on the grounds that this amounted to "non-consensual co-platforming"—a species of wrongdoing not usually recognized by the academy (Dembroff et al. 2019). About a month prior to the discovery of this grave transgression, philosopher Mark Lance, a colleague of Kukla's, cited more familiar sins, complaining in *Inside Higher Education* that "to produce arguments, in this context—that trans women are not women, or trans lesbians are not lesbians—is not just a view we can easily reject as confused and offensive. It is complicity with systemic violence and active encouragement of oppression" (Lance 2019). Lance left no room for doubt about how we ought to handle colleagues guilty of such complicity, such as Stock and Lawford-Smith, both of whom defend the suddenly-beyond-the-pale position that there is no such thing as a male lesbian (and who happen to be lesbians themselves). Rather than doing what philosophers typically do—give arguments, raise objections, offer rejoinders—Lance urged instead that "those who treat this like an intellectual game should not be engaged with. They should be told to [unprintable here]" (square brackets in original).

The climate does not seem to have become more inclusive, more accommodating of disagreement, in the intervening five years. A 2023 book, *What Even Is Gender?*, published by Routledge, a leading and well-respected imprint, characterizes Stock, the philosopher Rebecca Reilly-Cooper, and one of us (Byrne) as "transphobic," an astonishing accusation to find in an academic work (Briggs and George 2023: 28). ("Muddleheaded," "misguided," and "wrong" would have been completely acceptable alternatives!) If the authors had any evidence for this professionally damaging charge, they did not bother to disclose it. More importantly, the moral turpitude of one's opponents is irrelevant to the assessment of their arguments.

When organizers of the 2025 American Philosophical Association (APA) Eastern Division Meeting attempted to assemble a panel on philosophy and pediatric gender medicine, they were unable to find enough philosophers or bioethicists to participate. The explanation for this unusual state of affairs—there is typically no shortage of scholars keen to present at an APA meeting—was that a critic of the gender-affirming model (one of us—Gorin) had already agreed to sit on the panel. This inspired nearly all other invitees to decline the APA's invitation on explicitly "no platform"-style grounds. One opined that Gorin was an "anti-trans activist"—language likely borrowed directly from Transgender Map, a website that exhaustively catalogs the perceived enemies of the site's owner, trans activist Andrea James. James also posts information about her targets' family members, including their children (Singal 2023); the present authors each have dedicated web pages on James's website, complete with personal information, and where we are characterized as "anti-transgender activists." What would have been the APA's first panel on this politically urgent and philosophically rich topic never took place. To add insult to injury, the plans were dashed even before they had reached the stage at which some ambitious APA member might have accrued progressive credit by rallying for the panel's cancellation. A missed opportunity, no matter how one looks at it.

Such shenanigans are by no means restricted to philosophy or bioethics. This is relevant because academic fields are not separated by impermeable barriers: a chilly climate in one discipline can blow over to another. Philosophers or bioethicists wondering whether to dip their toes into pediatric gender medicine might well have second thoughts, not just because of the treatment of their own colleagues, but because of what they can see happening to colleagues elsewhere.

Recommendations

We have argued that the relative absence of debate and critical discussion within bioethics and philosophy on pediatric gender medicine and related topics is plausibly explained in large part by a censorious climate. Needless to say, this is not conducive to intellectually serious activity. It is also detrimental to vulnerable people such as psychologically distressed minors, who depend on responsible adults to act in their best interests. What can be done to improve matters? In this final section, we briefly offer a few recommendations; although they are couched as specific to philosophy, they have wider application.

One recommendation is worth stating only to point out that it is useless: philosophers should grow a backbone or—more appropriate in the present context—have some balls. (Some powerful male philosophers, it must be said, have been the least-willing to speak up.) But the incentives are not in place to solve this collective action problem: why take the risk of putting your head above the parapet first? Who in their right mind wants to be labeled "transphobic" in sober academic texts?

Another recommendation is more feasible to implement. We have noticed that an increasing number of publications (in both philosophy and other disciplines) include the author's autobiographical information or personal narrative within the text itself. Unless the personal experience or autobiographical details are clearly and substantively related to the argument of the paper or to its findings—for example, as they would be in a memoir or in autoethnographic research—such information can color the judgment of peer reviewers. In some cases, this may be detrimental to the author's prospects for publication, while in others, it may be beneficial. Bias should not play any role either way. For this reason, we think publishers should explicitly disallow the inclusion of superfluous personal information and narrative: "As trans philosophers, we…" and "As cis philosophers, we…" should be equally deprecated.[8] Knowing where to draw the line is sometimes difficult, but the default should favor the impersonal.

Another scholarly norm that journals should explicitly promote is the citation of and engagement with opponents. Normally in philosophy, this norm needs no encouragement at all: one standard template for a philosophy paper is "Philosopher X is seriously confused, and here's why." However, as noted above, there is a tendency for philosophers who endorse mainstream views on sex and gender to studiously ignore the "gender critical" side. Here's one more example from a paper on gender identity forthcoming in a respectable journal: "We have avoided citing other arguments here because we take them to be openly transphobic, and we resist giving them more uptake" (Hernandez and Bell forthcoming: 2, fn. 1). If a curious reader wonders what these "other arguments" might be, or who made them, the paper offers not the slightest clue. It would be easy enough for journals with some sense of self-respect to ensure that this kind of anti-intellectualism is disallowed. Enforcing engagement with critics might also help improve publication quality, which in this area of philosophy is depressingly low.

Peer review is another problem: philosophers working on sex and gender outside the mainstream have numerous stories of the corruption of the peer review process. Given the volume of submissions, journal editors welcome any excuse to reject a paper, and the "expert reviewers" may be those who think that gender-critical views do not deserve an airing any more than the musings of flat-earthers or 9/11-truthers. Conversely, expert reviewers may overlook glaring flaws or poor citation practices in papers defending approved conclusions. Open peer review—in particular, publishing referee reports—could raise standards by increasing transparency. Worth a try, in any case.

Although this doesn't amount to a recommendation, we should note that the *Journal of Controversial Ideas* (established in 2021) and external organizations such as the Academic Freedom Alliance (AFA), the Foundation for Individual Rights and Expression (FIRE), and the UK's Free Speech Union (FSU) have valuable roles to play. The JCI has been the home for gender-critical papers that otherwise would have died under the weight of rejections, and the FSU was instrumental in forcing Oxford University Press to honor its contract with Holly Lawford-Smith and publish her second gender-critical book, *Sex Matters*.[9]

Our final recommendation is that philosophers and bioethicists skeptical of current orthodoxies about sex and gender, who have already outed themselves, should try to engage in "public philosophy" in whatever medium suits them. Apart from reaching the public, this sends a signal to the younger generation of academics, including graduate students, that a successful career can include unpopular dissent and has its own rewards. As is sometimes remarked, the main problem with academic shunning and shaming isn't the tribulation visited upon the shunned and shamed, regrettable though that is. Rather, it's the collateral damage, the understandable reluctance of junior scholars to pursue research that may lead them to be eaten alive too. After a few iterations of this process, the purge is complete, and there are no heretics left to suppress.

The events of the last few years have shown that the scholarly norms of philosophy are much more fragile than one might have hoped. Even though this came as something of a shock to the present writers, perhaps it should not have. Spend time around professional philosophers, and you will realize that they have no special immunity to fashionable political trends, the latest un-replicable social science research, peer pressure, motivated reasoning, and the temptations of status. Threats to the integrity of the discipline of the sort chronicled above will always be present: when the temperature around sex and gender has decreased to livable levels, sooner or later, another controversy will take its place. It would be a mistake to think that a permanent solution is in the offing. The price of a healthy academy is eternal vigilance.[10]

The Gender Wars and Academic Freedom

JUDITH SUISSA AND ALICE SULLIVAN

Philosophical arguments regarding academic freedom can sometimes appear removed from the real conflicts playing out in contemporary universities. This chapter focuses on a set of issues at the front line of these conflicts, namely, questions regarding sex, gender, and gender identity. As a philosopher and a sociologist, we aim to elucidate the costs of curtailing discussion on fundamental demographic and conceptual categories. We argue that these costs are educational in the broadest sense: constricting the possibility of shared learning and knowledge production, which are vital to a functioning democracy. This chapter is an updated and abridged version of previous work, drawing on Suissa and Sullivan 2021b, 2022, 2021a.

Sex, Gender Identity, and "Transphobia"

For gender identity campaigners, simply asserting that sex exists as a meaningful category, distinct from people's self-declared "gender identity," is deemed transphobic. Lobby groups such as Stonewall demand affirmation of the mantra "Trans Women Are Women," with explicit and repeated calls for "No debate." The slogan functions as a demand to adhere to the ontological position that claims about people's "gender identity" trump claims about their biological sex. Gender identity ideology is, in this sense, absolutist, demanding that we ignore material evidence of the relevance of sex in any context. Repetition of the mantra "Trans Women Are Women" obstructs any attempt at a nuanced discussion about the circumstances under which sex might be relevant. The view that it is transphobic to acknowledge natal sex as even potentially relevant has led gender identity campaigners to demand that social and human scientists must not collect data on sex and that philosophers must not use sex as a conceptual category. Such demands are fundamentally antithetical to academic freedom.

In practice, the kinds of statements that routinely lead to people (overwhelmingly women) being denounced as transphobes include: that humans, like all mammals, have two sexes, male and female; that females are the sex that produces large immobile gametes called ova; that males are the sex that produces small mobile gametes called sperm; that women are adult human females; that women do not have penises; that homosexuality is same-sex attraction; that only women have cervixes; that a transwoman who transitions as an adult has not always been female; that non-gender-conforming young children should not be encouraged to believe that they may have been "born in the wrong body."

Most people could, in principle, fall foul of the charge of transphobia, but in practice, it is most commonly applied to women who have articulated and defended an account of women's rights that assumes the biological reality of the male/female distinction and, accordingly, defines women as a sex class. Many of these women are also feminists who believe that gender is a socially constructed system that maintains male privilege and oppresses females on the basis of their sexed bodies, primarily through controlling their reproductive capacity. This view of hierarchical systems of gender as historically and socially contingent is at odds with the view that everyone *has* a gender or gender identity and that this, rather than their sex, serves to categorize them for all purposes.

The denial that humans are sexually dimorphic mammals appears, at the very least, problematic for a range of scientific disciplines, and the belief that sex is not objectively real and determined at conception but merely "assigned" at birth as a social label has implications for a range of social and political questions. Yet, these beliefs are so fundamental to the orthodox gender identitarian position that merely to point out the contentious nature of the ontological claims on which they rest is to attract accusations of transphobia.

The absolutism of this position militates against reasoned debate. If the campaigning slogan "Trans Women Are Women" is taken as true in an absolute and literal sense, then there can be no scope for discussion of the ways in which the possession of a male body may be relevant in different ways in different contexts, from sex-segregated sports to changing rooms to prisons to lesbian relationships, and no scope for compromise regarding women's concerns and boundaries.

It is against this background that current concerns over academic freedom arise. Academics who have questioned the above orthodoxies have found themselves targeted in various ways. Such academics are often described using the loose term "gender critical," which we adopt in the following discussion while recognizing that it encompasses a broad range of views (Sullivan and Todd 2023).

The Suppression of Academic Freedom on Sex and Gender

This section documents some examples of the suppression of academic freedom on sex and gender. It is not intended to be exhaustive but to give a sense of the terrain. We focus primarily on instances in Britain. Our aim here is to document a range of examples and tactics, as advocates of gender identity ideology often deny that any silencing of opponents of their position is taking place or diminish its extent and significance. Most incidents do not receive publicity; thus, we are limited to discussing a small sample of incidents that have. The prevalence of practices such as these has an inevitable wider chilling effect on academic life.

Suppression of Research

The extreme tactics used by gender identity campaigners to suppress research, including the use of defamatory allegations against researchers, have been described by social historian Alice Dreger (Dreger 2016, Dreger 2008). Dreger documents the campaign against psychologist J. Michael Bailey, which included targeting his family and false allegations that he sexually abused his children. For exposing this abuse, Dreger was targeted by the same group of activists. She received threatening messages mentioning her family, vexatious ethics charges filed against her, and organized complaints directed at institutions that invited her to speak.

Whereas research on gender identity may have seemed a niche interest in 2003 when Bailey (Bailey 2003) was writing about adult male transsexuals, the stakes are now much higher, as the number of young people expressing trans identities has risen. The first research paper to examine the broader social and psychological reasons for the surge in gender dysphoria among teenage girls, Lisa Littman's "Rapid-Onset Gender Dysphoria in Adolescents and Young Adults" (Littman 2018), prompted protests from gender identity campaigners. Brown University bowed to pressure by removing publicity for the paper from its website, while the journal that had published the peer-reviewed paper, *PLOS One*, carried out a post-publication review. This vindicated the analysis and results, yet the journal insisted on some "reframing" of the paper in a corrected version (Heber 2019).

Sallie Baxendale has highlighted the paucity of evidence regarding the effects of puberty blockers on cognitive development (Baxendale 2024a). She has also raised concerns regarding the difficulty of publishing work that subjects the views of activists to evidence-based scrutiny (Baxendale 2024b). The politicization of academic publishing has similarly threatened scholarship on women's rights (Murray Blackburn Mackenzie 2020b; Murray and Hunter Blackburn 2019).

Ethics committees have also been implicated in blocking research on sex and gender. In the UK, proposed research on people who "de-transition" was blocked by Bath Spa University, apparently due to concerns about potential reputational damage to the university (Revesz 2017). Kings College London (KCL) ethics committee blocked a proposed survey of athletes on the question of whether males should be included in female sports categories on the grounds that using the terms male and female constitutes "misgendering" (Armstrong 2023).

Blacklisting, Harassment, and Smear Campaigns

Several academics have faced attempts to get them sacked for writing on sex, gender, and gender identity. For example, philosopher Kathleen Stock faced calls for her dismissal by student activists angered by her articulation of concerns about the conceptual assumptions behind the slogan "Trans Women Are Women" and about the potential effects of allowing males to claim the status of women based on self-declaration. This ongoing campaign of bullying and harassment escalated in October 2021, leading Stock to resign from Sussex University as she no longer felt safe on campus. In a parallel case in the US, evolutionary biologist Carole Hooven was hounded from her post at Harvard after stating on television that sex is biological and binary (Hooven 2023).

Attempts to remove academics from their posts can take the form of coordinated campaigns of (often anonymous) complaints to university administrators, which, though they may fail to get their target fired, often trigger a stressful and time-consuming administrative process. For example, a lecturer at Goldsmiths University, Natacha Kennedy, orchestrated a campaign to oust feminist academics from their jobs by accusing them of hate crimes. Kennedy was supported by some Goldsmiths students, who argued, in all earnestness, that their opponents should be sent to the Gulag for reeducation (Woolcock and Bannerman 2018).

Another tactic is to launch a petition calling for an academic with dissenting views to be fired. This technique was deployed against disabilities scholar Michele Moore in an attempt to remove her from the editorship of the journal *Disability and Society* for expressing concern about the narrative that children can be "born in the wrong body" and the fact that vulnerable and autistic children are disproportionately likely to be referred to gender identity services (Yeomans 2019).

Physical threats and intimidation are also common. The history faculty at the University of Oxford received credible threats against the historian Selina Todd, forcing them to provide security at her lectures. There are many lower-profile cases of (mainly) female academics facing smear campaigns and calls to have them sacked (Stock 2019). The personal costs of such processes, in terms of mental and emotional stress and financial insecurity, especially for those on precarious contracts, should not be underestimated.

Simply defending academic freedom is enough to draw accusations of transphobia. After becoming aware of how fellow academics—overwhelmingly women—were being harassed, bullied, verbally abused, and threatened for voicing a particular view on sex and gender, we published three short pieces expressing concern about the shutting down of academic freedom on these issues. Since doing so, we have had colleagues refuse to work with us, been disinvited from talks on topics that have nothing to do with sex and gender, had complaints about our views directed at our managers, been subject to calls for students to avoid our classes, and have had to report death threats to the police. A flyer featuring a photograph of Suissa, denouncing her as a fascist, was displayed in her faculty building. In Sullivan's case, advocating for accurate sex-based data collection has led to de-platforming (Griffiths 2019; Simons and Coen 2024).

"Cancel culture" on campus is often characterized as a conflict between students and academics. The truth is more complex. A small minority of students and university staff are active in the harassment of their peers, and students (Somerville 2020) and non-academic staff are also targeted. For example, Kevin Price, a college porter at Clare College Cambridge, resigned from his role as a Labour councillor rather than support a council motion containing the slogan "Trans Women Are Women." For this principled political action, entirely unrelated to his duties as a porter, the Students' Union called for him to be sacked (Watson 2020).

No-Platforming, Disinvitations, and Shutting Down of Events

Public attention is often focused on the no-platforming of speakers, particularly well-known public figures. But activists have also targeted events involving individuals with gender-critical views, even where these views are not the topic of the event, as in the case of a planned conference on prison reform, which was canceled after pressure from activists, a talk on women's art by the artist Rachel Ara, or a talk on the history of women's education by historian Selina Todd.

Events discussing the consequences of policy changes have been targeted. For example, the criminologist Jo Phoenix, then at the Open University, had a planned talk on trans rights in prisons canceled by the University of Essex following protests from activists who objected to her raising questions about possible tensions within the criminal justice system. An event at Edinburgh University to discuss women's sex-based rights in June 2019 was subject to a campaign of intimidation, including attempts to sabotage the booking system, defamatory allegations against the speakers, a petition to get the event shut down, and a rally with banners showing misogynistic slurs. The university was forced to provide a high level of security. Several subsequent attempts to hold gender-critical events at the University of Edinburgh have been threatened or sabotaged by activists (Benjamin 2023a).

These assaults on academic freedom do not operate within traditional intellectual and professional parameters. For example, academics who have never used population data have lobbied to prevent the UK Census from including data on sex (Sullivan, Murray, and Mackenzie 2023; Sullivan 2020b, 2021, 2020a). The interventions of university staff attempting to shut down events on women's rights are grounded in an ideological position, not any relevant disciplinary expertise. We have personal experience of the way in which the bureaucratic levers of "Equality, Diversity, and Inclusion" can be used by activists to generate barriers to research and discussion on sex and gender (Sullivan and Suissa 2022). Within such a climate, it is possible to be no-platformed and harassed for expressing views that are quietly shared by the majority of one's peers.

The University of Essex commissioned a report following two instances of no-platforming (Reindorf 2021). The above-mentioned cancellation of criminologist Jo Phoenix's talk on trans rights in prisons followed credible threats from students and a flyer circulating displaying a gun-toting figure captioned "SHUT THE FUCK UP, TERF." Reindorf notes that the initial cancellation was justified due to security concerns, but this was on the basis that the talk would be rescheduled. However, the sociology department subsequently decided not only not to reissue the invitation but to blacklist Phoenix from any future invitation. Reindorf states: "The later decision to exclude and blacklist Prof. Phoenix was also unlawful. There was no reasonable basis for thinking that Prof. Phoenix would engage in harassment or any other kind of unlawful speech. The decision was unnecessary and disproportionate. Moreover the violent flyer was wholly unacceptable and should have been the subject of a timely disciplinary investigation."

Professor Rosa Freedman of the University of Reading was invited to take part in a roundtable on antisemitism as part of the University of Essex's program for Holocaust Memorial Week. The invitation was rescinded after concerns were raised about her views on sex and gender. Freedman wrote to her MP and to the Universities Minister and spoke to the press, and as a result, the invitation was reinstated. A member of the academic staff at Essex responded with a tweet comparing Freedman, who is Jewish, to a Holocaust denier. Reindorf notes: "If the invitation had not been reinstated she would have been subjected to an interference with her right to freedom of expression. This would have been particularly egregious given that the topic on which she was due to speak was entirely unconnected to the question of gender identity and was a matter of academic expertise."

The Reindorf Report is a landmark document, providing vital insights into the processes through which universities fail to act in line with their stated commitment and policies on academic freedom. Notably, the evidence gathered by Reindorf pointed to a wider climate of fear for staff wishing to express views outside the orthodoxy of gender identity ideology. Reindorf stated that Essex may be in breach of its Public Sector Equality Duty to foster good relations between persons with particular protected characteristics. In addition, Reindorf noted that suppressing gender-critical views may constitute indirect sex discrimination, given that the academics targeted are overwhelmingly women.

What Is Academic Freedom, and Why Does It Matter?

While conceptual discussions of academic freedom and the related idea of free speech are complex, the legal framework governing practice in this area is clear.

In the UK, the main relevant legal context is the Education Reform Act 1988, section 202(2)(a) of which states: "[A]cademic staff have freedom within the law to question and test received wisdom, and to put forward new ideas and controversial or unpopular opinions, without placing themselves in jeopardy of losing their jobs or privileges they may have at their institutions." This formulation is included verbatim in the statutes and charters of many UK universities. The Education Act (No. 2) 1986 (Section 43) also enshrines a positive and proactive legal duty on universities to promote and protect freedom of speech on campus by requiring that universities "shall take such steps as are reasonably practicable to ensure that freedom of speech within the law is secured for members, students and employees of the establishment and for visiting speakers." This legislation has subsequently been strengthened by the Higher Education (Freedom of Speech) Act (2023), which requires universities in England to take steps to ensure freedom of speech on campus.

Academic Freedom and the Possibility of Learning

As an empirical social scientist and a philosopher, we rely on conceptual distinctions such as that between sex and gender in our teaching and research, whether in collecting data about sex differences in education or discussing the gendered division of labor in the context of theories of justice. Crucially, what we do when we employ such concepts and tools is not just design and carry out research, write papers, or present well-worked-out positions. Underpinning our activity is a form of thinking aloud, putting forward ideas that conversational companions—whether students, colleagues, or members of the public—engage with and may disagree with. In the course of such conversations, people may express ideas that are not fully developed or defended. They may say things that we disagree with, but we try to make sense of the disagreement, clarify what we mean by the terms and positions we describe, explore their implications, and reach towards a common understanding—or, at least, a shared view on what it is we disagree about and why.

This activity is precisely what is enabled when the university is really an environment bound not just by the principles of academic freedom but by a broader commitment to free speech. Recognizing this does not mean that we collapse the distinction between academic freedom and free speech, but it does require that we acknowledge that this distinction does not map neatly onto the reality of academic life. Nor does the insistence that the commitment to free speech is an integral part of academic life, rather than separate from it, commit us to positions invoking a "battle of ideas" or the "marketplace of ideas" envisaged by classic liberal theorists, where the best argument will win out in the march towards the truth. A somewhat different emphasis, based on an account of the essential pluralism of thought and action, comes from Hannah Arendt, who argued that freedom of speech, an essential element in political freedom, means that we will always hear other opinions, other perspectives, and other arguments than our own:

> If someone wants to see and experience the world as it 'really' is, he can do so only by understanding it as something that is shared by many people, lies between them, separates them, showing itself differently to each and comprehensible only to the extent that many people can talk about it and exchange their opinions and perspectives with one another, over against one another. Only in the freedom of our speaking with one another does the world, as that about which we speak, emerge in its objectivity and visibility from all sides (Arendt 2005: pp. 128–129).

On this view, it is not only the truth about the world that we are striving for, but the viability of the world as a shared place. As academics, we share this world with students and colleagues whose experience of it is often different from ours. In coming together in a spirit of intellectual inquiry, we are not only engaging in abstract theoretical debates or trying to win arguments, but trying to make sense of this world, to offer explanations that make sense of our lives within it and help us think about how we can change it for the better. It is this ability to conduct such forms of thinking aloud that is frozen out in the current climate. When students and staff have whispered exchanges in corridors rather than thinking out loud in seminar rooms and lecture halls, we all lose out, because these seminar rooms and lecture halls become places where "the world as that about which we speak" is less likely to emerge as a shared place.

In a context where a shared understanding of basic concepts such as sex and gender has substantive implications for a range of social issues, one might think that the existence of widespread and deep disagreement would call for more, not less, discussion. Yet the effective silencing of voices and self-censorship, as a result of tactics such as those described above, is now commonplace, as reflected in our own frequent experience of being contacted by students and colleagues who say they agree with us but are too frightened to express their views in class or in public. Often, these are junior staff on casualized contracts, members of minority groups, or young women at the start of their careers.

This refusal to engage with "offensive" views means that certain views are widely available only in a misrepresented form. The historian Mary Beard provides an example in a recent review of Germaine Greer's book on rape. Beard shows, with careful quotes from the book, how a lot of what Greer is accused of saying about rape (mostly based on a talk she gave at the Hay Festival) completely misrepresents her arguments. Beard notes that perhaps the anger directed at Greer for her much-quoted remarks on the trans community "has clouded fair judgement of her arguments on rape" (Beard 2019).

A vicious circle of ignorance and offense follows: once an individual has been denounced, her work can be freely misrepresented since her opponents will not give it a fair reading (or any reading at all). A similar process occurred in the case of Rebecca Tuvel, a scholar who was subject to ad hominem attacks and online shaming, accompanied by a striking failure to engage with what she had actually written (Singal 2017, Tuvel 2017). The book burnings and #RIPJKRowling hashtag provoked by J. K. Rowling's latest novel before its release exemplify the capacity for those so-minded to be outraged by words they have not read. Kathleen Stock has written of the way her views have been misrepresented in order to demonize her, including in an open letter that had to be corrected with an erratum because a central claim was patently false (Stock 2021). When intellectual engagement is replaced by denunciation, the possibility of learning is lost.

Overtly discriminatory or threatening statements that target individuals are rightly excluded from the realm of protected speech. But if statements, theoretical positions, and conceptual distinctions are denounced as transphobic by definition, irrespective of the speaker's actual views or arguments, these arguments are never heard and never engaged with on anything other than the most superficial level. Thus, the discursive realm in which anyone can contribute to social or political discussions of sexism, gender, or sexuality is narrowed. This state of affairs is not only profoundly anti-intellectual and anti-democratic, but educationally disastrous.

JUDITH SUISSA AND ALICE SULLIVAN

Academic Freedom and Democracy

The ability to engage beyond the university lies at the heart of the connection between academic freedom and democracy.

It should be a basic right for all workers to take part in the democratic process without fear of losing their livelihoods. But for academics, public engagement has a special importance, because it is essential that policy discussions are informed by reasoned argument and evidence. In a climate where discussion is being shut down and threats are used to silence opponents, it is particularly important that universities provide a space where discussion can occur without fear. Indeed, as universities are not the only organizations engaged in knowledge production and dissemination, there is a case for extending the rights and responsibilities entailed by academic freedom to research organizations outside academia.

These debates about sex and gender are not abstract. In the UK, they were triggered partly by proposed legislative changes to the 2004 Gender Recognition Act, which would allow individuals to change their legal sex on the basis of self-ID without meeting any diagnostic or other criteria. It is important to note that in the UK, transgender people are already protected from discrimination under the 2010 Equality Act, which lists "gender reassignment" as a protected characteristic as well as sex. Yet beyond this proposed legislative change, lobby groups have campaigned to remove the existing legal protections for single-sex spaces and for the effective erasure of sex as a category in language, law, and data (Jones and Mackenzie 2020). Lobbyists such as Stonewall have been highly effective in achieving "policy capture" of organizations, meaning that, even without legislative change, the status of the category of sex in policy and practice has been eroded with extraordinary rapidity and without proper democratic scrutiny (Biggs 2022, Murray and Hunter Blackburn 2019). Similar processes have occurred internationally (Burt 2020, 2023; Murray Blackburn Mackenzie 2020a).

The need for academic freedom to research and discuss sex and gender identity seems clear-cut, given the wide range of questions at stake and their implications for law, policy, and practice. Gender self-ID has implications for equalities monitoring (a public sector duty under the Equality Act), women's legal rights to sex-based protections (Asteriti and Bull 2020,

Auchmuty and Freedman 2023), single-sex services, and women's sports. The narrative that one can be "born in the wrong body" has implications for adolescent development and education.

Given that the rapid growth in the number of young people, especially girls, presenting with gender dysphoria (psychological distress relating to one's bodily sex) is not well understood, there is a prima facie public interest in facilitating scientific research in this area. These are not purely theoretical questions. Professionals working in this field have a duty to ensure that children turning to them for help and support receive the most appropriate treatment based on rigorous research and evidence. Yet there are serious concerns that experimental treatments are not receiving the scrutiny that one would expect (Biggs 2023, Barnes 2023).

This lack of normal scientific scrutiny, particularly when it comes to assessing medical interventions, carries serious risks for vulnerable people. An article raising concerns in the *British Medical Journal* notes, "We sought the views of methodologists and clinical trial statisticians, but few were prepared to speak publicly for fear of reprisal" (Cohen and Barnes 2019). Concerns about gender medicine were highlighted by the case of Keira Bell (*Bell v. Tavistock* 2020), a young woman who took action against the Tavistock clinic following her medical transition and subsequent regret and decision to "de-transition." Though the verdict was overturned on appeal, the case highlighted the lack of data collection and evidence for the treatments provided.

Dr. Hilary Cass's "'Independent review of gender identity services for children and young people" has subsequently decisively exposed this lack of evidence. Cass notes that "throughout the course of the Review, it has been evident that there has been a failure to reliably collect even the most basic data and information in a consistent and comprehensive manner; data have often not been shared or have been unavailable" (Cass 2024). There are clear dangers to the mental and physical health of vulnerable people when professional standards of scrutiny, safeguarding, and research ethics are abandoned in the face of ideological demands. Extensive evidence of how ideological concerns were prioritized over clinical practice in the UK's flagship gender service for children has been uncovered by investigative journalist Hannah Barnes (Barnes 2023).

Women who have attempted to discuss girls' and women's rights and their experience as a sex class in this context have faced concerted attempts to silence them. Woman's Place UK was formed after a meeting to discuss proposed legislative reform was targeted for harassment, and a sixty-year-old woman was battered by male gender identity activists. Yet women's organizations that campaign within the law to protect women's existing rights, such as Woman's Place UK and Fair Play for Women, as well as groups defending the rights of gays and lesbians, such as The LGB Alliance, are slandered and denounced as "hate groups." Accusations of fascism abound, directed at lifelong socialists and trade-union activists, in order to justify denying these women a platform. It is worth noting that the traditional Left basis for no-platforming fascists is often misunderstood. This rests on the view that fascists will shut down democratic debate and organizing through the use of violence against opponents. The argument was not that fascist speakers have dangerous ideas that might influence their audience, but that there is no sense in trying to reason with violent thugs. Careless use of the term "fascist" is far from new, and the parameters of the "no-platform for fascists" policy have been contested throughout its history. But it seems that no-platforming has now been turned on its head, as opponents of freedom of speech and association use no-platforming to silence dissent. None of the feminists who have been no-platformed for gender-critical views have committed or incited violent acts. Accusations of fascism and "literal violence" leveled against these women may appear comical but have real consequences in dehumanizing them, thereby justifying harassment and even violence against them.

Attempts to deny women's rights and to silence them with threats of violence and slanderous attacks on their reputations are as old as history. Yet we have been shocked by the outpouring of hatred directed at women, typically accompanied by the term "TERF," effectively used as a replacement for epithets such as "witch," "bitch," or "cunt." The treatment of J. K. Rowling, subjected to a tidal wave of verbal abuse in response to a strikingly thoughtful and empathetic essay, is simply the highest profile case of a commonplace phenomenon.

The need for academics to communicate evidence and rational analysis is all the more apparent when political discussion is constrained by fear and intimidation. Yet dehumanizing name-calling, mindless slogans, and associated threats are not restricted to social media, but appear in peer-reviewed journals (Allen et al. 2019) and in teaching materials. The lack of a vigilant and robust defense of a positive conception of academic freedom risks allowing those engaged in what amounts to bullying to set the parameters of what can and cannot be discussed.

Academics have both a right and a duty to engage in research and discussion that illuminates questions of public and policy importance and that enables students to approach contemporary issues equipped with a broad range of intellectual resources and critical capacities. To stifle such intellectual activity risks real harm, particularly in a climate of post-truth politics, polarization, and intolerance.

Academic Freedom and Acceptable Speech: Where Do the Boundaries Lie?

Academic freedom is often described as a "foundational value" in higher education. It is important to note that free speech and academic freedom are conceptually distinct yet interdependent values and that neither value translates into an unrestricted right of individuals to say whatever they like.

In the UK, the Education Act (No. 2) 1986 (Section 43) requires universities to "take such steps as are reasonably practicable to ensure that freedom of speech within the law is secured for members, students and employees of the establishment and for visiting speakers." The Higher Education (Freedom of Speech) Act 2023 strengthens the legal requirements on universities and colleges in England (but not the rest of the UK) in relation to free speech and academic freedom. But as the phrase "within the law" indicates, there are significant constraints on these freedoms, in line with existing legislation on the prevention of disorder or crime, protection of the reputation or rights of others, and protection of national security and public safety. The Criminal Justice and Public Order Act (1994) expressly forbids communication that is "threatening or abusive, and is intended to harass, alarm, or distress someone," and similarly, the Racial and Religious Hatred Act (2006) forbids the harassment of individuals and incitation to racial or religious hatred. But these unlawful acts are narrowly defined and require, in general, either "threatening, abusive or insulting words or behaviour" or conduct that "creates an intimidating, hostile, degrading, humiliating or offensive environment" for another individual, with particular reference to the protected characteristics under the Equality Act 2010.

Discrimination and harassment directed at trans students or staff simply for being trans should, of course, be treated with the relevant disciplinary procedures, and university disciplinary policies reference the 2010 Equality Act, where gender reassignment is one of the nine protected characteristics (see below). But while judgments about unlawful discrimination and harassment may not always be straightforward, the existence of legal restrictions on free speech cannot be allowed to undermine the fundamental commitment to its central value for universities.

Discrimination and Gender-Critical Beliefs

The Equality Act (2010) makes it unlawful to discriminate against people on the basis of nine protected characteristics: age, disability, gender reassignment, marriage and civil partnership, pregnancy and maternity, race, religion or belief, sex, and sexual orientation. To be protected, a belief must be worthy of respect in a democratic society and not affect other people's fundamental rights. So, for example, racist beliefs would not be protected.

Gender-critical beliefs are defined in UK law as the belief "that biological sex is real, important, immutable and not to be conflated with gender identity." The UK Employment Tribunal case of Maya Forstater, who lost her job at a think tank following tweets expressing gender-critical views, established that gender-critical beliefs are "worthy of respect in a democratic society" and therefore protected (*Forstater v. CGD Europe and Others* 2021). This means that it is illegal to discriminate against individuals because of their gender-critical beliefs.

The case of *Phoenix v. the Open University* (2024) applied the Forstater principle to the university sector. In a landmark judgment, the employment tribunal found that Professor Phoenix was discriminated against, harassed, constructively dismissed, and victimized by the Open University (OU) due to her gender-critical beliefs.

Phoenix, a criminologist, had expressed concerns about male-bodied offenders being housed in the female prison estate and about the suppression of speech on such issues. She set up a Gender Critical Research Network (GCRN) to create a space for research and discussion at the OU. In response, 368 OU staff and postgraduate students signed an open letter demanding that the OU withdraw its support from the network and denouncing it as transphobic. A further public statement accusing the network members of transphobia was published on the OU website.

The tribunal found that the OU failed to protect Phoenix from a targeted campaign of harassment and that the public denunciations of the GCRN contained untruths that were detrimental to the claimant's professional reputation. The OU had argued that they were obliged to allow the targeted harassment of Phoenix as an expression of academic freedom. The tribunal disagreed, stating that "upholding academic freedom did not prevent the Respondent from taking action to prohibit the harassment." As such, the judgment may be seen as drawing a line between the exercise of academic freedom on the one hand and campaigns of bullying intended to suppress academic research and debate on the other.

Conclusion: What Can Universities Do?

The defense of academic freedom is the collective responsibility of the academic community. Current challenges to upholding this value include a marketized system in which students are seen by university leaders primarily as customers rather than learners, encouraging an instrumentalism at odds with educational values. Increasing precarity among academic staff makes the exercise of academic freedom in teaching and research too risky for many to contemplate, given the frequent consequences, as documented above, of questioning ideological orthodoxies. The trend for university administrators to police the boundaries of academic freedom within the parameters of "risk assessments" and "reputational damage," rather than seeing academic freedom as a matter for the academic community, is central to the problem. There is a lot more that universities could do to ensure that an understanding of what academic freedom means and why it matters is well embedded throughout all institutional policies and procedures.

Academic research undertaken in good faith and by experienced researchers can be, and regularly is, criticized for its methodology, for its underpinning assumptions, and for what it does not say, as well as what it does say. But in an era of "post-truth" and "alternative facts," it is imperative to be careful and accurate in distinguishing rigorous academic research from dogma and ideology. The language of harm and safety must be treated critically and seriously. While we should all be vigilant in addressing the disadvantage and discrimination faced by various minority groups, students and staff should be able to distinguish between the expression of dissenting views and actions and speech that constitute overt forms of harassment, intimidation, and threats towards individuals.

A commitment to free speech and academic freedom does not and should not constitute a defense of harassment or attempts to close down the speech of others. Universities must take appropriate disciplinary action against students and staff who engage in campaigns of harassment against other students and staff.

Commentators in some sections of the Left increasingly assume that there is something right-wing or elitist about upholding the values of free speech and academic freedom or regard concerns about attacks on academic freedom as part of a manufactured, right-wing "culture war." Yet this is both historically illiterate and grossly short-sighted. It perversely ignores the power dynamics at play and the fact that abandoning academic freedom as a value to be upheld by the academic community means ceding decisions about what can and cannot be said to administrators who may equally be swayed by government, financial donors, or social media mobs.

We have focused here on academic freedom, highlighting cases involving university employees. However, there is a complementary case for strengthening free speech as an employment right for all workers, given that the absence of such protection tends to expose organizations to policy capture, weakens democratic discourse, and can only be detrimental to the ability of policymakers to know the views of the people they represent. Universities are not ivory towers, and our ability to deliver knowledge as a public good is undermined by a wider climate of censorship.

Many academics have only recently become aware of the political project to deny the material reality of sex and its implications. This chapter focuses on the threat to academic freedom in the case of sex and gender, not because it is a hard case, but because it is an easy one, with implications across the disciplines. If we cannot defend academic freedom in such a case, we cannot defend it at all.

The Desexing of Language in Women's Health Research and Care: A Story of Marginalization of Science, Cultural Imperialism, and Abuse of Power

KARLEEN GRIBBLE

As others in this volume have emphasized, in recent years, health research, policy, advocacy, and public communications have become fraught with controversy related to the rise of the concept of gender identity and associated activism to marginalize sex. Pregnancy, birth, and breastfeeding have become a crucial axis of this battle as there is not an area of human experience where sex difference is more obvious or important than reproduction. As a researcher in breastfeeding working in a Western high-income and global public health context, I've had a front-row seat to this. I am also the lead author of an influential paper on the importance of sexed language in maternity (Gribble et al. 2022), written in an attempt to counter the unthinking acceptance of desexed language as progressive and more inclusive. This chapter will provide an international perspective on how sexed use of words like "women" and "mothers" became controversial and the impact of advocacy against sexed language.

The Origins of Gender Identity and How and Why Language Is Desexed

It has been long understood that the sex of individuals should be made visible when sex is relevant to ensure sex-based needs and rights are not overlooked. Thus, the sexed terms of "women," meaning adult female people, and "mothers," meaning female parents, have been central to research, policy, advocacy, and communications regarding female reproduction. However, in recent years, those working in maternal and child health have experienced pressure to avoid these terms and desex our language. This pressure has come about due to the rise in cultural salience, personal application, and institutionalization of a concept developed in the United States of America (US) called gender identity (Money 1994).

When the term "gender identity" was first used, it described an individual's knowledge of their sex as male or female (Byrne 2023). However, the meaning soon changed so that gender identity delineated an internal sense of oneself as living in the social role of a man or a woman (Money 1994). In this way, gender identity became enmeshed with the concept of "gender" developed by feminist philosophers to describe the social expectations of men and women (Mikkola 2024). Current definitions of gender identity tend to be circular and retain reference to sex stereotypes, masculinity, and femininity (e.g. Chiniara 2023; Coleman et al. 2022). Further evolution of the concept has allowed for "non-binary" gender identities and the understanding that gender identity is not fixed but can change over time (Coleman et al. 2022). For those who subscribe to gender identity, it is their internal sense of self that makes them a man or a woman rather than biology. Recently, increasing numbers of individuals, particularly in the US but also in other Western nations and beyond, are asserting they have a gender identity in conflict with their sex and so are transgender (McKechnie et al. 2023; Wiepjes et al. 2018; Zhang et al. 2021).

Those identifying as transgender have higher rates of poor mental health and other adverse experiences than the general population and are considered a marginalized and suffering group (Farvid et al. 2021). Someone who is female but who has a gender identity of "man" may experience great distress if referred to as a woman. Thus, some promote desexed language in a desire to avoid distress and protect those they see as vulnerable. Other individuals and organizations appear to have a broader intent to marginalize sex and to replace it with gender identity across the board, including in policy and data collection (Murray and Hunter Blackburn 2019; Sullivan 2023). Of course, to do this, it is necessary to obscure sex in language.

Desexing language results in straightforward and clear terms such as "women" and "mothers" being replaced with descriptors referring to physiological processes or body parts or otherwise making sex invisible. These replacements may use technical language or are dehumanizing, such as "menstruators," "birthers," or "vagina owners," or inappropriately subsume women into "people" or "parents." Although both men and women have breast tissue, an association between femaleness and breasts means "breastfeeding" and "breastmilk" may be replaced with words like "chestfeeding" or "human milk." Finally, the phrases "assigned sex" or "sex assigned at birth" are used to communicate sex is not inherent to individuals but is arbitrarily allocated independent of biology. This decoupling of the word "sex" from sex as a biological variable effectively desexes it. Table 1 presents some desexed replacement terms used for women, mothers, breastfeeding, and breastmilk. For an ongoing documentation of desexed language in women's health, see the blog by writer Milli Hill, The Word is Woman (Hill 2024).

Table 1. Sexed terms and their desexed replacements in relation to female reproduction.

Sexed terms	Desexed replacement terms
Women	People, individuals, humans, patients, someone, uterus havers, people with uteri, cervix havers, people with vaginas, bodies with vaginas, birthing bodies, female-assigned bodies, menstruating personnel, menstruators, bleeders, menstruating individuals, bodies that bleed, menstruating bodies, people who experience periods, people who period, people who menstruate, vulva owners, vulva people, someone with a vulva, birth people, non-males, non-men, non-prostate owners, individuals who have receptive vaginal sex, womb carriers, those assigned female at birth, reproductive aged persons identified as female, person with reproductive potential, person of childbearing potential, people with reproductive organs, people with female-identified reproductive organs, individuals who have reproductive capacity.

Pregnant women	Pregnant people, pregnant families, birth givers, gestational carriers, gestational parents, carriers, gestators, expectant parents, expectant persons, people expecting, pregnant persons, pregnant patients, childbearing persons, childbearing individuals, childbearing communities, pregnancy termination seekers, maternal agent, service users, expectant birthing individuals, person pregnant, those who labor, people expecting.
Mothers	Parents, families, gestational parents, postpartum individuals, postpartum participant, birthers, birthing parents, postpartum individuals, postpartum people, caregivers, parentals.
Mother-infant dyad	Parent-infant pair, postpartum dyad, peripartum unit, lactating dyads.
Breastfeeding mothers	Breastfeeding parents, nursing parents, breastfeeding families, breastfeeders, chestfeeding parents, lactating parents, lactating families, lactators, lactating individuals, pumping parents, those who are latching their baby.
Breastfeeding	Human milk feeding, human milk provision, chestfeeding, lactating, body feeding, feeding from the body, mammal feeding, feeding, nursing.
Breastmilk	Human milk, parents' milk, chest milk.

Cultural Imperialism and Desexed Language

The pressure to desex language has occurred not just in the US but also in other Western and non-Western countries and in and via international organizations through US cultural power. The US origin of desexed language is supported by internet tools showing the uptake and spread of desexed terms. Google Books Ngram Viewer® shows trends in language in the Google Books® corpus. "Chestfeeding" first appeared in Ngram in 2010, and there has been a sharp increase in its use from 2012 until the present. Ngram shows the phrase "assigned sex" as far back as the mid-1800s and "sex assigned at birth" from the mid-1950s in small numbers. The long-term use of these terms relates to infants with disorders or differences of sexual development (DSDs), a small proportion of whom do not have a clear sex, so "assigning" them a sex is accurate (Markosyan and Ahmed 2017). However, since 2010, there has been a sharp increase in the frequency of "assigned sex" and "sex assigned at birth," reflecting the new use of these phrases to supplant "biological sex" or simply "sex." Similarly, "pregnant person" has been included sporadically in books since the 1800s, with more consistent use from the mid-twentieth century and a sharp increase in use since 2013.

The spread in the use of desexed language can be seen using Google Trends®, which analyzes random samples of internet searches to show the popularity of search terms across time and geography. Google Trends® consistently shows the geographic spread over time from the US to across the world of terms like "assigned sex," "menstruators," "birthing people," and "chest feeding" from about 2015–16 onwards.

Process of Growing Pressure to Desex Language

Recommendations to use desexed language started in the US and Canada around 2014–15. This was most visible in relation to the novel term "chestfeeding" and there were a number of papers aimed at health professionals promoting this word at that time (e.g. Farrow 2014; Wolfe-Roubatis and Spatz 2015). However, research suggests that chestfeeding was not then in widespread use amongst those who identified as transgender. MacDonald et al. (2016) considered "experiences with lactation, chestfeeding, and gender identity" of "transmasculine individuals," but of twenty-two research participants, only three said they preferred the term chestfeeding. Similarly, Walks (2018) wrote a book chapter, "Chestfeeding as a gender fluid practice," but all study participant quotations referred to breastfeeding and none to chestfeeding. The promotion of chestfeeding by these authors demonstrates how advocating for desexed language may come not just from a desire to support individuals but also to shift how societies conceptualize sex and gender identity and use language in relation to each.

When the idea of desexed language was first presented, it was understood that desexed terms should be used on a one-to-one basis with people who preferred them or in tailored resources for this group. Using preferred terms is an evidence-based tenant of person-centered health care, and tailored resources for groups with particular needs are also common, so this was not objectionable. However, it soon became clear that the desexing of all communications was the end goal. Advocacy for desexing language intensified during the COVID-19 pandemic and has continued since. Below, I describe the process and manifestation of desexing language with examples from journals, publishers, university ethics committees, and health organizations.

Professional Peer-Reviewed Journals, Publishers and Desexed Language

Desexed language became noticeably more common in academic publications during the COVID-19 pandemic and has continued to increase in use since. This can be seen in Google Scholar searches for "pregnant people" for 2018, 2019, 2020, 2021, 2022, and 2023, which yields 340, 439, 852, 1,720, 3,400, and 4,600 publications, respectively (a fourteen fold increase) and similarly for "birthing people" over the same years, which yields 23, 49, 119, 351, 738, and 1,190 publications (a fifty-two-fold increase).

Desexed language can be dehumanizing but nonetheless has been published unimpeded by journal review or editorial processes. *The Lancet* not only published a paper describing women as "bodies with vaginas" (Davis 2021) but quoted this phrase on the issue cover. Research published in *The Lancet Western Pacific* described the stigma of menstruation but nonetheless referred to study participants as "menstruators" (Wilbur et al. 2021). The *Harvard Journal of Law and Gender* published a paper discussing deficiencies in maternity care for black American women that described them as "birthing bodies" (Nash 2021).

Desexing language can introduce errors, and the most important publications in the most prestigious journals have not been exempt. For example, the groundbreaking paper on COVID-19 vaccination during pregnancy by Shimabukuro et al. (2021) in the *New England Journal of Medicine* miscited research as saying COVID-19 severity was greater in pregnant persons as compared to non-pregnant persons when the comparator was, in fact, non-pregnant women.

Some uses of desexed language in peer-reviewed papers are undoubtedly the authors' preference. For example, the reasoning for using "obstetric body" to mean women was described in a paper on female reproductive anatomy and clearly came from the authors' desire to promote gender identity over sex (VanSickle, Liese, and Rutherford 2022). However, journal reviewers and editors have also played an important role in the increasing use of desexed language. Reviewers not uncommonly ask authors to replace sexed terms like "women" and "mothers" with desexed terms like "birthing people" or "breastfeeding parents." Sometimes these requests are nonsensical; evolutionary biologist Natalie Dinsdale described how a reviewer for her paper on endometriosis and polycystic ovary syndrome requested "women" be replaced with "people with female-identified reproductive organs" (Dinsdale, Chizen, and Crespi 2023).

Some journals have implemented policies requiring desexed language. I was inside the process of such policy implementation in 2020 as a member of the editorial review board (ERB) of the *Journal of Human Lactation* (*JHL*). The *JHL* ERB was presented with a draft policy on "gendered and inclusive language," advising authors to prioritize gender identity over sex. The policy stated that authors should not use language implying there are two sexes and that "mother" should be understood and used only to apply to gender identity (Bamberger and Farrow 2021a). Guidance accompanying the policy also stated that "assigned sex" rather than "sex" should be used (Bamberger and Farrow 2021b). I provided feedback to the editor, saying that being able to clearly identify women and mothers in language was important when writing about breastfeeding. I argued some requirements were unscientific and inappropriate for an international journal as they imposed an American/Western philosophy on authors from diverse countries, thus presenting a type of US cultural imperialism. My protests and those of others made no difference, and the policy was implemented. I and two other ERB members who objected to the policy were shortly thereafter removed as members.

Other journals and publishers have instituted similar policies. Recently, *The Lancet* instructed, "Authors should use the term 'sex assigned at birth' rather than 'biological sex,' 'birth sex' or 'natal sex'" (*The Lancet* 2024). It is difficult to comprehend the imposition of the unscientific concept of "sex assignment" and requiring authors publishing in *The Lancet* to use this term as other than neocolonialism. Nonetheless, the editor of *The Lancet*,

Dr. Richard Horton, has written about the importance of decolonization, stating, "Decolonisation seems an urgent obligation if those who advocate for equity wish to be taken seriously and retain even a vestige of moral conscience" (Horton 2021). I cannot see how these conflicting positions can be reconciled.

The publisher Elsevier has produced a template of author instructions for their 2,900 journals using the phrase "sex assigned at birth" and stating sex is not binary. Authors are also required to complete a survey during manuscript submission that presumes gender identity is a universal characteristic and includes no option to indicate the concept does not apply. Thus, Elsevier institutionalizes not just desexed language but forces the concept of gender identity on all authors, irrespective of location, culture, or beliefs.

Authors may be asked to desex language by editors at any point of manuscript processing. A paper on pregnancy accepted by the *British Medical Journal* had all instances of "women" replaced at proofing with "people" by a member of the editorial staff, which was reverted after author objection (Goodfellow, Curran, and Care 2024). An editor at *The Lancet Global Health* requested that my coauthors and I add "chestfeeding parents" to a letter to the editor on breastfeeding women in the Democratic Republic of the Congo (Ververs et al. 2024) (we declined). Authors from Bangladesh were informed by an editor of a requirement to refer to the "assigned sex" of infants during the last round of review of their paper and felt they must comply (Sujon, Khatun, and Mumin 2023). It seems authors from non-Western or non-English speaking backgrounds may have more difficulty understanding and less confidence objecting to such requests.

University Ethics Committees

Some human research ethics committees ask researchers to desex the language used in their studies. The already mentioned Natalie Dinsdale was repeatedly requested by her university ethics committee to desex the language of her study on female reproductive anatomy and function (Dinsdale 2023). Over three rounds of revision, the ethics committee questioned the use of "women" and "female" in study materials, while Dr. Dinsdale defended her use of sexed language. In the final revision request, the direction was to "remove the references to women," which was rejected by Dr Dinsdale, who informed the ethics committee their requests were inappropriate, and the study was finally approved. In another case, a researcher was required by her ethics committee to replace "women" with a desexed alternative. The study subsequently had a poor response rate, and complaints were received about language. An ethics variation was sought and granted to reinstate "women," following which recruitment improved. Relatedly, some ethics committees are discouraging researchers from collecting data on sex (e.g. Longborough University 2024; Monash University), adversely affecting efforts toward closing the female data gap. In some cases, such guidance was developed by diversity and inclusion sections of universities.

Desexed Language within Health Organizations

Desexed language also became more commonly used by organizations over the COVID-19 pandemic. Many used sexed language at the beginning of 2020, but by the end of 2021, desexed language was their norm. The United States Centers for Disease Control and Prevention (USCDC) progressively desexed their COVID-19 breastfeeding guidance from May 2020 to January 2021 with the terms "woman," "women," "mother," and "mothers" replaced with "people," "person," "lactating person," "lactating parent," and "someone." At the time (and contrary to the evidence), the USCDC was also encouraging the separation of mothers with COVID-19 from their newborn infants, which impacted practice not just in the US but globally (Gribble et al. 2023). Alongside this, other health organizations, including the US-headquartered International Lactation Consultant Association, began using ambiguous desexed language, speaking of "breastfeeding/ chestfeeding families" and the need to keep "parents and infants together" (International Lactation Consultant Association 2021). Thus, in the midst of a pandemic, when mothers and newborn infants were being separated in unprecedented numbers and advocacy was urgently needed regarding the unique importance of mothers to their infants, the word "mother" was declining in use.

As with scientific publications, organizations introduced errors into communications when they desexed language. For example, the Australian

Department of Health desexed its COVID-19 vaccination information for pregnant and breastfeeding women in August 2021. A simple "find and replace" of "women" with "people" made statistics on the increased severity of COVID-19 during pregnancy incorrect (Australian Government Department of Health 2021). The UK's National Health Service began progressively desexing public communications on female reproductive health in 2020–21, including the blunder, "8 in 10 people under 40 will get pregnant within 1 year of trying by having regular sex without contraception" (National Health Service 2020).

Many changes made by organizations to public health communications also included technical or dehumanizing language. The health information service Healthline referred to women as "vulva owners" on multiple web pages (Catlet 2021; Kassel 2020; Litner 2021). Health Services Executive Ireland outlined that cervical cancer screening is needed "if you have a cervix" but did not explain that those who have cervixes are women (Health Services Executive Ireland 2023). Sometimes organizations simply avoided giving any indication a procedure might be sex-specific, such as the Canadian Cancer Society, which simply said, "If you've ever been sexually active, you should start having regular Pap tests by the time you're 21" (Canadian Cancer Society).

In grassroots women's health organizations, pressure to desex language has caused turmoil. In 2020, the US-headquartered breastfeeding support organization La Leche League International (LLLI) instituted a policy requiring their communications to include "a variety of terms," a euphemism for using phrases like "breastfeeding parents" or "breastfeeding/chestfeeding." In 2021, 250 LLLI breastfeeding counselors from forty-five countries protested against this requirement, calling it "oppressive and colonialist" (Lineham 2021). More recently, directors of the LLLI affiliate, La Leche League Great Britain (LLLGB), wrote to the LLLI board saying LLLGB would not use desexed language or provide services to males (LLLI had specified they should assist "everyone" who wants to breastfeed, irrespective of sex.) They pointed out that LLLGB is a UK charity with a registered mission "to help mothers to breastfeed through mother-to-mother support" (Charity Commission for England and Wales) and is a single-sex organization. The LLLI board responded by describing the word "mother" as a "roadblock" and suspending the accreditation of these directors (Martin 2024). The actions of LLLI in imposing a belief in, and prioritization of, gender identity across their whole organization is another clear example of cultural imperialism and abuse of power.

Insight into the process of advancing desexed language within health organizations is provided by Drysdale et al. (2024) who considered desexing language in health policy, promotion, education, and delivery in cervical cancer screening programs in Australia. Study participants working in policy, health promotion, health provision, and advocacy spoke in favor of desexing the language of cervical cancer screening. They described negotiation on language continually taking place within their organizations with the point of urgency for change to desexed terminology being reached after a long "slow drip" process. They viewed getting desexed language into guidelines and policy as a way of promoting "top-down" language changes throughout health systems and "pushing the envelope further and faster." Policy was viewed as a critical axis for change to enable the implementation of desexed language more broadly.

Desexed language in policy is particularly concerning as it might skew health priorities and resourcing, thus "pushing the envelope" from language to practice. One possible vector for change is through women's health strategies, some of which now use the term "women" to refer to gender identity rather than the female sex. The Women and Girl's Health Strategy for the state of Queensland, Australia, defines women and girls as "all people who identify as a women or girl" (Queensland Health 2024). In this way, males with a gender identity of woman or girl are included. The current strategy does not address male health conditions, but it is not difficult to imagine this may change. In fact, many initiatives previously focusing on women and girls have shifted to include both sexes. The organization UN Women was established to assist in the implementation of the United Nations Convention on the Elimination of All Forms of Discrimination Against Women (CEDAW) (United Nations 1981). CEDAW is an explicitly sex-based convention; however, UN Women now prioritizes gender identity over sex. It describes males with a woman gender identity as amongst the most marginalized women and advocates for their access to resources developed to promote women's equality, such as female-only spaces and women's sports (UN Women 2022).

What Are the Arguments Against Sexed Language?

I have already mentioned a desire to support individuals who identify as transgender is an important motivation for some who advocate against sexed language. Arguments against sexed language are also usually presented within a worldview where gender identity is a universal characteristic of humans through time, culture, and geography and wherein there have always been transgender people. Further underlying beliefs are that gender identity is of great consequence while sex is unimportant and the notion that there are two sexes is a colonial imposition.

These beliefs demonstrate a type of ethnocentrism (using beliefs of one's own culture as a frame of reference for other cultures) and presentism (application of present-day concepts and values to history) and are not only incorrect but offensive. This is particularly so in relation to the assertion that the fact there are two sexes was unknown outside of Europe until the fifteenth century (i.e., before colonialism) and also that gender identity is universal and transgender people have always existed. In regard to the latter, it is true that across time and cultures, there have always been individuals who don't conform to social expectations of their sex. However, it cannot be said such people experience a gender identity in conflict with their sex or are transgender. Maori researcher Elizabeth Kerekere (2017) pointed out just this, stating that in precolonial New Zealand, diversity of gender expression existed, but there is no evidence of individuals having diverse gender identities. In addition, while there are many cultures providing specific accommodations and status for people (usually androphilic males) who are non-conforming to sex stereotypes, such as the Fa'fa'fine of Samoa or Khawaja Sira of Pakistan, it is a mistake to consider these people transgender. To do so is, as anthropologist Paul Vasey described, "gender colonialism" (Vasey 2022) and also ignores the diminishment of such culture-bound understandings that is occurring as the concept of gender identity spreads

(Azhar 2017; "South Asia's non-binary communities worry about losing their identity" 2021).

Nonetheless, these beliefs strongly shape arguments made against sexed language, resulting in predominantly ideologically focused claims rather than evidence-based ones. Examples of ideologically focused claims in a recent publication by Pezaro et al. (2024) (as far as I can decipher or represent them) include:

- Making sex invisible in language will reduce sexism.
- Sexed language privileges the power of white men.
- Using sexed language is a politically conservative position.
- Sexed language perpetuates reproductive injustice.
- Recognizing in language that it is women who become pregnant and give birth is misogynistic.

Those who argue against sexed language also commonly utilize straw man arguments and may falsely say that those who assert sex is important view the social expectations of the sexes or sexism as immutable when the actual proposition is sex is immutable, not sex stereotypes (Pezaro et al. 2024). Finally, it is often claimed that sexed language was imposed on non-Western cultures via colonialism (e.g. Pezaro et al. 2024; United States Breastfeeding Committee 2022). This argument is without any basis. Where individuals and organizations from non-Western contexts promote desexed language, a link can usually be drawn to the US/the West. For example, a Nigerian health advocate who referred to women as "menstruators" informed me she used desexed language as she was *"privileged…[to] work with Western and "woke" organizations."* In another example, views in favor of desexed language by academics from non-Western backgrounds were put forward to argue sexed language has colonial origins (United States Breastfeeding Committee 2022). However, these academics were educated in and living in the US or other Western countries, and the underpinning of queer theory and ideas regarding gender identity in their views is unmistakable. In one case, the academic undertook her graduate studies in gender studies at the University of California, Berkeley, now home of the "godmother of queer theory" Judith Butler.

Importantly, those who argue against sexed language have generally not engaged with the challenges of desexed language flowing from less clear health communications. Pezaro et al. (2024) do not address this issue at all. The Equity and Inclusive Language Toolkit of the US Breastfeeding Public Health Partners advises "mothers" and "breastfeeding" could be replaced with "gestational parent" and "bodyfeeding" but also advises using plain language and avoiding technical terms (Breastfeeding Public Health Partners 2024). There is no discussion on how recommendations for desexed and plain language conflict with one another.

Why Has the Impact of Desexed Language Not been Researched?

Perhaps what is most extraordinary about the current situation is the lack of research supporting the implementation of desexed language or documenting the impact of these changes. In our 2022 paper on the importance of sexed language, we proposed that desexed language risks decreasing overall inclusivity, dehumanizes women, includes people who should be excluded, introduces inaccuracy, precludes precision, creates confusion, undermines breastfeeding, and works against recognition of the mother-infant relationship (Gribble et al. 2022). We provided examples where each of these detriments was evident and argued that desexed language would likely have the most impact on the most vulnerable women with low education, health literacy, or language skills or who are marginalized as mothers. We pointed out that desexing language works against well-established public health principles to prioritize plain language in health communications. This paper has now been read or downloaded by more than 175,000 people in over 160 countries, suggesting the concerns we described have resonated with many individuals in many contexts.

However, to date, there are just two studies considering the understandability and acceptability of desexed language; one from the US and the other from Uganda (Kinney et al. 2023; Nakijoba et al. 2025). The US study (sixteen participants) found that confusion resulted if "mother" was replaced with "parent" and that replacing "breastfeeding" with "chestfeeding" or "bodyfeeding" had low acceptability. One woman said she would exit a survey using "chestfeeding," while another distressingly connected "bodyfeeding" to her experience of rape and said she would reject a survey with this term. The Uganda study (146 participants) considered the understandability and cultural acceptability of terms including "person with a uterus," "person who menstruates," "birthing parent," and "chest feeding" (Nakijoba et al., 2024). This research also found confusion and offense; "birthing parent" was understood to refer to both mothers and fathers and the language itself was described as "disrespectful," "obscene," and "shameful" (Nakijoba et al., 2024). There is enormous potential for desexed language to contribute to poor individual and population health outcomes if women do not understand communications or are unwilling to engage with health services or research due to the language used.

Given the potential adverse impacts of desexed language, the lack of underpinning evidence leads to two questions: Why are changes being made without an evidence base, and why has research on this subject not been published? Regarding the former, this question was asked by women's health advocate Ciara Curran to a House of Lords Committee in the UK. Ciara is a bereaved mother who established an organization called Little Heartbeats to support women experiencing a serious pregnancy complication called preterm prelabor rupture of membranes (PPROM) ("Little Heartbeats). Ciara told the House of Lords, "I keep being told we need medical research to back any changes and polices, yet no research is in place to support removal of terms like women and mothers from health communications and policy and no one appears to have taken into account impact of the loss of these words on women who have lost babies" (House of Lords 2024). Ciara singled out "bodies with cervixes" as unacceptable and became emotional as she connected desexed terminology to dehumanization of women during pregnancy and childbirth and her own perception of being seen as "a body on the bed" rather than a woman needing health care when she experienced PPROM (Ely 2023). Ciara's words should be no surprise: researchers have repeatedly heard women who have experienced birth trauma describe feeling dehumanized and treated like "just a body" (Keedle, Keedle, and

Dahlen 2022; Walburg, Friederich, and Callahan 2014). Ciara's distress at desexed language being introduced without research is understandable.

Regarding the question of why research has not been published, first, the controversial nature of this subject has dissuaded researchers. Desexing language has been framed as unequivocally positive and "more inclusive," while questioning desexed language is bigoted and transphobic. I and other authors of the 2022 paper on sexed language (Gribble et al. 2022) experienced personal insults, attempts to have us deplatformed from speaking events (unsuccessful), and complaints to employers and organizations we work with (also unsuccessful). Two individuals who contributed to this paper withdrew before submission, being concerned about how their university would react, or because they thought being a named author might affect their funding prospects.

Second, research may have been undertaken but not published because negative outcomes to desexed language were identified. Non-publication of negative findings is a well-recognized phenomenon, particularly in pharmaceutical research (Khan et al. 2021), but this has also occurred in gender medicine. It was recently revealed that the World Professional Association for Transgender Health had suppressed the publication of papers reporting on commissioned systematic reviews of gender transition treatments because the findings were not as desired (Clinical Advisory Network on Sex and Gender 2024).

Third, researchers may have been unable to obtain funding to research desexed language due to ideological capture of funding bodies. I was a coinvestigator with very senior and experienced colleagues on two grant applications to the UK's National Institute for Health and Care Research to consider the impact of desexed health communications on women. Our applications were supported by letters from eleven women's health organizations outlining the dire need for this research. Both applications were prioritized highly in the initial review stage. However, both were rejected at the funding committee stage. With the first application, reviewers were reportedly "horrified" this research was being proposed; it was "dreadful research"; and the topic itself was offensive. With the second application, reviewers made a series of clearly ideologically based complaints, including that we had used the terms "female" and "woman" interchangeably, had not referred to queer theory, and the research question was not presented in neutral language. For the record, the research question was: "What are the extent, impacts and implications of de-sexed language in reproductive health communications for women in the UK?" It is hard to imagine a more neutral framing, and indeed, we were approaching this research with genuine curiosity as to what we would find, both positive and negative, about desexed language.

In the absence of research, we are left in a place of unknowing regarding the impact of desexed language. Anecdotal evidence of misunderstanding, aversion, and miscommunication supports adverse consequences. It is likely, for example, that cervical cancer screening promotions describing those eligible for screening as "anyone with a cervix" are not understood by some women. But which and how many women? What are the morbidity and mortality costs resulting from missed screening for women who do not understand these health communications or reject the message? How might misunderstanding or rejection be mitigated? Who benefits from desexed language, and under what circumstances? We do not know the answers to any of these questions. And yet journals, publishers, university ethics committees, health organizations, and others are moving full steam ahead using, promoting, and sometimes enforcing desexed language. This is a completely unacceptable situation. Desexing the language of women's health is a major intervention, one that should not be made without a strong evidence base.

Taking Action to Uphold Science and Evidence-Based Health Care

In order to protect individual and population health, increased understanding, appropriate policies, and research on sexed and desexed language are required. Those in science and health need to appreciate and communicate the known and potential detriments of desexed language and ensure they are properly considered by decision-makers. Providing concrete examples of problems with desexed language can enable understanding that this is not a simple, risk-free change. In international settings, making clear the difficulties for non-native English speakers and the cultural imperialism of using desexed language can be helpful. I am aware of many meetings where desexed language has been proposed but where individuals have suggested caution, referred to the paper on the importance of sexed language (Gribble et al. 2022), and sexed language has been retained.

Working together with colleagues can also increase confidence and provide protection. Recently, I collaborated with others in writing and gathering support for responses to two Australian government consultations addressing sexed and desexed language in research and data collection on sex (Gribble et al. 2025). While the responses were written by a small number, they were endorsed by many (over two hundred). Numerous signatories were cognizant of the challenges of desexed language, but for others, just being asked to consider supporting the responses also performed an educative role.

Journal publishers and editors should implement appropriate policies in relation to sexed and desexed language to guide authors and reviewers. Journals should not permit papers to be published using dehumanizing, desexed language. They should have policies preventing authors who have used sexed language in their recruitment and study instruments from using desexed language in their papers as a matter of ethics and respect for research participants. They should not require terms like "assigned sex" or universalize gender identity. They should ensure authors do not feel pressured by reviewers to use desexed language. Where desexed language is used by authors, journals must ensure accuracy is not compromised, nor research of others misrepresented. They should make certain that where "women" has been used in a gender-identity-based meaning (including males), data is nonetheless disaggregated by sex so that sex differences are visible (e.g. Thornhill et al. 2022). If disaggregation by sex is not possible because sex data has not been collected, this should be noted as a limitation. The UK Network of Professors of Midwifery and Maternal and Newborn Health has produced a position statement on sexed language which includes recommendations that "journal editors should not seek to influence authors to use de-sexed language" and that "authors of journal articles, policy, health information should avoid language that disembodies and dehumanises people by reducing them to body parts or functions" (UK Network of Professors of Midwifery and Maternal and Newborn Health 2023). A number of women's health journals are using this position statement to guide their work.

Ethics committees should not outsource guidance development to diversity, equity, and inclusion bodies. Rather, they should develop their own policies prioritizing research quality and ethical conduct, including the avoidance of dehumanizing language. They should ensure sex data is collected where any of the following may be relevant:

- Physiological/anatomical differences between the sexes
- Social expectations of the sexes or sexism
- Women's reproductive capacity
- Women's reproductive work

Ethics committees should also ensure study participants are able to indicate gender identity is not a characteristic applicable to them. (This is also a recommendation of the UK Network of Professors of Midwifery and Maternal and Newborn Health.)

Organizations must ensure they do not impose desexed language upon individuals or other organizations and show cultural respect. Those holding positions of power have a particular responsibility not to act in a culturally imperialistic manner, but this is not occurring. For example, the HIV Language Guide of the research funder the US National Institute of Allergy and Infectious Disease describes the terms "mothers," "women," "girls," and "breastfeeding" as "stigmatizing terms to avoid" (National Institute of Allergy and Infectious Disease 2021). This guide also states that stigmatizing language (including these terms) should not be used in any communications, including grant applications and publications, with no exemption for non-US grant applicants or recipients. This is completely unacceptable. Nakijoba et al. (2024) noted their motivation to consider the acceptability of desexed language in Uganda being in response to "pressure from journals and some partner organizations" to use desexed terminology. Their research concluded that "the shift towards de-sexed language may be seen as an imposition of foreign values rather than a progressive step toward inclusivity."

Organizations generally should be extremely cautious about desexing language absent research to evaluate its impact. Applying established health communication principles and using plain and clear language for the majority population and targeted materials for those with particular needs around sex and gender identity is a defensible position.

Those determining research priorities must urgently act to encourage research on the impact of desexed language and ways in which communication regarding sex and gender identity can be optimized. In addition, any ideological capture of funding organizations needs to be undone. The chilling effect of real or perceived ideological capture of universities and funders on the willingness of researchers to undertake controversial research or state unpopular opinions must be recognized, and a culture encouraging dissent cultivated (Nemeth 2018).

We must act with integrity to advocate for a reinstatement of evidence over ideology in science and health care.

SECTION 6
What Can Be Done?

The concluding section of this volume focuses on the more upbeat, and yet challenging, question of what can be done to restore scientific integrity, open inquiry, rational debate, and free speech in academia and in society more generally. Needless to say, there is no silver bullet, and different authors reflect different perspectives on both the challenges involved and the opportunities that may arise. As several of the authors of this section emphasize, the challenge of how to return to the previous norms of academic discourse and scholarly inquiry will involve undoing past damage and removing the institutional infrastructures (and often the individuals at their helm) that now provide impediments to progress.

Dorian Abbot begins this section with an essay describing a set of principles he views as essential to ensure the protection of science from politics. He views self-censorship by many academics as a key problem and describes his own voyage to become an outspoken advocate for action, urging more faculty to follow his lead. He describes three different parables reflecting different aspects of the current problems in the academy and what actions might be taken to avoid each one.

The need for faculty to take the lead in reforming higher education is further explored by Richard Redding, who argues that the only way to depoliticize science is to ensure a more intellectually diverse professoriate.

Note that the diversity he describes has nothing to do with race or gender. Rather, he argues that an intellectually homogenous faculty, currently overwhelmingly liberal, does not allow for the scholarly dialectic that is an essential feature of true academic freedom and progress. He suggests, as a result, that current DEI efforts should move from their misplaced fixation on identity and victimization toward working to encourage true viewpoint diversity.

Steven Pinker reflects on the damage that has already occurred at major research universities, including his own institution, Harvard, and argues that a long-term concerted plan to systematically undo the self-inflicted damage is necessary by adhering to five commitments: Free Speech, Institutional Neutrality, Nonviolence, Viewpoint Diversity, and Disempowering DEI.

Nicholas Christakis, in a short, second contribution to this volume, argues that one must move beyond faculty to teach students to embrace true inclusion, namely inclusion of viewpoint diversity. If faculty do not demonstrate this in their teaching, then arguing "do as I say, but not as I do" is bound to be ineffective and will do little to change the real future of higher education.

Finally, Jordan Peterson does a deep dive into what he thinks are the underlying long-term psychological and historical allegorical cultural factors that have resulted in the current ideological capture of much of Western academia. He sees deep-seated fundamental patterns of archetypal narratives as illustrative of the problems that must be addressed. In the end, he sees it as incumbent on every one of us to risk being a redeemer. As he puts it in his concluding remarks: "The true spirit of the academy is the prophetic voice that redeems. We abandon that at our great peril and to the detriment of all that depends on education itself."

Science and Politics: Three Principles, Three Fables

DORIAN ABBOT

Note: This essay appeared in Liberties Journal *in April of 2022.*

1. Introduction

Science is a creative endeavor that requires the free and open exchange of ideas to thrive. Society has benefited immensely from scientific progress, and in order for science to continue to pay dividends, scientific work must be judged on the basis of scientific merit alone. Over the past decade, scientific departments and organizations have become increasingly politicized, to the point that it is now significantly impeding scientific development. This time, the assault originated from the radical left, but conservatives have done their share of meddling in science in the past and are likely to again in the future. Keeping politics out of science is something that all people of goodwill, both Democrats and Republicans, should be able to agree on.

How can we ensure political neutrality in science? I want to propose three critical principles for the protection of science from politics and to illustrate them with three playful, slightly naughty fables about what has been happening when they are violated. The three principles are: (1) all scientists need to be able to say and argue whatever they want, even if it offends someone else; (2) universities and academic societies need to maintain strict neutrality on all social and political issues; and (3) hiring needs to be done on the basis of scientific merit alone. These principles have been lucidly outlined in three important documents at the University of Chicago, where

I teach geophysical science: the Chicago Principles, which were issued by the university in 2014; the Kalven Report "on the university's role in social and political action" from 1967; and the Shils Report on the "criteria for academic appointments" from 1970. All these reports assume, as the Kalven Report puts it, that "the mission of the university is the discovery, improvement, and dissemination of knowledge." This sounds prosaic, but the definition is important to emphasize because some people are now challenging it. They argue that faculty members should be activists who promote certain political positions and agendas rather than pursue truth wherever it may lead. I should add that we are not perfect at the University of Chicago, and I sometimes fear that we honor these principles more in the breach than in the observance. And yet, these principles are important goals for every scientific institution to at least aim for.

I started my journey in this area simply by self-censoring—for no less than five years. I stayed away from campus whenever possible and avoided departmental gatherings. At first, I thought that the problem was a few bad apples in my department yelling at everyone who disagreed with them and accusing people of being various types of witches. I only slowly learned that I was observing just a small part of a national movement in favor of censorship and the suppression of alternative viewpoints. It is absolutely essential that we resist this movement and encourage students and faculty to speak freely about whatever they want on campus: we all lose when people self-censor.

Unfortunately, students and faculty are now self-censoring at alarming rates, in part as a result of high-profile cancellations of academics guilty of wrongthink. For example, FIRE has documented 471 attempts to get professors fired or punished for their speech over the past six years, the vast majority of which resulted in an official sanction. In a recent report for CSPI, Eric Kauffman estimates that three in ten thousand faculty experience such an attack each year, which corresponds to about one every three years at a large university with one thousand faculty. Because these cancellations are so public and potentially harmful to the victim's career, a small number can have an outsized impact on free expression. According to the same report, 70 percent of US centrist and conservative faculty report a climate hostile to their beliefs, and 91 percent of Trump-voting faculty say that a Trump voter would not express his/her views or are unsure. Similarly, after a major academic freedom incident in the fall of 2021, MIT polled faculty at two faculty forums and found that approximately 80 percent are "worried given the current atmosphere in society that your voice or your colleagues' voices are increasingly in jeopardy" and more than 50 percent "feel on an everyday basis that your voice, or the voices of your colleagues are constrained at MIT." The problem extends to students, more than 80 percent of whom self-censor on campus, according to a 2021 FIRE survey. To get a sense of the magnitude of the self-censorship problem at universities, contrast these numbers with the fact that, according to a recent paper in SSRN by James Gibson and Joseph Sutherland, only 13 percent of American respondents did not feel free to speak their mind in 1954, at the height of McCarthyism. The discovery and transmission of knowledge are severely hindered under these conditions.

It was my wife who inspired me to start making my views known. Her reason was that she was born in Ukraine at the tail end of the Soviet Union. When I told her what was happening on campus, she had a question: "If you speak out, will you have trouble?" I said, "Why?" And she replied: "It sounds like what my mother told me about Soviet times, and people who spoke out had serious trouble then." That was enough to convince me. Not in this country, not on my watch. My wife's mother is a teacher. In the aftermath of communism, she brought home the old propaganda books from school about Lenin and his pals. For fond memories? To teach her children about the old ways? Not at all. She brought them home to burn them to stay warm in the winter. Lenin's system failed so utterly that people had to burn the old propaganda just to stay warm. And no small part of this failure was due to the fact that they had allowed science to become politicized.

2. A Speech Crisis at AIT

Now for the first story. Consider a situation at the Awesome Institute of Technology, or AIT. It involves two main characters. The first character is Dr. Centrist, a professor at AIT. Dr. Centrist has devoted the past thirty years of his life to developing his biomedical skills and is now at the top of his field. He is working in cancer research and is close to finding a cure. No one questions his ability in the lab, or his scientific honesty, or his devotion to science and his students. Moreover, Dr. Centrist has advised people of both sexes, all races, all sexual orientations, all religions, and all nationalities, and treated them all with equal respect. Dr. Centrist is a political centrist, smack in the middle of mainstream American viewpoints. Some of his views are center-left—for example, he supports broad access to education and health care as well as protecting the environment for everyone. But some are center-right, and these will end up being considered "provocative" and "controversial" at AIT and cause him quite a bit of trouble.

The second character in my story is Mr. Woke (he/they), an undergraduate. Mr. Woke entered AIT as a physics major, but physics just "wasn't a good fit," so he switched to anthropology, which leaves him/them much more time for other interests such as Twitter.

At some point, it slips out that in his personal life, Dr. Centrist "clings to his guns and Bible." Upon hearing of this, Mr. Woke looks up in shock from his/their Macbook Pro with all the appropriate political decal stickers, spits out a mouthful of his/their venti soy chai latte, and declares that this is highly "problematic." According to Mr. Woke, supporting gun rights is a "dog whistle" or "coded language" for white supremacist vigilantism, and therefore, minoritized people at AIT cannot feel safe and will be irreparably harmed if Dr. Centrist is allowed to continue his scientific research there. Moreover, Christianity is an exertion of power used by the cis-hetero-patriarchy to oppress gender and sexual minorities. The LGBTQIA2S+ community at AIT will not tolerate this type of bigotry on campus, according to Mr. Woke, who has apparently appointed himself/themselves spokesfolx for the entire community.

But it gets worse. In a conversation at lunch about a recent Supreme Court case, Dr. Centrist lets it slip that he is pro-life. Mr. Woke declares that this is a blatant "war-on-women" position that cannot be tolerated. AIT needs to be a safe space where no gender minority ever has to hear, and thereby be harmed by, a viewpoint with which she disagrees. Mr. Woke says inclusivity dictates that this sort of violent hate speech must be restricted, and he/they takes to Twitter to demand a "Speech Code" and a "Code of Conduct" to ensure that the "climate" at AIT is made safe and inclusive for everyone—by inhibiting and silencing anyone who disagrees with him/them, of course.

I have saved the worst for last. Eventually it comes out that, horror of horrors, Dr. Centrist is a "deplorable" who actually voted for Donald Trump. Mr. Woke scrambles into action. He/they organizes a letter of denunciation of Dr. Centrist, demanding that he be fired in order to protect minoritized people on campus who have been threatened by Dr. Centrist's violent and aggressive racist hate vote. Mr. Woke has plenty of "allies" who sign, and then he/they threatens everyone else with a similar denunciation if they refuse to sign, declaring: "Silence is violence, and if you don't sign now you will be tarnished as a racist for the rest of your career. I will make sure of it!" Eventually most of the students at AIT sign.

AIT President Craven is interrupted from a busy schedule of meetings on inclusive pronoun usage, equitable landscaping, and bathroom diversity to deal with the latest campus controversy. President Craven is presented with a real conundrum: should he defend the fundamental purpose of AIT, which is the unfettered pursuit of truth, and risk being called a scary name by Mr. Woke, or should he panic and do whatever it takes to make his anxiety go away quickly so that he can return to attending his important meetings and enjoying his outrageous salary in peace. For President Craven, the choice is easy. He fires Dr. Centrist and returns to his pronoun and landscaping meetings.

An unfortunate result of his decision is that we never get the cure for cancer that Dr. Centrist was close to discovering—will another university or another laboratory hire such a disgraced individual?—but this was not high on President Craven's list of priorities. Afterwards Mr. Woke insists that Dr. Centrist was not "canceled." He was, rather, "held accountable" for his hateful, bigoted, and generally "problematic" views. Mr. Woke is fine with this result, because he/they believes that cancer is a social construct caused by systemic racism, and Dr. Centrist's racist methods of science are useless compared to the medicinal benefits of other ways of knowing. But what does the public, who funded Dr. Centrist's research and pays for most of the tuition at AIT through federal grants, think about losing out on progress toward a cure for cancer because of someone's disapproval of Dr. Centrist's political views, which many people also hold?

What would it take to avoid this disaster at AIT? As annoying as Mr. Woke is, I think the real villain is President Craven. In order to prevent the terrible outcome that we have just described, President Craven doesn't exactly have to turn into Churchill, he just needs to turn into President Not-A-Complete-Dingleberry. He just needs a tiny bit of spine. All he has to do is say, "Sorry, Mr. Woke. That's not how we do things around here. You are free to express your opinions, and Dr. Centrist is free to express his opinions. You don't get to silence people you disagree with at AIT." This idea is described in the Chicago Principles as follows:

It is not the proper role of the University to attempt to shield individuals from ideas and opinions they find unwelcome, disagreeable, or even deeply offensive...concerns about civility and mutual respect can never be used as a justification for closing off discussion of ideas, however offensive or disagreeable those ideas may be to some members of our community.

It is important to get the Chicago Principles adopted on your own campus before a crisis occurs. Even President Craven might have been brave enough to stand up to Mr. Woke if he had an official policy to point to as an excuse. A standard part of the orientation for new students and faculty should be to explain the moral and intellectual foundations of these principles and, more generally, of academic freedom: both why they are important and how they will be enforced. (There also must be real penalties for violations.) This would help Mr. Woke understand that illiberal tactics will not work.

I should emphasize that the right of *everyone* to speak on campus needs to be defended. Let me give you an example. I have a colleague who has tiled images of Karl Marx on his website and a Soviet flag in his office. He is actively introducing his Marxist-Communist views into campus settings in a way that Dr. Centrist did not introduce his own. This is deeply offensive, of course, to anyone who has taken ten minutes to study the history of the twentieth century, let alone actually suffered under communism. But the fact that my colleague openly advocates for what I consider to be an indefensible political position has absolutely no bearing on his scientific and mathematical ability. No matter how extreme, immoral, and offensive his or anyone else's political views may be, we need to defend his right to express them freely without letting them hinder his scientific career. Throughout history, many famous scientists have been highly eccentric and held weird and even repulsive social and political views. So what? Should we, therefore, renounce the fundamental and critical science that they produced?

3. The Global Social Justice Forum

Now a story in the Swiss Alps. Professor Right and Professor Left are attending the Global Economic Forum. They have been appearing at the forum for decades, and they agree on almost nothing. Whenever an economic issue arises, Professor Right argues for less government, and Professor Left argues for more government. But they listen to each other's lectures seriously and they respond to each other's arguments. Sometimes, Professor Left gets excited and makes a hasty comment on Professor Right's lecture implying that she is stupid, but he never calls her evil. Even though Professor Right disagrees with Professor Left, she modifies her perspective when Professor Left shows data that contradicts a claim that she is making. In the end, after some back-and-forth, they tend to reach some sort of conclusion about the matter in question that both can agree is empirically justified. They do not like each other, but each understands that the other is necessary for the critical examination of his or her own views and can lead him or her to better economic research.

Enter a graduate student. Let us call her Ms. Oppressed. Ms. Oppressed doesn't think that Professor Right is merely wrong; she believes that she is morally corrupt. How else could someone argue for small government, when big government is clearly what is needed to fix the obvious systemic problems in our society that are oppressing women and marginalized people? Ms. Oppressed starts a Twitter campaign to force the Global Economic Forum to issue an official statement acknowledging that increasing the size of government is the only solution to all social and economic problems, as well as add a condition that in order to present a paper at the forum, every participant must sign a pledge of agreement with this statement.

Professor Left finds himself in a bit of a bind. On the one hand, Ms. Oppressed seems to agree with him on most policy issues, and he has never been terribly fond of Professor Right. On the other hand, banning speakers rather than contending with their arguments seems to go against the liberal tradition in which Professor Left usually locates himself. While he is considering this, Ms. Oppressed tells Professor Left that "silence is violence" and that he had better get on board with the program or she will turn to Twitter. Professor Left decides that the best course of action is to declare "no enemies to the Left" and go along with Ms. Oppressed. The statement and the pledge are instituted, Professor Right refuses to sign, and she is henceforth banned from the Global Economic Forum.

Professor Left finds the next forum meeting quite exhilarating. He can expound all his wildest ideas without the annoying Professor Right demanding evidence or logic. Yet he starts to get an uneasy feeling when he attends Ms. Oppressed's lecture, which is titled, "Data: Research or Oppression?" In it, Ms. Oppressed argues that data and the ideal of disinterested methodological rigor should no longer be used in economic research because "this idea of intellectual debate and rigor as the pinnacle of intellectualism comes from a world in which white men dominated." (These deathless words were uttered in October 2021 by Phoebe Cohen, chair of geosciences at Williams College.) In fact, at the end of her lecture, Ms. Oppressed actually attacks Professor Left because he refused to ban Professor Right for so many years. Silence is violence, after all, and by allowing that sort of hate think at the Global Economic Forum, Professor Left has actively participated in a horrible system of oppression.

After the forum, Ms. Oppressed organizes a letter to have Professor Left and anyone else over the age of thirty-five banned from all future forums for past collaboration with evil right-wingers, which, in addition to being essential for social justice, will have the added benefit of opening up lots of career opportunities for Ms. Oppressed and her allies. It works, naturally, and Professor Left soon finds himself banned from the most important meeting in the field, suffering the same fate that was visited upon Professor

Right only a year before. Meanwhile, at the next meeting of the Global Economic Forum, all the presentations have titles such as "Indigenous Ways of Managing Global Economies," "Feminist Perspectives on Inflation," and "Intersectional Debt Management." No one dares to present data or make a rational argument, for fear of being labeled a white supremacist. Needless to say, the discussions of economics at the event quickly lose their previous influence upon business leaders and policymakers, whose job it is to make actual decisions in the actual world, but it becomes very popular with journalists at the prestigious journal of record, the *New York Spaces*, who write favorable pieces about the exciting new developments in a field that they used to treat with a mixture of confusion and disgust. It is not long before Ms. Oppressed is rewarded for speaking truth to power with a FitzArthur "Genius" Award.

The key error here was that Professor Left compromised on the principle that universities and societies should never take positions on social and political issues. He did this because he tended to agree with the political positions that were proposed. Doing so makes universities and societies political entities rather than scientific ones and has the effect of restricting free expression by members of the university community who disagree with the official position. It is particularly important right now that professors on the left do not fall for this trap. Aside from the principled reason for this, there is a practical reason: they will never be revolutionary enough, and the revolution is sure to eat them next if they fail to stop it now. In the words of the Kalven Report,

> The instrument of dissent and criticism is the individual faculty member or the individual student. The university is the home and sponsor of critics; it is not itself the critic. To perform its mission in the society, a university must maintain an extraordinary environment of freedom of inquiry and maintain an independence from political fashions, passions, and pressures. A university, if it is to be true to its faith in intellectual inquiry, must embrace, be hospitable to, and encourage the widest diversity of views within its own community. It is a community, but only for the

limited, albeit great, purposes of teaching and research. It is not a club, it is not a trade association, it is not a lobby.

Since the university is a community only for these limited and distinctive purposes, it is a community which cannot take collective action on the issues of the day without endangering the conditions for its existence and effectiveness. There is no mechanism by which it can reach a collective position without inhibiting the full freedom of dissent on which it thrives. It cannot insist that all members favor a given view on social policy; if it takes collective action, therefore, it does so at the price of censuring any minority who do not agree with the view adopted. In brief, it is a community which cannot resort to majority vote to reach positions on public issues.

The principle of political neutrality is extremely important for a university, though it is often neglected relative to the principle of free expression. But you cannot have the latter without the former. Free expression is not possible in practice at universities that release statements on social and political issues. Consider 2020 as an example of how this is *not* supposed to work: universities and societies across the country issued statements on social and political issues, and faculty members who disagreed with them publicly were attacked, silenced, and sometimes even fired. The attackers felt justified by the official statements.

4. Dean Shifty Pulls a Fast One

Finally, there is the situation developing in the job search at the physics department at Winthrop University. Winthrop has had an aggressive DEI (diversity, equity, and inclusion) program that has been in place for more than a decade, and it has already hired dozens of DEI deans and deanlets to implement and promote it. Yet Winthrop Physical Sciences Dean Shifty has recently received word from the president of the Henry Foundation that the foundation is not happy with the numbers that Winthrop's DEI program has produced. In particular, the foundation is expressing concern with the slow progress in appointing an appropriate number of underrepresented faculty in the physical sciences.

Given the Henry Foundation's deep pockets and cultural influence, Dean Shifty can see her dreams of a nice presidency at a liberal arts college with a fat paycheck slipping away. And so, she takes immediate action. Although the advertisement for the physics faculty search explicitly says that there will be no discrimination on the basis of race or sex, Dean Shifty slyly sends the message to the chair of the physics department and the members of the search committee that she will not consider a nomination for the faculty position if it is an Asian or white man. She does this orally and through an intermediary because she knows that it is a violation of Titles VI and IX of the Civil Rights Act. The members of the faculty search committee are uncomfortable, but they feel that they have no choice but to comply. They do not actually know what is in the Civil Rights Act—they are physicists, not lawyers; but they assume that Dean Shifty would not do anything illegal.

A fierce debate soon emerges on the hiring committee. It turns out that half of the committee thinks the department needs a woman, and half of the committee thinks that the department needs an underrepresented minority. Instead of debating the scientific merit of the candidates, the committee spends its time debating which type of underrepresented person should be recruited. In the end, they settle on hiring a woman, because there are more women than underrepresented minorities among the graduate students, and the students need more faculty members who "look like them." They hire a woman, and both Dean Shifty and the president of the Henry Foundation are thrilled. Of course, the entirety of the faculty is vaguely aware of what happened, which leads to a strange and uncomfortable situation for the new member of the department. Meanwhile, similar hiring shenanigans have been implemented at universities across the country, so the male Asian and white candidates find it extremely difficult to get faculty jobs, and many end up leaving the field.

Notice what happened when hiring criteria other than scientific merit were introduced: it immediately made the process political. Whether to hire a woman or an underrepresented minority is a political question, not a scientific question. In order to avoid the politicization of science, therefore, it is absolutely essential that all admission, hiring, promotion, and honors be awarded on the basis of scientific merit alone. Politics is automatically introduced when purely merit-based decisions are abandoned. Moreover, ideological purity tests called DEI statements are now often used as a gate-keeping mechanism to ensure political uniformity in faculty hiring, and this quite obviously violates the principle of political neutrality. Also, note that the purpose of the university and of science is being violated if criteria other than merit are used for hiring: in such cases, we are no longer pursuing truth to the best of our ability. We have instead substituted some other goal.

These matters were presciently discussed in the Shils Report:

> The conception of the proper tasks of the University determines the criteria which should govern the appointment, retention, and promotion of members of the academic staff. The criteria which are to be applied in the case of appointments to the University of Chicago should, therefore, be criteria which give preference above all to actual and prospective scholarly and scientific accomplishment of the highest order, actual and prospective teaching accomplishment of the highest order, and actual and prospective contribution to the intellectual quality of the University through critical stimulation of others within the University to produce work of the highest quality.

Note that the last clause refers strictly to stimulating others "to produce work of the highest quality" and should not be interpreted as a way to sneak other criteria into consideration. And later:

> There must be no consideration of sex, ethnic or national characteristics, or political or religious beliefs or affiliations, in any decision regarding appointment, promotion, or reappointment at any level of the academic staff.

The objective of this rule is simple: *fairness*.

5. Conclusion

The principles of academic freedom confer not only a right but also a duty. Some people think that the duty of academic freedom is to restrict your speech in certain cases, but this is incorrect. The duty of academic freedom is to use it. My obligation as a professor and a scientist is to say what I really think in public while, of course, focusing my teaching on the subject I was hired to teach, not least because so many people in society cannot: that is the whole point of the professional protection known as tenure. Too often, tenure is wasted on the timid. Anyway, they can't cancel all of us.

A Five-Point Plan to Save Universities from Themselves

STEVEN PINKER

For universities to have a leg to stand on when they try to stand on principle, they must embark on a long-term plan to undo the damage they have inflicted on themselves. This includes Harvard.

For almost four centuries, Harvard University, my employer, has amassed a reputation as one of the country's most eminent universities. But it has spent the past year divesting itself of tranches of this endowment. Notorious incidents of cancellation and censorship have contributed to a plunge in confidence in institutions of higher education,[1] prompting me and more than one hundred colleagues to found a new Council on Academic Freedom at Harvard.[2] That was before Harvard came in at last place in the Foundation for Individual Rights and Expression's Free Speech ranking of 248 colleges, with a score of zero out of one hundred[3]—originally less than zero, but Harvard benefited from a bit of grade inflation. (I'm a FIRE adviser but had no role in the rankings.)

Then in June, the Supreme Court ruled against Harvard in a suit claiming it had discriminated against Asian American applicants. And in October, after the massacre of 1,200 Israelis by Hamas, thirty-four student organizations calling themselves the Harvard College Palestine Solidarity Committee blamed the pogrom "entirely" on the victims'[4] own government. Harvard's newly installed president, Claudine Gay, issued a muted, both-sidesy statement.[5] Following an outcry, with headlines like "Harvard's

Horror"[6] and "Harvard Is a National Disgrace,"[7] she followed up with a second statement and then a third, pleasing no one.

Capping off the annus horribilis, last week, Gay was grilled on antisemitism in the most-watched hearing in the history of the US Congress.[8] In response to the question of whether a call by students for genocide of Jews violated university policies, she gave the inadvertently Bartlett's-worthy answer, "It depends on the context." Her other responses struck viewers as evasive, formulaic, and lawyer-coached.

The fury was white-hot. Harvard is now the place where using the wrong pronoun is a hanging offense, but calling for another Holocaust depends on context. Gay was excoriated not only by conservative politicians but by liberal alumni, donors, and faculty, by pundits across the spectrum, even by a White House spokesperson, and by the second gentleman of the United States. Petitions demanding her resignation have circulated in Congress, X, and factions of the Harvard community, and a prediction market[9] is posted 1.2:1 odds that she would be ousted in 2023 (she resigned in January 2024).

I don't believe that firing Gay is the appropriate response to the fiasco. It wasn't just Gay who fumbled the genocide question but two other elite university presidents—Sally Kornbluth of MIT (my former employer) and Elizabeth Magill of the University of Pennsylvania, who resigned following her testimony[10]—which suggests that the problem with Gay's performance[11] betrays a deeper problem in American universities.

Congressional inquiries are often televised ambushes, and as Gay walked into the line of fire, she had been rendered defenseless by decades of rot in campus policies. In the exchange that went viral, Republican Representative Elise Stefanik of New York asked Gay whether "calling for the genocide of Jews violate Harvard's rules on bullying and harassment."[12]

Gay interpreted the question not at face value but as pertaining to whether Harvard students who had brandished slogans like "Globalize the intifada" and "From the river to the sea," which many people interpret as tantamount to a call for genocide, could be prosecuted under Harvard's policies. Though the slogans are simplistic and reprehensible, they are not calls for genocide in so many words. So even if a university could punish direct calls for genocide as some form of harassment, it might justifiably choose not to prosecute students for an interpretation of their words they did not intend.

Nor can a university with a commitment to academic freedom prohibit all calls for political violence. That would require it to punish, say, students who express support for the invasion of Gaza knowing that it must result in the deaths of thousands of civilians. Thus, Gay was correct in saying that students' political slogans are not punishable by Harvard's rules on harassment and bullying unless they cross over into intimidation, personal threats, or direct incitement of violence.

Gay was correct yet again in replying to Stefanik's insistent demand, "What action has been taken against students who are harassing Jews on campus?" by noting that no action can be taken until an investigation has been completed. Harvard should not mete out summary justice like the Queen of Hearts in *Alice in Wonderland*: Sentence first, verdict afterward.

The real problem with Gay's testimony was that she could not clearly and credibly invoke those principles because they either have never been explicitly adopted by Harvard or they have been flagrantly flouted in the past (as Stefanik was quick to point out).[13] Harvard has persecuted scholars who said there are two sexes[14] or who signed an amicus brief taking the conservative side[15] in a Supreme Court deliberation. It has retracted acceptances from students who were outed by jealous peers for having used racist trash talk on social media when they were teens. Harvard's subzero FIRE rating reveals many other punishments of politically incorrect peccadillos.

So, for the president of Harvard to suddenly come out as a born-again free-speech absolutist, disapproving of what genocidaires say but defending to the death their right to say it, struck onlookers as disingenuous or worse.

In the wake of this debacle, the natural defense mechanism of a modern university is to expand the category of forbidden speech to include antisemitism (and as night follows day, Islamophobia). Bad idea. A history of punishing speech is what sapped the presidents' credibility in the first place, and a promise to double down on it did not save Magill. Deplorable speech should be refuted, not criminalized. Outlawing hate speech would only result in students calling anything they didn't want to hear "hate speech."

Even the apparent no-brainer of prohibiting calls for genocide would backfire. Trans activists would say that opponents of transgender women in women's sports were advocating genocide,[16] and Palestinian activists would use the ban to keep Israeli officials from speaking on campus.

For universities to have a leg to stand on when they try to stand on principle, they must embark on a long-term plan to undo the damage they have inflicted on themselves. This requires five commitments.

Free speech. Universities should adopt a clear and conspicuous policy on academic freedom. It might start with the First Amendment, which binds public universities, and which has been refined over the decades with carefully justified exceptions. These include crimes that by their very nature are committed with speech, like extortion, bribery, libel, and threats, together with incitement of imminent lawless action. It also permits restraints on the time, place, and manner of expression. The First Amendment does not entitle someone to blare propaganda from a sound truck in a residential neighborhood at 3 a.m. or to set up a soapbox in the middle of a busy freeway.

Since universities are institutions with a mission of research and education, they are also entitled to controls on speech that are necessary to fulfill that mission. These include standards of quality and relevance: you can't teach anything you want at Harvard, just like you can't publish anything you want in the *Boston Globe*. And it includes an environment conducive to learning. Though a university should not punish a student for holding up a placard, it has a legitimate interest in preventing a group from permanently repurposing its walls as political billboards or from forcing students to walk through a gauntlet of intimidating slogan-chanters on their way to class every day.

Institutional neutrality. A university does not need a foreign policy, and it does not need to issue pronouncements on the controversies and events of the day. It is a forum for debate, not a protagonist in debates. When a university takes a public stand, it either puts words in the mouths of faculty and students who can speak for themselves or unfairly pits them against their own employer. It's even worse when individual departments take positions, because it sets up a conflict of interest with any dissenting students and faculty whose fates they control.

The events of this autumn also show that university pronouncements are an invitation to rancor and distraction. Inevitably there will be constituencies who feel a statement is too strong, too weak, too late, or wrongheaded. The resulting apologies and backtracking compromise the reputation of the university and interfere with the task of administering it. For this reason, a stated policy of institutional neutrality would be a godsend to university administrators. Such a policy would still allow them to comment on issues that directly affect university business, just like any institution.

Nonviolence. Some students think it is a legitimate form of political expression to drown out a speaker, block the audience's view with a screen, obstruct public passageways, invade a lecture hall chanting slogans over bullhorns, force administrators out of their offices and occupy the building, or get in the faces of other students.

Universities should not indulge acts of vandalism, trespassing, and extortion. Free speech does not include a heckler's veto, which blocks the speech of others. These goon tactics also violate the deepest value of a university, which is that opinions are advanced by reason and persuasion, not by force. And they bring further discredit to the institution: Parents and taxpayers wonder why they should support, at fantastic expense, students being forced to listen to political propaganda from other students when they should be learning math and history from their professors.

Viewpoint diversity. Universities have become intellectual and political monocultures. Seventy-seven percent of the professors in Harvard's Faculty of Arts and Sciences describe themselves[17] as liberal, and fewer than 3 percent as conservative. Many university programs have been monopolized by extreme ideologies, such as the conspiracy theory that the world's problems are the deliberate designs of a white heterosexual male colonialist oppressor class. (The appalling antisemitism infesting college campuses grew out of the corollary that Israelis, and by extension Jews who support them, are a party to this conspiracy.) Vast regions in the landscape of ideas are no-go zones, and dissenting ideas are greeted with incomprehension, outrage, and censorship.

The entrenchment of dogma is a hazard of policies that hire and promote on the say-so of faculty backed by peer evaluations. Though intended to protect departments from outside interference, the policies can devolve into a network of like-minded cronies conferring prestige on each other. Universities should incentivize departments to diversify their ideologies, and they should find ways of opening up their programs to sanity checks from the world outside.

Disempowering DEI. Many of the assaults on academic freedom (not to mention common sense) come from a burgeoning bureaucracy that calls itself diversity, equity, and inclusion while enforcing a uniformity of opinion, a hierarchy of victim groups, and the exclusion of freethinkers. Often hastily appointed by deans as expiation for some gaffe or outrage, these officers stealthily implement policies that were never approved in faculty deliberations or by university leaders willing to take responsibility for them.

An infamous example is the freshman training sessions that terrify students with warnings of all the ways they can be racist (such as asking, "Where are you from?"). Another is the mandatory diversity statements for job applicants, which purge the next generation of scholars of anyone who isn't a woke ideologue or a skilled liar. And since overt bigotry is, in fact, rare in elite universities, bureaucrats whose job depends on rooting out instances of it are incentivized to hone their Rorschach skills to discern ever-more-subtle forms of "systemic" or "implicit" bias.

Universities should stanch the flood of DEI officials, expose their policies to the light of day, and repeal the ones that cannot be publicly justified.

A fivefold way of free speech, institutional neutrality, nonviolence, viewpoint diversity, and DEI disempowerment will not be a quick fix for universities. But it's necessary to reverse their tanking credibility and better than the alternatives of firing the coach or deepening the hole they have dug for themselves.

Teaching Inclusion in a Divided World

NICHOLAS A. CHRISTAKIS

One of the most difficult intellectual and emotional challenges I faced earlier this year at Yale was finding an answer to a Native American student's poignant question: Why should she put any faith in institutions in our society—including our judicial system and universities—when those institutions had clearly betrayed her people in generations past?

"The same Constitution with its protection of the rights to free expression and assembly that you revere," she said, "was previously of no use to people like me."

She was right, of course. So why should she and other young people place trust in systems that can perennially fail us?

I wish I had told her that the way out of this conundrum is to make these institutions her own. I wish I had told her that these institutions are worth respecting and preserving for their (albeit imperfect) embodiment of Enlightenment values, that she surely should want to embrace those values, and that her generation could make those values more true, not less. These institutions could be hers, and I believe she should *want* them to be hers.

Students are demanding greater inclusion, and they are absolutely right. But inclusion in what? At our universities, students of all kinds are joining traditions that revere free expression, wide engagement, open assembly, rational debate, and civil discourse. These things are worth defending. In fact, they are the predicates for the very demands the students have been making across the United States.

Conversely, it is entirely illiberal (even if permissible) to use these traditions to demand the censorship of others, to besmirch fellow students rather than refute the ideas that they express, and to treat ideological claims as if they were perforce facts. When students (and faculty) do this, they are burning the furniture to heat the house.

Open, extended conversations among students themselves are essential not only to the pursuit of truth but also to deep moral learning and to righteous social progress. The faculty must step up and show students a way forward: to learn to be harder on the problems we face in our society but easier on each other. We must demonstrate that we cannot be a community of searchers and learners if we do not share the same principles at the core of our universities.

And so the faculty must cut at the root of a set of ideas that are wholly illiberal. Disagreement is not oppression. Argument is not assault. Words—even provocative or repugnant ones—are not violence. The answer to speech we do not like is more speech.

If we fail to see this, we risk confirming for our students the old joke that we wouldn't want to join a club that would have us.

Depoliticizing Science Requires Revolutionary Reform of the Professoriate

RICHARD E. REDDING

Scientists now confront an ideologically censorious culture, often coming from academia (al-Gharbri and Clark 2024, Clark et al. 2024, Honeycutt et al. 2023, Jussim et al. 2024), which threatens open inquiry in the sciences and social sciences (which I refer to collectively as "science"). In the past, it was mostly the "softer" social sciences that were subject to ideological influences and the censorious practices that often accompany them (e.g., Redding 2013), but nowadays, even the natural sciences are impacted. Consider, for example, the vitriolic controversies surrounding the definitions and relevance of gender, sexuality, and race in biology, psychology, and medicine and the censorship of ideologically unpopular research in these areas and the denouncements and cancellations of researchers (e.g., Coyne and Maroja 2023, Satel 2021).

The fundamental problem is that science has become highly politicized. To be sure, such politicization comes from the Right as well as the Left. But today, it is primarily the radical left that is able to censor the science they find to be ideologically unappealing while at the same time privileging and promoting those ideas and scientists they prefer (see Clark et al. 2024, Frisby et al. 2023). Not only does this chill the open inquiry necessary for scientific progress, but it has diminished public confidence in science, particularly among conservatives (NORC 2023). When people perceive that a scientific topic (e.g., climate change, COVID-19) has been politicized in ways inconsistent with their own ideologies, they are more likely to discount it (Van Boven et al. 2018).

Unlike the radical right, the radical left has been able to censor science because progressives have captured the academy (Redding 2024), which is where the future scientists are trained and where most of the scientific research in the United States is conducted. The academy has always tilted decidedly to the left, but starting about twenty-five years ago, radically progressive, Marxist, and critical ideologies (which tend to be both inherently censorious and philosophically anti-scientific; see Mills, Maranto, and Redding 2024) began dominating the academic landscape. Numerous surveys (see Redding 2023b) show that the cohort of young professors and administrators (most of whom are tenured professors) include virtually no conservative, libertarian, or even centrist or apolitical professors. About 80–90 percent of professors in the physical and biological sciences are leftist, and in the social sciences, that number approaches 100 percent, with many departments and entire colleges not having a single non-leftist on the faculty (Redding 2023). Ten years ago, for which the most recent such data is available, 25 percent of professors considered themselves to be "radical" or Marxist, but that percentage is undoubtedly higher today.

After the 2020 murder of George Floyd, this ideologically captured professoriate instituted diversity, equity, and inclusion (DEI) policies and practices throughout university life. The new DEI regime instantiates the academy's longstanding diversity policies and practices on steroids, solidifying the ideological hegemony in teaching and research (Redding 2024). It often comes with a left-wing authoritarianism (Jussim et al. 2023) that restricts free speech and academic freedom, denounces and cancels speakers and professors, and discriminates against non-progressive faculty, prospective faculty, and students. Recent surveys of the professoriate have found that many self-censor on controversial topics, while others (especially liberals) censor, sanction, or discriminate against colleagues who conduct ideologically unpopular research (e.g., Clark et al. 2024, Honeycutt et al. 2023, Kauffman 2021). The professoriate also sought to ideologically indoctrinate students, who would be the future scientists, policymakers, and professors.

The professoriate will continue the groupthink that gives it license to enforce its ideological hegemony so long as its ranks are ideologically monolithic. Accompanying such groupthink is a culture wherein dissidents feel that they must self-censor while those in the majority compete to out-virtue signal their colleagues by censoring and marginalizing alternative viewpoints (Redding 2024). By contrast, when there is a critical mass of professors having minority ideological perspectives, they are no longer a marginalized outgroup, and it becomes acceptable to question the prevailing paradigms and explore alternatives in research and teaching (Redding 2023a).

Thus, to depoliticize science and cancel the censorship culture, we must reform university faculties to make them more ideologically diverse (Redding 2023a), and not only science faculties but those in other disciplines as well since they also exert censorious influences on science. The importance of reforming the professoriate lies in the fact that it produces most of the scientific research in the United States, is often relied upon by policymakers for its expertise, has a major influence on major scientific organizations such as the American Medical Association and the American Psychological Association (which have a significant influence on public policy and promulgate practice guidelines and educational accreditation standards), produces the next generation of K–12 science teachers, and trains our next generation of scientists.

We must attract more ideologically diverse students to enter the professoriate pipeline, and universities must hire them to join their faculties. There are significant barriers to achieving this, however. As the academy became increasingly progressive over the last twenty-five years, this created a snowball effect whereby leftist faculties replicated themselves more and more while non-leftists increasingly shied away from academic careers (Redding 2023b). Many prospective doctoral students who are conservative, libertarian, centrist, and even apolitical end up self-selecting out (during college or early in their graduate school career) of doctoral programs and academic careers because they discover the academic culture to be hostile to their worldviews—being a toxic educational and work environment for them. They also do not look forward to a career of kowtowing to the academy's ideological hegemony (Redding 2023b). Progressives are drawn to the academy precisely because academic culture is consistent with their values, with many seeing it as an ideal place to be politically active in scholarship, teaching, or advocacy. In addition, as studies have documented, there is discrimination and bias against conservatives and libertarians in graduate school admissions, faculty hiring and promotion/tenure decisions, and peer reviews of colleagues' work (Redding 2023b).

To be sure, it will be an uphill battle to overcome these obstacles. Yet whenever a societal institution's ideological policies and practices swing too far in one direction, self-correcting forces almost always eventually swing it back toward the center. I am pessimistic in the short term but optimistic in the long term. As academics, public intellectuals, politicians, and commentators begin to speak out against ideologically driven censorship, as well as the DEI regimes that tend to promote it, we may be seeing the pendulum's long swing back to sanity. Some university alumni, trustees, and state legislatures are now pushing back, with the former withholding their financial and other support and the latter passing legislation prohibiting or curtailing DEI regimes at state universities as well as initiating policies and initiatives aimed at fostering intellectual diversity (Redding 2024).

Although an important part of the solution, such pressures from outside the academy will have limited effect, given the academy's relative insularity and the fact that it is university faculties that control which students and faculty they will recruit, admit, and hire. Unfortunately, a commitment to even a modicum of ideological diversity in graduate student and faculty recruitment, as well as faculty hiring, will be a tough sell to academics, many of whom have the self-serving and tautological belief that the academy is hegemonically progressive because such ideas are the correct, moral ones (and that prospective students and faculty having different views are immoral, anti-intellectual, unintelligent, or uninformed) (Redding, 2023a, b). Accompanying such beliefs is a sense of self-righteousness and intellectual superiority, which many like to virtue signal to their colleagues via supporting and enacting censorious practices against those not conforming to their ideological worldview.

Perhaps the best way to persuade university faculties to diversify is to convince them that it is in their professional and personal interest to do so. We should emphasize the ways that an ideologically diverse faculty will improve their credibility with the public and policymakers (leading to more grant funding!), help guarantee open science (including their own free inquiry!), and enhance the ability of those in the applied sciences to serve diverse clients and communities (see Redding and Cobb 2023). There is now a rich theoretical, empirical, and applied literature on how ideological diversity benefits a discipline, those working it, and the communities they serve (for the discipline of psychology, see Frisby et al. 2023; Redding 2001, 2023b; Redding and Cobb 2023).

Indeed, ideological diversity should be an important component of professors' cherished DEI projects (Redding 2024). As I describe in detail elsewhere (Redding and Cobb 2023; Redding 2024), recent empirical research (conducted no less by science professors) strongly suggests that the values and goals of DEI are furthered by including ideological diversity within it. The proponents of DEI focus on demographic diversity because they believe four things to be true about people's demographic characteristics: (1) that they are central to identity, (2) that people are discriminated against based on such characteristics, (3) that accompanying demographic diversity is a diversity of knowledge, values, and life experiences, and (4) that members of minority demographic groups are victims of societal oppression (Redding 2024). However, recent research in behavior genetics as well as social and personality psychology shows that these things are also true—equally, if not more so—with respect to people's ideological and sociopolitical views. People's ideologies are as important to their identity as their demographic characteristics; their ideologies affect how they relate to others and how others relate to them; people are often discriminated against due to their ideologies; ideological diversity produces substantial educational and workplace benefits; and ideological minorities in a setting are often marginalized and oppressed (Redding and Cobb 2023).

Thus, universities should include viewpoint and ideological diversity within their DEI policies and initiatives. Academic departments should affirmatively recruit ideologically diverse professors and graduate students, implement ideologically inclusive admissions and hiring practices, and foster an ideologically inclusive environment (Redding 2023a).

I have focused on the revolutionary reform of academia—by diversifying university faculties ideologically—as the only long-term solution to politicized science. Organizations and think tanks advocating for higher education reform (e.g., the American Council of Trustees and Alumni, Manhattan Institute) are playing an important role. In addition, the recent establishment of new academic and scientific organizations whose mission is to encourage open inquiry and push back against explicit and implicit censorship (e.g., Academic Freedom Alliance, Foundation for Individual Rights in Expression, Heterodox Academy, the Society for Open Inquiry in Behavioral Science, National Association of Scholars) can play an important role in supporting such reform. In various ways, they provide resources and professional support for academics and fellow travelers who are heterodox against prevailing ideologies to conduct research and teach from divergent perspectives.

Indeed, we ought to encourage academics to be courageous in their research, teaching, and university administrative work. Change will only come about when university trustees, alumni, donors, and policymakers—*but especially professors and university administrators*—speak up and act in ways that defend and promote open scientific inquiry and resist ideological control and censorship.

Dear Reader, will *you* be brave enough to do so?

The Archetypal Fate of the Universities

JORDAN B. PETERSON

It is inevitable that any major cultural shift emerging as a consequence of the universal play of motivating ideas will take an archetypal form. This is a matter of definition, as much as anything, as ideas that strike people deeply and produce a sense of conviction or devotion must appeal to something akin to basic instinct for them to have a universal perception-shaping and action-directing effect.[1] This is what accounts for their reversion to the prototypical or archetypal pattern. In the absence of such a fundamental call to interest and action, no union, no true mass movement, is possible, even in principle, as conceptions must be deeply held to motivate and widely shared to unite. Thus, an ideological shift must in some way, however unsatisfactorily, satisfy a longing that might be inchoate but is still widely felt—and is therefore instinctive in its essence. Such idea-systems must as well offer a solution to such a broadly perceived problem that is equally emotionally or motivationally attractive. This means an explanation or answer that quells anxiety in relation to a shared and valued goal and that simultaneously and for the same reason instills a sense of purpose or hope.[2] That provision of security and purpose optimizes the function of the two fundamental axes of emotion, negative and positive, which are always experienced in relationship to some vision of purpose[3] or, even more fundamentally, higher or sacred purpose (however couched in secular language that aim, that distant star, might be).[4]

The ideology that has taken hold of the universities over the last six decades, accelerating mightily in the latter stages of that process, proves no exception to the rule. The investigation of the at least two fundamental patterns of archetypal narrative apparently at play shines a clarifying light on the terribly dangerous situation at hand. The first pattern is encapsulated in the *Enuma Elish*, "When on High," a Babylonian myth of creation dating from the late second millennium BC.[5] That great and ancient story puts forth a characterization of the structure of the world that is typical of foundational narratives, which invariably portray existence itself as the consequence of the interplay of two principles, often characterized, and perhaps most simply understood as order and chaos (alternatively: masculine and feminine, light and darkness, or known and unknown).[6] In the Babylonian accounting, the former is given the form of the divinity Apsu, the male principle of being and becoming, and the latter Tiamat, a female dragon, who is both the matter-like origin of all things (her husband playing the spiritual role) and the force that eternally threatens, like the biblical flood, when order itself decays.[7] Apsu might well be regarded as something akin to what is referred to so ungratefully, blindly, and derisively in modern parlance as "the patriarchy," and which can be understood more deeply as the *a priori* structure of perception that allows the world of possibility to be apprehended intelligibly—in some manner, the very "rationality" of the rationalists. Tiamat, for her part, is the characterization of what is there for perception—the unpredictable, dangerous, and promising possibility of the world, "there in the beginning," but without form and void, in the famous terminology (Genesis 1:2), requiring the structured capacity for perception for its realization.

Why bother with the analysis of such portrayals? It is critical to understand in that regard that the founding myths of entire cultures are not mere archaic superstitions, easily, casually, and rationally dispensed with by the truly and genuinely educated and enlightened. They are, in the same wise, not "inventions" or "fabrications" or "accounts of objective reality," nor ploys by the power-elite to maintain social control in the Marxist sense—not "opiates of the masses." Finally, they are also not childish delusions, protecting the naive and immature against the terrible burden of death anxiety, as the Freudians and neo-Freudians claimed.[8] They are instead communicable representations of the implicit structures, both psychological and social, that shape perception itself.[9] They are the net result, it might be said, of the competition between stories across vast spans of time to grip attention and embed themselves in memory—the very competition of the gods in heaven whose battles make up the cosmogonic narratives of cultures everywhere, as the Romanian historian of religions Mircea Eliade has so carefully documented.[10] Thus, the analysis of foundational myths offers the opportunity to get to the bottom of things in a manner that can possibly be challenged only—or better, abetted—by a deep analysis of biological structure and function itself.

How might such dovetailing of story and psychophysiology be conceptualized? It is perhaps simplest to begin by making a case that is simultaneously neuropsychological and phenomenological. The brain itself (and this is the case for all twinned-hemisphere brains) appears to consist of two primordially fundamental systems of perception, motivation, emotion, cognition, and action—one specialized for operation in territory that is established, safe, understood, and predictably desirable (that is the domain of order); the other for what is dangerous, rife with opportunity, unfamiliar—even predatory.[11] The world of experience (this is the phenomenological analysis) lays itself out in accordance with this twinned vision: we are stable, secure, and hopeful when what we desire and can predict in relationship to that desire is laying itself out, in the old and familiar way, and on alert and apprehensive or even paralyzed by terror when the domain of the unknown and terrifying makes itself known. It is for this reason that reality has a strongly bipartisan structure—the known, with its security and proclivity for stagnation, and the unknown, with its terrors and possibilities.[12] What does this mean about reality itself, in the "objective" sense? This is a very difficult question to answer, given that any account of reality requires a subjective observer, with the biases that a subjectivity hell-bent on surviving is likely to manifest—but it is clearly the case that organisms with nervous systems that do survive construe the natural order in a manner that appears to require some perception and conceptualization of what is and what has not yet come to be—of actuality and possibility, of masculine and feminine, of light and darkness; in the Taoist framing, of yang and yin.

The Mesopotamian account characterizes this dynamic interplay as the union of the two primordial deities, father and mother, order and chaos—the aforementioned Apsu and Tiamat. It is their creative interaction, portrayed sexually, which produces the second tier of the gods, which might be regarded both as the primal or not-so-noble savage state of man, and the first multiplicitous society of the cosmos. This account of psychological differentiation and emergent social order can also be read neuropsychologically, or neurodevelopmentally: the first "spirits" that arise when being itself is established might well be considered the most basic of instincts, present even at birth—even forming part of the animal world, far down the phylogenetic chain—that rule, regulate, and govern perception and action "in the beginning."[13] The social structure ruled by these deities of basic instinct is far from optimal, however, according to the authors of the *Enuma Elish*. The polity or psyche produced by their emergence and interaction is noisy, fractious, careless, hedonistic, derisive, arrogant, self-centered—radically immature, and prone to all of the catastrophes that the immature are destined to encounter and to produce. The primal children of Apsu and Tiamat go about their business, creating, producing, fighting, and multiplying—but they have no respect for the past and no integrated, sacrificial view of the future. The consequence of this essentially immature and narcissistic heedlessness? The first descendants of order and chaos kill their father in a fit of chaotic, arrogant ingratitude. Furthermore, and equally importantly, they attempt, in the aftermath of his death, to subsist on his corpse.

The sub-deities that arise in the wake of such death can be many, but they take the primary form of the desire for power, the hedonism that is the purpose of power (including the sadistic delight that inevitably accompanies such will to pleasure), or the nihilistic elevation and worship of nothing itself.[14]

And what of the corpse itself that is now the temporary dwelling place of the usurpers and assassins of culture? A society—a civilization, a culture—can well and truly be considered a storehouse of value, a hoard of treasure, a place of plenty. When the spirit of culture is truly alive, that storehouse is self-replacing, as it contains not only the accumulated material value bequeathed to those who currently inhabit the earth, in the practical form of technology and practically useful infrastructure, but the representations of the patterns of perception and action that originally gave rise to such accumulation and representation. When Apsu is killed—when God dies; when the power-mad Luciferian usurper arises; when the hedonists rule; when the moral responsibility for productive activity is foregone (all of these are the same thing)—then the murderous, fractious, and faithless denizens of the degenerate society who remain can feast for a while on his corpse. After that? Resources diminish, as they fail to be replenished. Afterward? The inevitable reversion to unformed possibility. That is the return of the eternal dragon, the rearing by the chaos of its ugly head.

This is an account reflective, as it must inevitably be, of something else that is deep and permanent: the eternal biological conflict between host and parasite. The parasite problem is essentially this: how can a complex organism sustain and reproduce itself, across time, when simpler and faster-reproducing parasites inevitably emerge to take advantage of the resources such organisms store or, more profoundly, constitute, in their embodied form? This is no trivial matter: the problem of parasitism is certainly one of the deepest and most permanently challenging or even insoluble biological and, therefore, existential problems.[15] Sex itself evolved, hundreds of millions of years ago, at least in part to mix genes in a manner that allows the sexually reproducing host to stay a jump ahead of the organisms that are competing for its own resources. We have stored up such resources, in the form of giant pools of immense wealth, everywhere in the West, particularly since World War II.

What is (or was) the direct and incontrovertible and all-too-easily purloined dollar value of the endowments of the Ivy Leagues? To say nothing of the now somewhat damaged but still kingly value of their brands? The same metaphor of treasure-house can be appropriately applied, in general, to all the great institutions of the West, which are all arguably buckling under the assault that has already placed the universities beyond the pale. Has it not been said that a corpse rots from the head down? And what is at the head, in the West, other than the universities? It has also been said that culture is upstream of politics. What is upstream of culture other than education? The great storage places of value that derived their material wealth and brand status from the actions of those imbued with the true redeeming spirit of civilization have been invaded by the most anti-meritocratic, morally relativistic, cynical, nihilistic, power-mad, and bitter immature hedonists imaginable. Chaos, therefore, looms, if it has not arrived already.

The past is indeed a great leviathan, washed up, undeserved on shore. It is entirely possible to live inside that storehouse, that treasury of value, and to feast off its remains without acknowledging the fate that delivered it, without shouldering any of the responsibility for its maintenance or replenishment, without undertaking any of the sacrifices necessary to ensure its deliverance to the generations that follow. This is precisely the tactic of scavengers and parasites, biologically speaking, hell-bent on stripping whatever flesh magically presents itself to the bone—all the while chasing their own narrowly self-centered, hedonistic, and even psychopathic indulgences, domineering aims, and imprudent and infantile purposes. Is Harvard, for example, not reasonably construed as a whale carcass, ripe for the plundering—with its $50 billion endowment, with what was once a university somewhat carelessly and secondarily attached to it? In its prime, it was certainly a genuine treasure house, built in reality and reputation upon the truly diligent work of those who labored upward and properly within its confines in the past. Like its sister institutions—MIT, UPenn, Yale, Princeton, Stanford, and the rest—it became, as a consequence of that extreme care, a place of brilliant lecturers, creative and honest scientists, and highly qualified, ambitious, and disciplined students. In consequence, perceiving their opportunity, the parasites arrived, plotting their takeover during the Harvard-psychology-department-Timothy-Leary-inspired tune-in, drop-out hippie extravaganza of the 1960s.

All the virtues that made these places what they most truly were—the disciplined striving, the upward aiming, the long apprenticeship, the concern for the truth, the high and discerning intelligence—were disdainfully dismissed by the parasitical invaders. All that "patriarchal hierarchy" deemed nothing but the expression of arbitrary power and the manifestation of economic, sexist, racist, and then, of course, the metastasis—"intersectional"[16] oppression. This was an accusation that had just enough truth to stick, as all complex institutions are both tainted and threatened by the actions of those who seek to force and control—but "tainted" and "threatened" are words that are by no means synonymous with "founded or predicated upon."

More truly, however, the artful, treacherous, and self-serving reduction of the very spirit of complex hierarchies of competence and productivity to nothing but the motivations of power and oppression was and is something best viewed as an example of the secondary gain of psychoanalytic fame. The purveyors and wielders of such theories have, apart from the obvious unearned psychological advantages of demonizing the host, the great opportunity to employ force, compulsion, and power themselves—as no other motivations exist within the purview of their conceptualizations. Thus, we have the current Machiavellian justification for the battle against true ability: "If those in positions of so-called 'authority' obtained everything they have in consequence of their theft and misuse of power, it is only natural justice that those they victimized and oppressed turn the tables and steal from them (particularly given that, in the final analysis, there are is no other true motivation but that of power)."

The very reductive and simple-minded but certainly canny ideologues whose victimized proclaiming was so cowardly accepted, abetted, and even admired ("they're so brave") by the very targets of their shaming very quickly had the doors to the warehouse flung open to them. This was the assault on merit by those who either had none or who had rejected even what little they had—accepted as meaningful critique by those who were immorally indiscriminate in their self-flagellation and guilt. The parasites came flooding in, consequentially, in droves, multiplying the number of sham academic positions and entire disciplines and positions, demolishing one by one everyone who opposed their profoundly self-serving acting or indicated its essentially narcissistic and shallow nature. This was the invasion of academia by the poseurs and charlatans of the permissive, infantilizing, accusatory mob; the radical feminists and pseudo-lesbians, the grievance studies "scholars," the postmodernists allied so corruptly and unforgivably with the vengeful and murderous Marxists, the do-gooding activists, the rainbow-banner bearing hedonists of pride, even the outright communists—the whole bloody mess of those claiming no other motivation but compassion and care for the marginalized but actually doing nothing other than ravaging the corpse of their forebears.

Thus, we in the West came to subsidize three or four generations of the absolute foes even of the culture that made even their very existence possible. Thus, we enabled the mass action and even the thriving of those who, without such largesse, would have had difficulty providing for themselves even the basic needs and relationships of life. How did this happen, at the level of detail? Consider this: How much trouble could one person hell-bent on producing nothing but such trouble manage to cause, in a lifetime, if freed from the necessities of genuine productive and generous travail and provided with inexhaustible means to do so, all the while? Then multiply that error by a thousand—or by the tens of thousands—in the case of the faux-academics and "researchers" who emerged in the wake of the capitulation of the true university faculties and managers. Imagine, in addition, that something rotting from the head down does so in the most complete and poisonous manner possible. That is exactly what took place in the West—particularly in the decades following the degenerate 1960s.

The consequences of all that now stare us in the face, as we stare back, in the hapless paralysis of the prey animal, shocked to be in such a position. When the source is polluted, everything downstream inevitably becomes poisoned—everything is corrupting simultaneously, as is only too evident, under the current circumstances. Hence, the increasing unreliableness of professional organizations and governing bodies everywhere and the sad and degenerating state of the legacy media, which, with few exceptions, does nothing but lie. Hence, the pathetic virtue signaling and apocalypse-mongering hypocrisy and betrayal of the international political elite. The heads of the hydra that have always been said to envelop the world during its decline are prone to appear, magically and uncannily, everywhere at once.[17] That is the death of the center and the victory of the margin—the very victory that the barbarians who stormed the gates promised so forthrightly and colorfully to deliver. The generations who lived immediately prior to us built up tremendous storehouses of value in the peaceful and productive aftermath of World War II. They left those places of value unguarded. In consequence, the vandals and the Visigoths came to occupy and will continue to do so in an ever-worsening and more comprehensive manner unless they are attended to and forthrightly stopped.

What, then, has happened to the universities? The Luciferian intellect has usurped the place properly reserved for the highest of unifying spirits. It has thereby killed the goose that for so long laid the truly golden eggs. The remnant carcass makes for a short-term feast but ensures that poverty and hunger, not least of spirit, will more permanently rule. The postmodernists famously insist on the skepticism toward meta-narratives that denies the very existence of a highest or most fundamental value.[18] This is nothing more or less than an extension of the idea of the death of God. What dread spirit necessarily arises to take the place of what was once set in the highest place? That same question can be rephrased usefully: What rules in the absence of a unity? Obviously, a plurality, if not nothing—even, shall we, a "diverse" and "inclusive" plurality. There is, therefore, little difference, even at the onset of such a movement, between the rule of diversity and chaos itself—and it is certain that this "little difference, even to begin with" will soon and inevitably degenerate into the very culture-destroying emergence of the goddess of Chaos herself. Diversity is indistinguishable, on the shadow side, from aimlessness, confusion, and conflict—the cause of war in the pluralistic world at large, as systems of value inevitably compete for dominance and the introduction of precisely such strife and disarray into the smaller subsidiary systems of any given culture, prone to an inevitable hopelessness, anomie, and disunity as they lose their center. "Mere anarchy is loosed upon the world," indeed.[19] The Mesopotamian story, the culmination of a much older oral tradition, took untold millennia to develop. The wisdom it contains is eternal and is ignored at our great peril. The same can be said for the story of Cain and Abel. The parasitic spirit is prideful, bitter, usurping, vengeful, chaotic—then murderous; then genocidal. Ungrateful children attempt to live on the corpse of their father. Fate unfolds quite terribly from that starting position.

Is there a way out? Practically, it looks very difficult. The institutions of higher education in the West are administratively top-heavy, in precisely the manner allowing that very administration to prey upon the future earnings of the students they hypothetically serve. Furthermore, the faculty itself leans very heavily in the radical direction despite being the dubious beneficiaries of administrative expansion. Student-to-faculty ratios have increased dramatically. The status and autonomy of the faculty have been very badly compromised. None of this seems to have made the professors and researchers any more forthcoming in their resistance or any more skeptical about the self-righteous "progressive" ideals they advocate. Finally, the students themselves have been infantilized, as well as taught outright that progressive radicalism is identical to virtue. It is, therefore, very difficult to imagine anything approximating an across-the-board improvement in behavior at the universities, in consequence. There are no imaginable circumstances under which the extant Ivy League institutions, for example, would be likely to dispense with the 80 percent of administrative staff and a somewhat lesser but still high proportion of the faculty that would have to disappear for any real change in orientation to occur.

There have been recent initiatives, most famously in the states of Texas and Florida, to push back against the diversity, inclusivity, and equity bureaucracy-mob, but a simple name change is no trouble for the serpentine Machiavellians of the radical fringe to manage. This means that the same old corrupt players can just change their ideational camouflage, and at the surface level, just as they did when the revelations of Communist excess became too self-evident to deny in the 1970s, and Marxism metastasized into the intersectional oppression idiocy of today. It is, therefore, highly possible that the academy in the West is rotting because it is, in fact, dead and cannot be revived in anything resembling its current form. In that case, the flame will have to be kept burning at new institutions, whether in the classic bricks-and-mortar and in-person guise, if that proves possible (and I doubt it), or electronically, if the major changes in the conceptualization of education and the technology necessary to offer mass higher learning prove capable of delivering on their obvious but by no means certain promise.

Is there a way out personally, psychologically, or conceptually? Or, more specifically, what is presently incumbent upon those who are currently or will in the future be engaged in the university teaching or research enterprise? We can turn, once again, to the *Enuma Elish* account to elucidate the archetypal pattern. It is always in the very darkest hour that the redeeming hero emerges. Deliverance comes to the faithless murderer-children of chaos and order in the story in question in the hero-form of Marduk, the

Mesopotamian savior-son-of-God. After Cain and Abel, we are presented with the cataclysm of the flood, as well as the erection of the Tower of Babel, the totalitarian edifice which is a structure of value with the wrong value at its pinnacle. Those are the twin fates of the descendants of Cain, the parasites who will not sacrifice properly and who undermine what even they themselves most truly value and depend upon. What are these fates? The descent of the world into the primordial waters of chaos that reigned before the beginning; alternatively, the worship of the false idols of the Luciferian intellect that likewise produce disunity and confusion—in the latter case, the eventual inability of those who aim wrong to even understand or utter the same words. "What is a woman?"—indeed. What is the way forward in the biblical narrative—perhaps running parallel to the Mesopotamian story?

Careless pride eternally comes before a fall, but the humility that is its medicament is the attention that leaves nothing out of its purview and can, therefore, perceive the proper course forward. We could consider the account of Jonah, the everyman alerted by his conscience to speak the words of truth that set society straight. He flees, initially, from the requirement—who wants to risk standing against the mob?—but upon encountering the abysmal consequences of maintaining silence when there is something that demands to be said, he reconsiders his cowardice, turns his voice over to the control of the divine, and redeems the corrupt city of his enemies. What is the moral of such stories for the timorous, ideology-addled professors of today—or for those who enable them through their silence or even "allyship"? What might be done by forthright and faithful believers in the redeeming quality of truth and sacrifice when the walls we relied on to protect us have been breached and the barbarians have entered the gates? And have they? Here's a telling data point: it is none other than the despicable Michel Foucault, darling of the Luciferian radicals, who is by some accounts the most cited "scholar" in the world.[20]

What might be done, indeed? We professors, scholars, and researchers could open the eyes and the mouths God has given us instead of continuing to take each of the thousands of small steps backward that have enabled the devouring ideologues at our institutions. We could forthrightly note the chaos that looms and identify and hold to account its purveyors and acolytes. We could do the duty required of us by the people who have charged the inhabitants of the academy with the responsibility of preserving and seeking the truth, and we could call out the frauds, hucksters, Machiavellians, and outright charlatan-psychopaths of the faux disciplines and idiot theories of oppressor and victim. We could, by so doing, speak the words of truthful magic that redeem the fragmented world and position the foundation back in its place. Alternatively, we could remain frozen in the corner like the petrified rabbits we have far too often revealed ourselves to be and let the parasitical rabble attain their temporary and inevitably self and society-destroying "victory." What did the great Aleksandr Solzhenitsyn say so effectively in his all-out and eventually successful assault on the predatory tyrants and parasites of the envious and ideology-possessed radical world? ONE WORD OF TRUTH SHALL OUTWEIGH THE WHOLE WORLD.[21] The protected professoriate pays for the opportunity and protection historically granted by the community to the academic endeavor with the vow to protect what is true, beautiful, and good. If we once again found the courage to be the standard-bearers of such value, the mad and ideological mob would be rapidly defeated.

Will we do what needs to be done? We have failed so far, and it would take a mad bit of optimism to suggest that the course could be reversed. Nonetheless, the situation is far from hopeless. Here is another bit of traditional mythology to consider with regard to such a situation. When Abraham is informed by God of the latter's decision to destroy the sinful cities of Sodom and Gomorrah, the great prophet objects, making the case that wholesale destruction is no temperate and proper response to the sins even of the many (Genesis 18: 22–33). God agrees to spare the threatened cities if even fifty righteous people might still be found dwelling within their walls. Abraham persists in his objections, even after receiving such consideration from God, bargaining the divine down to forty-five, then forty, then thirty, then twenty—then ten. What does this strange story indicate? Nothing less than this: an entire polity, and all of its denizens, can be saved even from an impending, dire, and deserved fate if even a tiny number of its inhabitants remain willing to risk the great adventure of the truth.

The moral of the story for the current denizens of the academies of higher education? It would not take a great number of professors, researchers, and students even now to speak and turn the tide. The moral, even more specifically, even more personally? It is incumbent on each and every one of us engaged in such enterprise to risk being one of the redeeming ten. We should remember, too, when considering such things: it is indeed a risk to see what is in front of our eyes and to say what needs to be said. It is also true, however, that willful blindness and failure to speak when there is something to say are ventures with perils of their own. The ten in question may find themselves in trouble when they speak, but they are simultaneously saving themselves and all those around them from the invisible hand of doom, threatening the future. This is an existential fact, alluded to in the great story of Jonah. It is not by mere chance that the fleeing prophet finds himself drowning when he holds his tongue after being commanded to speak—and then, worse, devoured by the beast of the abyss. There is indeed a descent into hell that threatens, both psychologically and socially, when conscience is ignored. It is the wisdom of the ages that the risk of seeing and speaking, all things considered, is much less than the risk of hiding and silence. There may be a price that is to be paid for the former—but the price of the latter is both soul and society. The true spirit of the academy is the prophetic voice that redeems. We abandon that at our great peril and to the detriment of all that depends on education itself.

See, therefore, and speak—or pay the inevitable price.

Epilogue

I am enormously grateful to the many colleagues here who devoted their time and energy to help this volume comprehensively describe the current challenge we face if we are to restore scholarship and an open scientific process—so necessary to deal with the real challenges of the twenty-first century—to academia and professional societies.

After reading all of their contributions, it is hard to know whether to come out as optimistic or pessimistic. On the one hand, pressures to regulate speech and limit open inquiry persist in many areas. On the other hand, more and more faculty are speaking out against these threats, and society in general seems to be becoming more suspicious of the real politics and impact of so-called diversity, equity, and inclusion initiatives.

There is an old saying that if the only tool you have is a hammer, everything looks like a nail. I have been an educator for much of my life, and it is therefore not surprising that I think that education is the key to surmounting the current war on science. In my experience, once the public becomes truly aware of initiatives that limit free inquiry and merit-based progress, they respond appropriately. It is my earnest hope that this volume can play a role in this effort.

Politics should play no role in governing the process of science and scholarship. It is for this reason that the authors here, all of whom have worked within the academic system, represent the full spectrum of political leanings. In spite of what might be our different views about the best role of government in public education, for example, we have come together because of our inherent respect for the process of education and research itself.

In the end, I hope the articles in this book serve to do more than depress you. Rather, I hope they energize you to speak out, talk to your friends, demand better of your governmental representatives, and hold the leaders of academic institutions and research organizations accountable for their tendency to virtue signal rather than demonstrate some spine when confronted with social media mobs demanding someone's head or complaining of perceived offenses.

It is hard to know what will cause the current social panic to subside and reverse and when it might happen. Numerous times over the past year, I have heard friends suggest the tide is turning. I am more wary. I think it is equally likely that it will get worse before it gets better. I hope I am wrong, and I hope the impact of the arguments presented here by my distinguished colleagues will help ensure that I am. Ultimately, this, too, shall pass, of course. But the longer it takes, the more opportunities for progress will have been squandered, and the greater the challenge will be for meaningful recovery. All of us need to work to ensure that the road to recovery begins here and now.

Lawrence M. Krauss, PEI 2024

References and Footnotes

To the reader: All footnotes and citations listed in the text can be viewed online at the following address. https://lawrencemkrauss.com/the-war-on-science-references/ Or use the QR Code below:

Authors

Dorian Abbot is a professor in the Department of the Geophysical Sciences at the University of Chicago. He is a member of the Committee of the Council of the University of Chicago Senate, the Board of Trustees of Florida Polytechnic University, and the Board of Advisors of the University of Austin. Abbot was awarded the Hero of Intellectual Freedom Award by the American Council of Trustees and Alumni.

John Armstrong is a Reader in Financial Mathematics at King's College London. His mathematical research includes pension investment, rough-path theory, and applications of geometry to stochastic processes. His educational research champions a scientific approach to equality, diversity, and inclusion. He is a cofounder of the London Universities' Council for Academic Freedom.

Peter Boghossian is the director of the National Progress Alliance and a founding faculty advisor for the University of Austin. His fundamental objective is to teach people how to think through what often seem to be intractable problems. @peterboghossian.

Dr. Maarten Boudry is a philosopher of science and the first holder of the Etienne Vermeersch Chair of Critical Thinking at Ghent University. He has published over fifty academic papers and two edited volumes: *Science Unlimited?* (2018) and *Philosophy of Pseudoscience* (2013). He also wrote six trade books in Dutch on science and philosophy, the latest one being *The Betrayal of Enlightenment (Het verraad aan de verlichting,* Prometheus Amsterdam, 2025). Substack: maartenboudry.substack.com.

Alex Byrne is a Laurence S. Rockefeller Professor of Philosophy at MIT. His most recent book is *Trouble with Gender: Sex Facts, Gender Fictions* (Polity, 2024).

Nicholas A. Christakis, MD, PhD, MPH, is the Sterling Professor of Social and Natural Science at Yale University, where he also directs the Human Nature Lab. A member of the National Academy of Sciences, he is the author of the *New York Times* bestseller *Blueprint: The Evolutionary Origins of a Good Society* (2019).

Roger B. Cohen, MD, is a professor of medicine at the Perelman School of Medicine at the University of Pennsylvania.

Jerry Coyne is a professor emeritus in the Department of Ecology and Evolution at the University of Chicago and works on the origin of species in the fruit fly Drosophila. He has published more than 125 scientific articles and 180 articles in popular magazines, books, and newspapers, and is the author of *Speciation* (with Allen Orr), *Why Evolution Is True,* and *Faith Versus Fact.*

Richard Dawkins is English. He was the Charles Simonyi Professor of the Public Understanding of Science at Oxford University. His eighteen books include *The Selfish Gene, Unweaving the Rainbow,* and, most recently, *The Genetic Book of the Dead.*

Sir Niall Ferguson, MA, DPhil, FRSE, is the Milbank Family Senior Fellow at the Hoover Institution, Stanford University, and a senior faculty fellow of the Belfer Center for Science and International Affairs at Harvard. He is the author of sixteen books, including *The Pity of War*, *The House of Rothschild*, and *Kissinger, 1923–1968: The Idealist*, which won the Council on Foreign Relations Arthur Ross Prize. His latest book, *Doom: The Politics of Catastrophe*, was published in 2021 by Penguin and was shortlisted for the Lionel Gelber Prize.

Janice Fiamengo retired in 2019 from her position as a professor of English at the University of Ottawa, where she taught for sixteen years. Before that, she had taught full-time at the University of Saskatchewan for four years. In retirement, she is at work on a history of feminist thought in English (from a non-feminist, non-adulatory perspective) and is the author of the *Fiamengo File 2.0*, a YouTube series showcasing the dark side of feminism.

Solveig Lucia Gold writes about good and bad ideas in the ancient and modern worlds. The Senior Fellow in Education and Society at the American Council of Trustees and Alumni, she has a PhD and an MPhil in classics from the University of Cambridge and an AB in classics from Princeton University.

Moti Gorin is a philosopher and bioethicist. He is an associate professor of philosophy at Colorado State University.

Karleen Gribble, PhD, BRurSc (Hons), is an adjunct professor in the School of Nursing and Midwifery at Western Sydney University, Australia. Her research focuses on breastfeeding in situations of adversity, including in emergencies and in the context of intergenerational trauma and child protection concerns. Karleen advocates for recognition of the importance of mothers to their infants and works to create environments that support breastfeeding and the mother-infant relationship.

Dr. Carole K. Hooven is the author of *T: The Story of Testosterone, the Hormone that Dominates and Divides Us*. She taught about hormones and behavior in the Department of Human Evolutionary Biology at Harvard University for over two decades and is now an associate in the Department of Psychology at Harvard and a non-resident senior fellow at the American Enterprise Institute.

Geoff Horsman is an associate professor of chemistry and biochemistry at Wilfrid Laurier University in Waterloo, Canada, where his research focuses on how microorganisms make small molecules like antibiotics. He is also active in the Society for Academic Freedom and Scholarship, is cochair of the Laurier Heterodox Academy Campus Community, and contributes to local school reform efforts through the grassroots organization Educating Minds, Parents of Waterloo Region (EMPOWR).

Joshua T. Katz is a senior fellow at the American Enterprise Institute, where he focuses on higher education, language and culture, the classical tradition, and the humanities broadly defined. Before taking up his position at AEI, he was the Cotsen Professor in the Humanities and a professor of classics at Princeton. He has a BA from Yale, an MPhil from Oxford, and a PhD from Harvard, all in linguistics.

Sergiu Klainerman, the Higgins Professor of Mathematics at Princeton University, received his undergraduate degree from Bucharest University and his PhD from New York University (1978). He was a Miller fellow at Berkeley, an assistant professor at NYU, and joined the Princeton mathematics department as a professor in 1987. He is a member of the US National Academy of Sciences, a foreign member of the French Academy of Sciences, the American Academy of Arts and Sciences, and on the board of the recently formed American Academy of Sciences and Letters. He is the recipient of the MacArthur Fellowship award (1991), Bocher prize of AMS (1999), and Le Conte Prize of the French Academy of Sciences (1996).

Lawrence M. Krauss is currently president of The Origins Project Foundation, chair of the board of the Free Speech Union, Canada, CEO of Gus What Productions, and host of *The Origins Podcast*. He is a theoretical physicist and the author of over three hundred scientific publications. He has held professorships at Yale University, Case Western Reserve University, Australian National University, New College of the Humanities, and Arizona State University, and he is the recipient of numerous international awards and fellowships for his research and writing. His popular books include the *NYT* bestsellers *The Physics of Star Trek* and *A Universe from Nothing*. Krauss also appears regularly on radio, TV, and films.

Anna I. Krylov is the University of Southern California Associates Chair in Natural Sciences and a professor of chemistry. Krylov's work, which includes more than three hundred papers and three hundred invited talks, has received numerous international awards. She is a fellow of the American Chemical Society, the American Physical Society, the Royal Society of Chemistry, and the American Association for the Advancement of Science, and she is an elected member of the American Academy of Sciences and Letters, the International Academy of Quantum Molecular Science, and the Academia Europaea. An outspoken advocate of freedom of speech and academic freedom, Krylov is a founding member of the Academic Freedom Alliance.

Luana S. Maroja is an evolutionary biologist and professor at Williams College. She got her undergraduate and master's degree from the Federal University of Rio de Janeiro, Brazil, and her PhD from Cornell University. She is interested in speciation and population genetics and has worked on a variety of organisms.

Christian Ott received his doctorate in astrophysics from Potsdam University and the Max Planck Institute for Gravitational Physics in 2007. He joined the faculty at the California Institution of Technology in 2009 and was a professor of astrophysics until resigning at the end of 2017. He is a cofounder of Pasadena Consulting Group LLC, which offers software, machine learning, and computational science consulting services.

Bruce Pardy is a professor of law at Queen's University in Canada and the executive director of Rights Probe, a law and liberty think tank. Bruce is an academic, lawyer, columnist, and outspoken critic of the illiberal managerial state. On X: @PardyBruce https://x.com/PardyBruce.

Dr. Jordan B. Peterson is a bestselling author, clinical psychologist, prominent public intellectual, professor emeritus at the University of Toronto, and host of the *Jordan B. Peterson* podcast. He has written three books, *Maps of Meaning*, the bestseller *12 Rules for Life*, and *Beyond Order*, which have sold more than fourteen million copies. His fourth book, *We Who Wrestle with God*, will be released in November. Dr. Peterson's international lecture tour has sold out more than eight hundred venues. Dr. Peterson cofounded the Peterson Academy, and over thirty thousand students have enrolled since the beta launched publicly in September 2024.

Steven Pinker is an experimental psychologist who does research on language, cognition, and social relations and writes on all aspects of human nature. He is the Johnstone Family Professor of Psychology at Harvard University and the author of twelve books, including, most recently, *Rationality: What It Is, Why It Seems Scarce, Why It Matters*.

Richard E. Redding is the Ronald D. Rotunda Distinguished Professor of Jurisprudence, associate dean for academic affairs, and professor of education at Chapman University. He received his PhD in psychology from the University of Virginia and his JD from Washington and Lee University. He is an elected member of the American Law Institute and an elected fellow of the American Psychological Association and the Association for Psychological Science. He has published widely on the issue of viewpoint diversity in the academy, including the recent book *Ideological and Political Bias in Psychology: Nature, Scope, and Solutions*.

Arthur Rousseau is board-certified in psychiatry by the American Board of Psychiatry and Neurology and is a distinguished life fellow of the American Psychiatric Association. He was in private practice for over forty years as a psychodynamic psychiatrist, which included forensic work, teaching as a clinical professor at the University of Oklahoma Health Sciences Center, and advising legislative policies at a state and national level. After retiring in 2022, he has dedicated his time to continued patient advocacy, writing, and family.

Gad Saad is a professor of marketing at Concordia University (Montreal, Canada) and a former holder of the Concordia University Research Chair in Evolutionary Behavioral Sciences and Darwinian Consumption (2008–2018). In 2024–2025, he is serving as a visiting professor and global ambassador at Northwood University. He has pioneered the use of evolutionary psychology in consumer behavior. His 2020 book *The Parasitic Mind: How Infectious Ideas Are Killing Common Sense* was an international bestseller.

Sally Satel is a psychiatrist, a senior fellow at the American Enterprise Institute, a lecturer at Yale University School of Medicine, and the medical director of a methadone clinic in Washington, DC. She is the author of *PC, M.D., How Political Correctness Is Corrupting Medicine* (Basic Books, 2000).

Lauren Schwartz is a board-certified psychiatrist by the American Board of Psychiatry and Neurology and a fellow of the American Psychiatric Association. She has been practicing psychodynamic psychiatry in private practice for the past eleven years. Alongside her full-time clinical work, she is a public speaker, author, wife, and mother of three children.

Alan Sokal is a professor of mathematics at the University College London and professor emeritus of physics at New York University. He is the coauthor (with Jean Bricmont) of *Fashionable Nonsense: Postmodern Intellectuals' Abuse of Science* and author of *Beyond the Hoax: Science, Philosophy and Culture.*

Alessandro Strumia is a physicist based in Italy who used to collaborate with CERN. He is the author of more than two hundred publications on theoretical physics, plus a few about Higgs detection, bibliometrics, and gender and physics.

Judith Suissa is an emerita professor of philosophy of education at the UCL Institute of Education in London. She has published widely on topics including anarchist and libertarian education, parenting, the politics of schooling, and academic freedom. She left academia in 2023 to try and find some breathing space.

Alice Sullivan is a professor of sociology at University College London. Her research interests include social and educational inequalities, quantitative methods, sex and gender, and academic freedom.

Jay Tanzman is a statistical consultant working in biostatistics, epidemiology, social sciences, and educational research. He has published criticisms of the statistical practices in the social sciences and has been an outspoken advocate for improving the statistical methods employed in experimental psychology. He uses statistical insights to assist scientists in fighting false claims of discrimination in STEM fields. Previously, he held academic positions at Loma Linda University and the University of California, Los Angeles.

Abigail Thompson received her PhD in mathematics from Rutgers University in 1986 and joined the UC Davis faculty in 1988, where she is currently a distinguished professor of mathematics. She is a fellow of the American Mathematical Society, where she served as vice president from 2019 to 2022. She is a founding member and member of the Academic Council of the Academic Freedom Alliance and a founding member and secretary of the Association for Mathematical Research.

Amy L. Wax is the Robert Mundheim Professor of Law at the University of Pennsylvania Law School, where she teaches courses in Remedies, Conservative Political and Legal Thought, and Law, Neuroscience, and Responsibility. Her scholarship addresses issues in social welfare law and public policy, as well as the relationship of the family, the workplace, and labor markets. She received an MD from Harvard Medical School and a JD from Columbia Law School. She served in the Solicitor General's Office of the Department of Justice during the Reagan and George H. W. Bush administrations. She is the author of *Race, Wrongs, and Remedies* (2009).

Elizabeth Weiss is a professor emeritus of anthropology at San José State University, where she curated the largest single-site collection of precontact Native American skeletal remains west of the Mississippi. She is the author of several books, including *Reading the Bones: Activity, Biology, and Culture* (University of Press Florida, 2017), *Repatriation and Erasing the Past* (University of Florida Press, 2020) with coauthor James W. Springer, and *On the Warpath: My Battles with Indians, Pretendians, and Woke Warriors* (Academica Press, 2024). From 2023 to 2024, she was a faculty fellow at Heterodox Academy's Center for Academic Pluralism, and she currently sits on the board of directors of the National Association of Scholars.

Frances Widdowson is a political scientist who was fired from Mount Royal University in Calgary for satirizing "woke" ideas. Widdowson was found to have been wrongly terminated in arbitration, but her lack of reinstatement is being appealed. Her talks have been canceled by the University of Lethbridge and the University of Regina, and she is currently suing the former for violating her charter rights. Details about Widdowson's battles with cancel culture can be found on the website www.wokeacademy.info.